Protein Bioinformatics:
From Sequence to Function

Praise from the Field for
Protein Bioinformatics: From Sequence to Function

A protein structure oriented bioinformatics book has been long overdue and I would like to congratulate Dr. Gromiha for his efforts to fill this gap.

Dr. Sandeep Kumar, Principal Scientist,
Pharmaceutical Sciences, Research and Development,
Global Biologics, Pfizer, Inc.

The book has excellent coverage for students wishing to practice bioinformatics.

Dr. Debnath Pal, Indian Institute of Science,
Bangalore, India

The author introduces the concepts in a simple manner so that even people with a non-mathematical background will find it very useful. The literature covered is quite extensive so that the beginner will find it as a very good platform to begin one's work.

Dr. S. Selvaraj, Department of Bioinformatics,
Bharathidasan University, Tamil Nadu

To my knowledge, there is no source currently available that solidifies protein bioinformatics the way this book does. The book tries to provide the broadest possible coverage of protein bioinformatics; it has practice exercises, and it links and discusses online databases/servers that the reader can actually use to analyze their own protein datasets.

Dr. Eric Gaucher, Georgia Institute of Technology,
Atlanta, GA USA

This book is a "must have" for my Protein Engineering students who need to have easy access to the most important bioinformatics tools for analyzing protein structures and predicting how mutations will affect their structure and/or function.

Dr. Gerard Pujadas Departament de Bioquímica i
Biotecnologia, Universitat Rovira i Virgili, Catalonia

Protein Bioinformatics: From Sequence to Function

M. Michael Gromiha

*Computational Biology Research Center (CBRC),
National Institute of Advanced Industrial
Science and Technology (AIST), Tokyo, Japan*

AMSTERDAM • BOSTON • HEIDELBERG • LONDON
NEW DELHI • NEW YORK • OXFORD • PARIS • SAN DIEGO
SAN FRANCISCO • SINGAPORE • SYDNEY • TOKYO
Academic Press is an imprint of Elsevier

Cover credits:

The figure is a portion of Color Insert 12 which was adapted from Gromiha MM, Pujadas G, Magyar C, Selvaraj S, Simon I. "Locating the stabilizing residues in (alpha/beta) 8 barrel proteins based on hydrophobicity, long-range interactions, and sequence conservation." *Proteins: Structure, Function and Bioinformatics.* 2004;55:316–329.

Academic Press is an imprint of Elsevier,
A Division of Reed Elsevier India Pvt. Ltd.
Registered Office: Gate No. 3, Building No. A-1, 2 Industrial Area, Kalkaji,
New Delhi – 110019

Library of Congress Cataloging-in-Publication Data
Gromiha, M. Michael.
 Protein bioinformatics : from sequence to function.
 1. Proteins–Structure–Data processing.   2. Proteins–
 Metabolism–Data processing.   3. Bioinformatics.
 572.6'0285-dc22

British Library Cataloguing in Publication Data
A catalogue record for this book is available from the British Library

ISBN-13: 978-81-312-2297-3

Printed and bound by CPI Group (UK) Ltd, Croydon, CR0 4YY
Transferred to digital print 2013

Published by Elsevier, A Division of Reed Elsevier India Pvt. Ltd.
10 11 12   5 4 3 2 1

Dedicated to the memory of my beloved father

Contents

Foreword

Bioinformatics is a broad field of research that escapes easy definition. As in all natural sciences, progress in the life sciences depends on the interplay of empirical research and data interpretation (theory building), as well as modeling (validation). Because of the vast improvements in data collection, we have to deal with very large numbers of highly complex empirical data—in particular molecular and genetic data—while our ability to understand the data in terms of new concepts and theoretical frameworks often lags behind. The gap between data and scientific understanding cannot be bridged without computational tools, and this is where bioinformatics comes into the picture. Since all steps of data collection, preprocessing, storage, and analysis have their particular computational tools, novices trying to understand "bioinformatics" in general feel inevitably lost in the jungle of often loosely defined terms and concepts.

Choosing protein bioinformatics as the subject of this book is highly justified for two particular reasons. First, protein research is a well-defined subfield both within the molecular sciences and within the industrial sector where protein products (pharmaceuticals, industrial enzymes, etc.) play dominant roles. Second, while genome biologists deal with large amounts of lightly annotated data, protein researchers often need to deal with a few, very well-characterized molecules, using a variety of methods. The book seeks to offer an introduction of these methods and tools.

The purpose of the book is to provide a resource to help researchers and students to understand the structure, function, folding, stability, and interactions of proteins through bioinformatics approaches. The book is meant for students and researchers who need a first-time introduction to protein bioinformatics and is meant to serve as a textbook. In addition, it will be an excellent reference book with up-to-date information for protein researchers who want to apply computational tools in their research projects.

The book is divided into seven chapters discussing protein fundamentals, sequence analysis, structure analysis, structure prediction, folding, stability, and protein interactions. Each chapter presents an overview of the main problem-solving approaches and tools. All chapters provide an overview of the currently available on-line tools, and many of them are illustrated with examples. The methods include both basic and advanced methodologies, which help the reader become familiar with the manifold approaches that characterize this varied and interdisciplinary field. At the end of each chapter, there is a list of exercises and problems that can be directly used in teaching courses.

Bioinformatics is a fast-evolving field. The author, Michael Gromiha, chose a well-defined target within this changing scenery and made an excellent summary of the computational tools of protein bioinformatics as we see it today. The approach is unique in the sense that it covers all aspects of proteins and their interactions in a coherent manner, and it provides a concise but comprehensive overview of the pertinent databases and Web servers. Students and researchers in protein research, bioinformatics, biophysics, computational biology, molecular modeling, and drug design will find this easy-to-understand book a ready reference for staying current and productive in this changing, interdisciplinary field.

It is with these thoughts that I recommend this well-written book to the reader.

Sándor Pongor
International Centre for Genetic engineering and Biotechnology
Trieste, Italy

Preface

Proteins are one of the most versatile molecules in living organisms. The structural and functional roles of proteins have been investigated through several approaches and different points of view. It ranges from the experimental studies such as the characterization of proteins using biochemical and molecular biological aspects, determination of protein structures, understanding the stability of proteins at extreme conditions and mutations, and designing drugs using the concept of protein–ligand interactions to numerous computational techniques for the analysis of protein sequences and structures, folding behavior, stabilizing interactions, prediction of protein structures, and docking studies.

The recent advances in information technology have enhanced the computational power as well as information exchange with rapid progress. This changed the trend of developing simple methods to complicated machine learning techniques to attain the highest accuracy between the data obtained from experiments and computational models. The availability of high-speed computers with extreme storage capacity has reduced the gap between experiment and theory.

The development of databases on various aspects and tools for different types of analysis, and the availability of both of them on the Web considerably reduced the work of experimentalists to search for the suitable sample and opened a way to the bioinformatists to deepen the study. Based on the variety of investigations on proteins and their importance on clinical aspects, it is necessary to compile most of the bioinformatics-related works on proteins, which will be useful to both the experimentalists as well as the bioinformatics/computational biology researchers. This book will be of immense use and a valuable guide to the students and researchers working on proteins and to those who have the interest to work on proteins. Furthermore, it provides a comprehensive survey of literature and applications to the students who are studying/interested on bioinformatics.

The main feature of the book is the coverage of all aspects of proteins, such as sequence and structure analysis, prediction of protein structures, protein folding, protein stability, and protein interactions. All the sections have the illustrations and Web addresses of available databases and online tools, and the details to construct the data and derive the important parameters. In addition, the literature has been thoroughly surveyed from the beginning, and important methods have been highlighted in detail.

Chapter 1 provides the general introduction of proteins, structural organizations, databases for protein sequences and structures, and their applications. This will give an overall idea about the availability of public domain resources to the students and researchers.

Chapter 2 illustrates the works that can be done with protein sequences. This includes the alignment of sequences, position specific scoring matrices, and the delineation of several amino acid properties based on just sequence. Furthermore, the available servers for analyzing protein sequences have been discussed.

Chapter 3 demonstrates the plenty of analyses carried out with known three-dimensional structures of proteins. It includes the assignment of secondary structures, computation of solvent accessibility to obtain the information about the residues that are in the interior of the protein and at the surface, contacts between amino acid residues in protein structures, development of contact potentials, clusters of amino acid residues in folded environment, and free energy calculations. In addition, the comparison of protein structures using different methods has been outlined with illustrated examples.

Chapter 4 deals with protein folding kinetics with the introduction of Φ-value analysis and the relationship between Φ values and amino acid properties. In addition, the importance of protein folding rates and the parameters that are influencing the folding rates of proteins have been surveyed. Furthermore, the relationship between Φ values and folding rates has been outlined.

Chapter 5 focuses on the prediction of protein structures, that includes the discrimination of different structural classes of globular and membrane proteins, secondary structure content, secondary structural regions, contacts between amino acid residues, solvent accessibility, and the three-dimensional structures of proteins.

Chapter 6 is devoted to protein stability beginning with the determination of protein stability using experimental techniques and the availability of databases for protein stability. The relative contributions of noncovalent interactions to protein stability and the factors influencing the stability of thermophilic proteins have been surveyed. The development of different methods for discriminating the stabilizing and destabilizing effect of proteins upon amino acid substitutions and predicting the stability change upon mutations has been elaborated.

Chapter 7 enlightens the importance of protein interactions, such as protein–protein, protein–nucleic acid, and protein–ligand interactions. The availability of databases for the binding of proteins with other molecules has been listed, and the analysis carried out to extract the principles governing the interactions has been included. Furthermore, the development of Web servers for predicting the binding sites in protein and their binding affinities has been explained.

Appendix A includes the collective lists of databases and Web servers, which are important in the field of protein bioinformatics. The exercises given in each chapter will be an added value to the book. In essence, this book would be a valuable resource to students and researchers to deepen the knowledge about the studies on proteins.

M. Michael Gromiha

Acknowledgments

I am deeply indebted to Professor P.K. Ponnuswamy, who introduced me to the field of proteins, and encouraged to deepen the knowledge in protein research.

I am grateful to Dr. S. Pongor, Dr. A. Sarai, Dr. Y. Akiyama, Dr. R. Majumdar, Dr. M. Lakshmanan, and Dr. K. Asai for their continuous support and advice.

My sincere thanks to my collaborators Dr. S. Ahmad, Dr. M. Babu, Dr. K. Fukui, Dr. L-T. Huang, Dr. D. Jacobs, Dr. H. Kono, Dr. S. Kumar, Dr. T.S. Kumarevel, Dr. C. Magyar, Dr. M. Oobatake, Dr. Y-Y. Ou, Dr. D.A.D. Parry, Dr. V. Parthiban, Dr. M.N. Ponnuswamy, Dr. G. Pujadas, Dr. G.P.S. Raghava, Dr. S. Selvaraj, Dr. I. Simon, Dr. M. Suwa, Dr. Y-h. Taguchi, Dr. H. Uedaira, and Dr. Y. Yabuki, who helped me in several ways at different stages of the work.

I extend my gratitude to Dr. Dr. P. Chakrabarti, Dr. E. Gaucher, Dr. I. Ghosh, Dr. E. Ortlund, Dr. Pal, Dr. K. Veluraja, and Dr. D. Velmurugan as well as my collaborators for their comments on various chapters in the book.

It is my pleasure to acknowledge my wife A. Mary Thangakani, son's Michael Mozim, and Michael Abejo, and daughter Angela Shalom, who have spared their committed time and allowed me to work on the book. I also immensely acknowledge my mother for her valuable guidance.

Finally, I thank all my well wishers and friends who encouraged me to write this book.

CHAPTER 1

Proteins

Proteins perform a variety of functions, including enzymatic catalysis, transporting ions and molecules from one organ to another, nutrients, contractile system of muscles, tendons, cartilage, antibodies, and regulating cellular and physiological activities. The functional properties of proteins depend on their three-dimensional structures. The native structure of a protein can be experimentally determined using X-ray crystallography, nuclear magnetic resonance (NMR) spectroscopy, electron microscopy, etc. Over the past 40 years, the structures of more than 53,000 proteins (as of May 12, 2009) have been determined. On the other hand, the amino acid sequences are determined for more than eight million proteins (as of May 5, 2009). The specific sequence of amino acids in a polypeptide chain folds to generate compact domains with a particular three-dimensional structure. Anfinsen (1973) stated that the polypeptide chain itself contains all the information necessary to specify its three-dimensional structure. Deciphering the three-dimensional structure of a protein from its amino acid sequence is a long-standing goal in molecular and computational biology.

1.1 Building blocks

Protein sequences consist of 20 different kinds of chemical compounds, known as amino acids, and they serve as building blocks of proteins. Amino acids contain a central carbon atom (C_α), which is attached to a hydrogen atom, an amino group (NH_2), and a carboxyl group (COOH) as shown in **Figure 1.1**. The letter R in **Figure 1.1** indicates the presence of a side chain, which distinguishes each amino acid.

1.1.1 Amino acids

Amino acids are naturally of 20 different types as specified by the genetic code emerged from DNA sequences. Furthermore, nonnatural amino acids occur, in rare cases, as the products of enzymatic modifications after translocation. The major difference among the 20 amino acids is the *side chain* attached to the C_α through its fourth valance. The variation of side chains in 20 amino acids is shown in **Figure 1.2**. These residues are represented by conventional three- and one-letter codes. Most of the databases use single-letter codes.

The amino acids are broadly divided into two groups, hydrophobic and hydrophilic, based on the tendency of their interactions in the presence of water

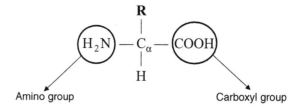

Figure 1.1   Representation of amino acids. R is the side chain that varies for the 20 amino acids.

molecule. The hydrophobic residues have the tendency of adhering to one another in aqueous environment. Generally, amino acids, Ala (A), Cys (C), Phe (F), Gly (G), Ile (I), Leu (L), Met (M), Val (V), Trp (W), and Tyr (Y), are considered as hydrophobic residues. In this category, Ala, Ile, Leu, and Val contain aliphatic side chains; Phe, Trp, and Tyr contain aromatic side chains; and Cys and Met contain sulfur atom. Gly has no side chain, and it has hydrogen (H) at the fourth position. Two Cys residues in different parts of the polypeptide chain but adjacent to each other in the three-dimensional structure of a protein can be oxidized to form a disulfide bridge. The

Hydrophobic residues

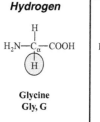

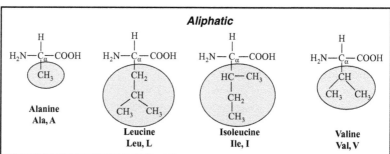

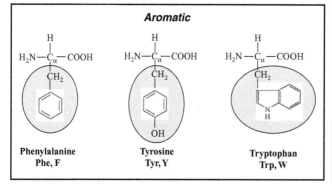

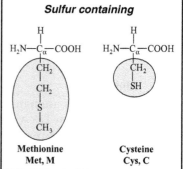

Figure 1.2   The common 20 amino acids in proteins. The three- and one-letter codes for the amino acids are also given. The amino acids are classified into hydrophobic (hydrogen, aliphatic, aromatic, and sulfur containing) and hydrophilic (negatively charged, positively charged, and polar). The side chains are marked with oval boxes.

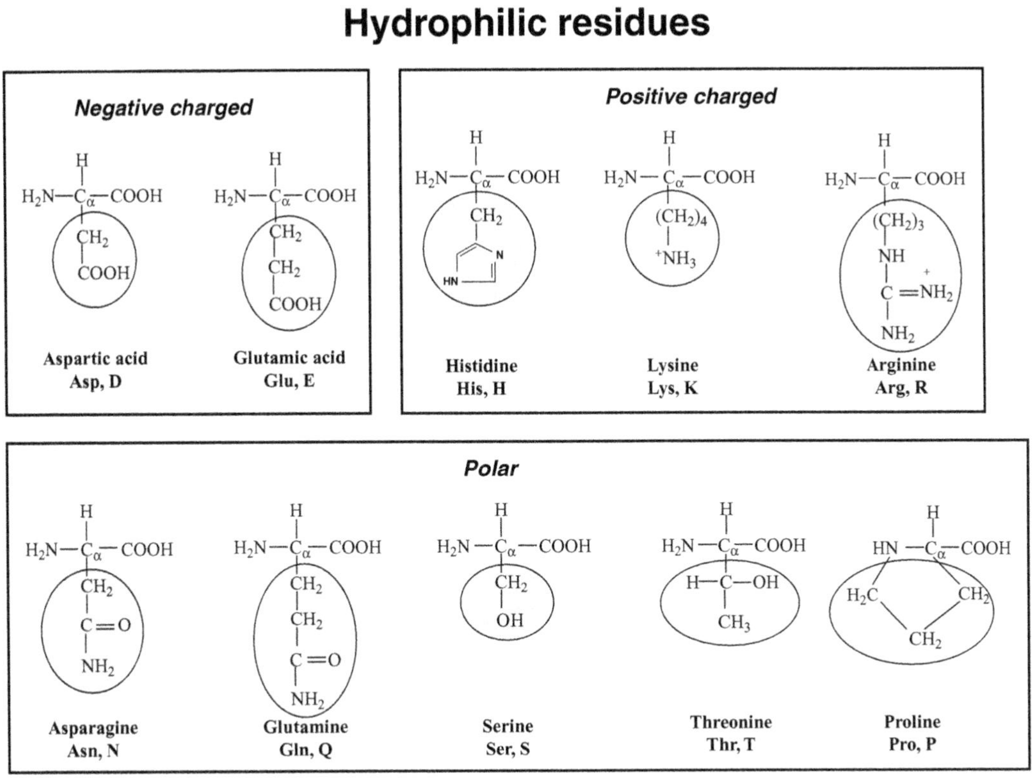

Hydrophilic residues

Negative charged

Aspartic acid
Asp, D

Glutamic acid
Glu, E

Positive charged

Histidine
His, H

Lysine
Lys, K

Arginine
Arg, R

Polar

Asparagine
Asn, N

Glutamine
Gln, Q

Serine
Ser, S

Threonine
Thr, T

Proline
Pro, P

FIGURE 1.2 (Continued)

formation of disulfide bridges in protein structures stabilizes the protein, making it less susceptible to degradation.

Amino acids, Asp (D), Glu (E), His (H), Lys (K), Asn (N), Pro (P), Gln (Q), Arg (R), Ser (S), and Thr (T), are classified as hydrophilic residues. In this category, Asp and Glu are negatively charged; His, Lys, and Arg are positively charged; and others are polar and uncharged.

1.1.2 Formation of peptide bonds

The carboxyl group of one amino acid interacts with the amino group of another to form a peptide bond by the elimination of water (**Figure 1.3**). Amino acids are joined end-to-end during protein synthesis by the formation of such peptide bonds. The peptide bond (C—N) has a partial double-bond character due to resonance, and hence there is no rotation about the peptide bond. In **Figure 1.3**, the peptide is represented as a planar unit with the C=O and N—H groups positioning in opposite directions in the plane. This is called *trans*-peptide. There is another form, *cis*-peptide in which the C=O and N—H groups point in the same direction. To avoid steric hindrance, the *trans* form is frequently presented in protein structures for all amino acids except Pro, which has both *trans* and *cis* forms. The *cis* prolines are found in bends of the polypeptide chains.

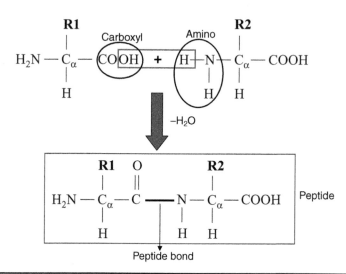

FIGURE **1.3**   Formation of a peptide bond by the elimination of a water molecule.

A protein chain is formed by several amino acids in which the amino group of the first amino acid and the carboxyl group of the last amino acid remain intact, and the chain is said to extend from the amino (N) to the carboxyl (C) terminus. This chain of amino acids is called a polypeptide chain, main chain, or backbone. Amino acids in a polypeptide chain lack a hydrogen atom at the amino terminal and an OH group at the carboxyl terminal (except at the ends), and hence amino acids are also called **amino acid residues** (simply residues). Nature selects the combination of amino acid residues to form polypeptide chains for their function, similar to the combination of alphabets to form meaningful words and sentences. These polypeptide chains that have specific functions are called **proteins**.

1.2   Hierarchical representation of proteins

Depending on their complexity, protein molecules may be described by four levels of structure (Nelson and Cox, 2005): primary, secondary, tertiary, and quaternary (**Figure 1.4**). Because of the advancements in the understanding of protein structures, two additional levels such as supersecondary and domain have been proposed between secondary and tertiary structures. A stable clustering of several elements of secondary structures is referred to as a supersecondary structure. A somewhat higher level of structure is the domain, which refers to a compact region and distinct structural unit within a large polypeptide chain.

1.2.1   Primary structure

Primary structure describes the linear sequence of amino acid residues in a protein. It includes all the covalent bonds between amino acids. The relative spatial arrangement of the linked amino acids is unspecified.

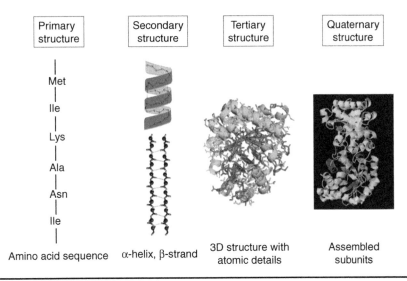

FIGURE **1.4**   Structural organization of proteins.

1.2.2   Secondary structure

Secondary structure refers to regular, recurring arrangements in space of adjacent amino acid residues in a polypeptide chain. It is maintained by hydrogen bonds between amide hydrogens and carbonyl oxygens of the peptide backbone. The major secondary structures are α-helices and β-structures.

The α-helical conformation was first proposed by Linus Pauling and co-workers (1951), and a typical α-helix is shown in **Figure 1.5**. In this structure, the polypeptide backbone is tightly wound around the long axis of the molecule, and R groups of the amino acid residues protrude outward from the helical backbone. The repeating

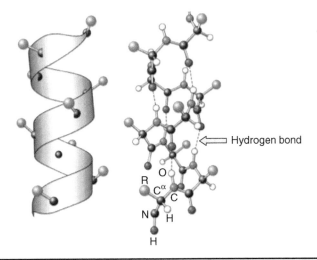

FIGURE **1.5**   Structure of a typical α-helix. The hydrogen bonds between the residues n and $n + 4$ are shown as dotted lines. Figure was taken as a screenshot from the Web, http://www.food-info. net/uk/protein/structure.htm

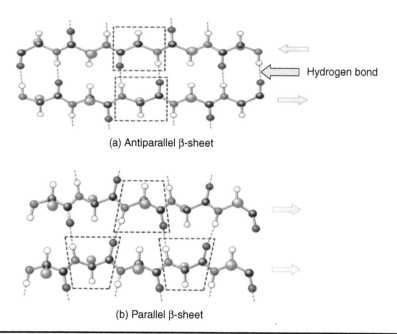

(a) Antiparallel β-sheet

(b) Parallel β-sheet

Figure 1.6   Structures of (a) antiparallel and (b) parallel β-sheets. The dotted lines show the hydrogen bonds between amino acid residues. The arrows indicate the directions of the polypeptide chain, from N- to C-terminal. Figure was taken as a screenshot from the Web, http://www.food-info.net/uk/protein/structure.htm.

unit is a single turn of a helix, which extends about 0.54 nm along the axis, and the number of amino acid residues required for one complete turn is 3.6. In an α-helix, each carbonyl oxygen (residue, n) of the polypeptide backbone is hydrogen bonded to the backbone amide hydrogen of the fourth residue further toward the C-terminus (residue, $n + 4$). The hydrogen bonds, which stabilize the helix, are nearly parallel to the long axis of the helix.

The other common secondary structure is β-structure that includes β-strands and β-sheets. β-strands are portions of the polypeptide chain that are almost fully extended, and several β-strands constitute β-sheets. β-sheets are stabilized by hydrogen bonds between carbonyl oxygens and amide hydrogens on adjacent β-strands (**Figure 1.6**). In β-sheets, the hydrogen bonds are nearly perpendicular to the extended polypeptide chains. The β-strands may be either parallel (running in the same N- to C-terminal) or antiparrallel (running in opposite N- to C-terminal directions).

In a polypeptide chain, the α-carbon atoms of adjacent amino acids are separated by three covalent bonds arranged as C_α–C–N–C_α. In these bonds, rotation is permitted about the N–C_α and C_α–C bonds, and the torsional angles are conventionally denoted as φ and ψ, respectively. Every secondary structure is described completely by these two torsional angles that are repeated at each residue. The allowed values for φ and ψ can be shown graphically by simply plotting these values known as Ramachandran plot (Ramachandran et al. 1963). **Figure 1.7** shows the conformations that are permitted for most amino acid residues in Ramachandran plot.

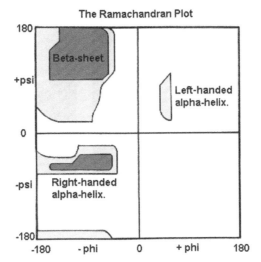

FIGURE 1.7   Ramachandran plot showing the allowed regions of α-helical and β-strand conformations. Figure was taken as a screenshot from the Web, http://swissmodel.expasy.org/course/text/chapter1.htm.

1.2.3   Tertiary structure

Tertiary structure refers to the spatial relationship among all amino acids in a polypeptide; it is the complete three-dimensional structure of the polypeptide with atomic details. Tertiary structures are stabilized by interactions of side chains of nonneighboring amino acid residues and primarily by noncovalent interactions. The formation of tertiary structure brings the amino acid residues that are far apart in the primary structure close together.

1.2.4   Quaternary structure

Quaternary structure refers to the spatial relationship of the polypeptides or subunits within the protein. It is the association of two or more polypeptide chains into a multisubunit or oligomeric protein. The polypeptide chains of an oligomeric protein may be identical or different. The quaternary structure also includes the cofactor and other metals, which form the catalytic unit and functional proteins.

1.3   Structural classification of proteins

Proteins are broadly classified into two major groups: fibrous proteins, having polypeptide chains arranged in long strands, and globular proteins, with polypeptide chains folded into a spherical or globular shape.

1.3.1   Fibrous proteins

Fibrous proteins are usually static molecules and play important structural roles in the anatomy and physiology of vertebrates, providing external protection, support, shape, and form. They are water insoluble and are typically built upon a single, repetitive structure assembled into cables or threads. Examples of fibrous proteins

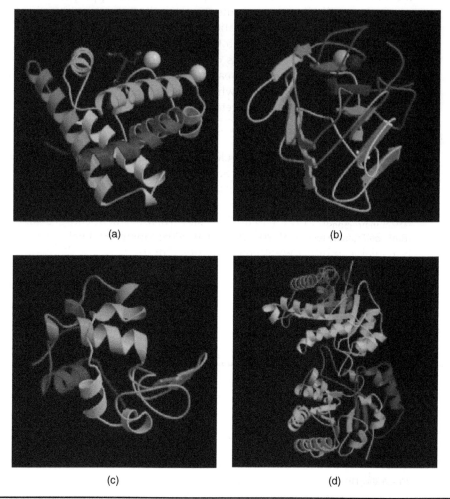

(a)   (b)

(c)   (d)

FIGURE **1.8**   Ribbon diagram for four typical protein structures in different structural classes (a) all-α (4MBN), (b) all-β (3CNA), (c) α + β (4LYZ), and (d) α/β (1TIM). Figure was adapted from Gromiha and Selvaraj (2004).

are α-keratin, the major component of hair and nails, and collagen, the major protein component of tendons, skin, bones, and teeth.

1.3.2   Classification of globular proteins

Globular proteins are categorized into four structural classes: all-α, all-β, α + β, and α/β (Levitt and Chothia, 1976). The ribbon diagrams illustrating the structures in each class are shown in **Figure 1.8**.

The all-α and all-β classes are dominated by α-helices (α > 40% and β < 5%) and by β-strands (β > 40% and α < 5%), respectively (**Figures 1.8a** and **b**). The α + β class contains both α-helices (>15%) and antiparallel β-strands (>10%) that do not mix but tend to segregate along the polypeptide chain (**Figure 1.8c**). The α/β class proteins (**Figure 1.8d**) have mixed or approximately alternating segments of α-helical (>15%) and parallel β-strands (>10%).

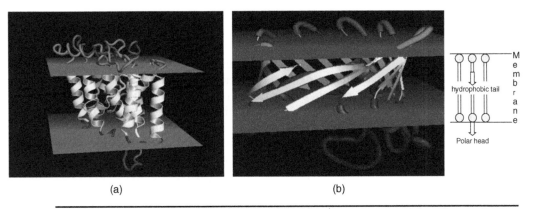

(a)   (b)

FIGURE 1.9   Representation of (a) α-helical and (b) β-barrel membrane proteins. The membrane spanning regions are shown within the disc. Protein structures were taken from Protein Data Bank of Transmembrane Proteins (http://pdbtm.enzim.hu/).

1.3.3   Membrane proteins

Membrane proteins, which require embedding into the lipid bilayers, have evolved to have amino acid sequences that will fold with a hydrophobic surface in contact with the alkane chains of the lipids and polar surface in contact with the aqueous phases on both sides of the membrane and the polar head groups of the lipids (**Figure 1.9**). In genomes, 30% of the proteins are suggested to be membrane proteins, and most of the transmembrane helical and strand proteins are identified as targets for drug design. Membrane proteins perform a variety of functions, including cell–cell signaling and mediating the transport of ions and solutes across the membrane. They are of two kinds: (i) transmembrane helical proteins in which they span the cytoplasmic membrane with α-helices (White and Wimley, 1999) and (ii) transmembrane β-barrel proteins that traverse the outer membranes of gram-negative bacteria with β-strands (Schulz, 2003). **Figure 1.9** shows the structures of membrane proteins with these two different motifs, α-helices and β-strands.

1.4   Databases for protein sequences

Recombinant DNA techniques have provided tools for the rapid determination of DNA sequences and, by inference, the amino acid sequences of proteins from structural genes. The number of such sequences is increasing exponentially, and these sequences have been deposited in the form of database, generally, known as protein sequence databases. Specifically, Georgetown University, Washington, D.C., USA, developed the database, Protein Information Resource (PIR). The Swiss Institute of Bioinformatics and European Bioinformatics Institute developed SWISS-PROT and TrEMBL databases. Recently, progress has been made to set up a single worldwide database of protein sequence and function, UniProt, by unifying PIR, SWISS-PROT, and TrEMBL database activities.

1.4.1   Protein Information Resource

PIR has evolved from the Atlas of Protein Sequence and Structure established in the early 1960s by Margaret O. Dayhoff (Dayhoff et al. 1965). It produces the largest, most comprehensive, annotated protein sequence database in the public domain, the PIR International Protein Sequence Database, in collaboration with the Munich Information Center for Protein Sequences and the Japan International Protein Sequence Database (Barker et al. 2000). It is freely available at http://pir.georgetown.edu/. PIR offers a wide variety of resources mainly oriented to assist the propagation and standardization of protein annotation on three major aspects: (i) PIRSF, protein family classification system; (ii) iProClass, integrated protein knowledgebase; and (iii) iProLink, literature, information, and knowledge. The iProClass database provides value-added information reports on protein sequences, structures, families, functions, interactions, expressions, and modifications. The sequence information of a specific protein can be searched with simple "Text search" in iProClass **(Figure 1.10a)**. The search yielded 10 proteins, and the correct one has been selected with a click on the left-side box. It is also possible to save the results as a table or in FASTA format. The result obtained for the search with "Human lysozyme" is shown in **Figure 1.10b**. It includes general information

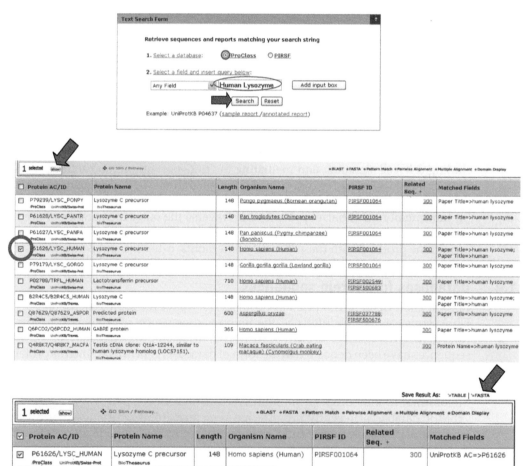

(a)

FIGURE 1.10   Text search in iProClass of PIR database: (a) the search with "Human lysozyme," along with intermediate steps, and (b) the information provided at the result page are shown.

/ProClass Summary Report for UniProtKB Entry: P61626

Related Sequences   *BioThesaurus*   *ID Mapping*

GENERAL INFORMATION			
Protein Name and ID	UniProtKB ID	UniProtKB Accession	Protein Name
	LYSC_HUMAN	P61626; P00695; Q13170; Q9UCF8	Lysozyme C precursor
	PIR-PSD: LZHU RefSeq: NP_000230.1 GenPept: AAA59535.1 ; AAC63078.1 ; EAW97222.1 ; AAA59536.1 ; CAA32175.1 ; AAH04147.1 ; EAW97221.1 ; AAA36188.1 IPI: IPI00019038		
Taxonomy	*Source Organism:* Homo sapiens (Human) *Taxon Group:* Euk/mammal *NCBI Taxon:* 9606 *Lineage:* Eukaryota; Metazoa; Chordata; Craniata; Vertebrata; Euteleostomi; Mammalia; Eutheria; Euarchontoglires; Primates; Haplorrhini; Catarrhini; Hominidae; Homo.		
Gene Name	LYZ; LZM		
Keywords	3d-structure; amyloid; amyloidosis; antimicrobial; bacteriolytic enzyme; complete proteome; direct protein sequencing; disease mutation; disulfide bond; glycosidase; hydrolase; polymorphism; polysaccharide degradation; signal		
Function	Lysozymes have primarily a bacteriolytic function; those in tissues and body fluids are associated with the monocyte-macrophage system and enhance the activity of immunoagents.		
Subunit	Monomer.		

CROSS-REFERENCES	
Bibliography	►View Bibliography Information ►Submit Bibliography *Annotated references:* PMID: 8105095; 10350481; 10469827; 10561612; 11887182; 11927576; 11986950 [PDB/GeneRIF] More *Other references:* PMID: 11849445; 12675840; 15745733; 8765309; 9659355; 9745729; 18391951; 9359845; 8566845; 17353931; 9883972; 366724; 10534505; 12477932; 10558865; 18591461
DNA Sequence	GenBank/EMBL/DDBJ: M21119; J03801; M19045; X14008; U25677; BC004147
Structure	1B5V: SCOP CATH FSSP MMDB PDBsum 1B5W: SCOP CATH FSSP MMDB PDBsum 1B5X: SCOP CATH FSSP MMDB PDBsum 1B5Y: SCOP CATH FSSP MMDB PDBsum 1B5Z: SCOP CATH FSSP MMDB PDBsum 1B7L: SCOP CATH FSSP MMDB PDBsum 1B7M: SCOP CATH FSSP MMDB PDBsum 1B7N: SCOP CATH FSSP MMDB PDBsum 1B7O: SCOP CATH FSSP MMDB PDBsum 1B7P: SCOP CATH FSSP MMDB PDBsum 1B7Q: SCOP CATH FSSP MMDB PDBsum 1B7R: SCOP CATH FSSP MMDB PDBsum 1B7S: SCOP CATH FSSP MMDB PDBsum 1BB3: SCOP CATH FSSP MMDB PDBsum 1BB4: SCOP CATH FSSP MMDB PDBsum More
PIR Feature & Post Translational Modifications	FEAT1; active site: Glu, Asp (53,71) [predicted] FEAT2; binding site: substrate (Asp) (120) [predicted] FEAT3; disulfide bonds: (24-146,48-134,83-99,95-113) [experimental] FEAT4; domain: signal sequence (1 18) [predicted] FEAT5; product: lysozyme (19-148) [experimental] Phosphosite: P61626

FAMILY CLASSIFICATION	
UniRef	UniRef100_P61626; UniRef90_P61626; UniRef50_P61626
PIRSF	PIRSF001064 lysozyme c
Pfam Domain	Pfam: PF00062: C-type lysozyme/alpha-lactalbumin family (19-146)
Prosite Motif	Prosite: PS00128; PDOC00119: Alpha-lactalbumin / lysozyme C signature. Prosite: PS51348; PDOC00119: Alpha-lactalbumin / lysozyme C family profile.
InterPro	InterPro: LYSC_HUMAN IPR001916: Glycoside hydrolase, family 22 IPR019799: Glycoside hydrolase, family 22, conserved site IPR000974: Glycoside hydrolase, family 22, lysozyme
SCOP Fold	►Class: Alpha and beta proteins (a+b); *Fold:* Lysozyme-like; *Superfamily:* Lysozyme-like; *Family:* C-type lysozyme [133L:A; 134L:A; 1B5U:A; 1B5V:A; 1B5W:A; 1B5X:A; 1B5Y:A; 1B5Z:A; 1B5Z:B; 1B7L:A; 1B7M:A; 1B7N:A; 1B7O:A; 1B7P:A; 1B7Q:A; 1B7R:A; 1B7S:A; 1BB3:A; 1BB3:B; 1BB4:A; 1BB4:A; 1BB5:A; 1BB5:B; 1C43:A; 1C45:A; 1C46:A; 1C7P:A; 1C3S:A; 1CJ7:A; 1CJ8:A; 1CJ9:A; 1CKC:A; 1CKD:A; 1CKF:A; 1CKG:A; 1CKG:B; 1CKH:A; 1D6P:A; 1D6Q:A; 1D13:A; 1D14:A; 1D15:A; 1EQ4:A; 1EQ5:A; 1EQE:A; 1GAY:A; 13AZ:A; 1OB0:A; 1GB2:A; More]
Other Classification	BLOCKS: IP6000974 Lysozyme signature PRINTS: PR00137 LYSOZYME PRINTS: PR00135 LYZLACT SMART: SM00263 LYZ1 HomoloGene: 37278

FEATURE & SEQUENCE DISPLAY	

Length = 148   Click on a bar to show its sequence; to copy and paste it, press ctlr then ctlr v.

```
            1    MKALIVLGLVLLSVTVQGKVFERCELARTLKRLGMDGYRGISLANWMCLAKWESGYNTRA
           61    TNYNAGDRSTDYGIFQINSRYWCNDGKTPGAVNACHLSCSALLQDNIADAVACAKRVVRD
          121    PQGIRAWVAWRNRCQNRDVRQYVQGCGV
```

(b)

FIGURE 1.10   *(Continued)*

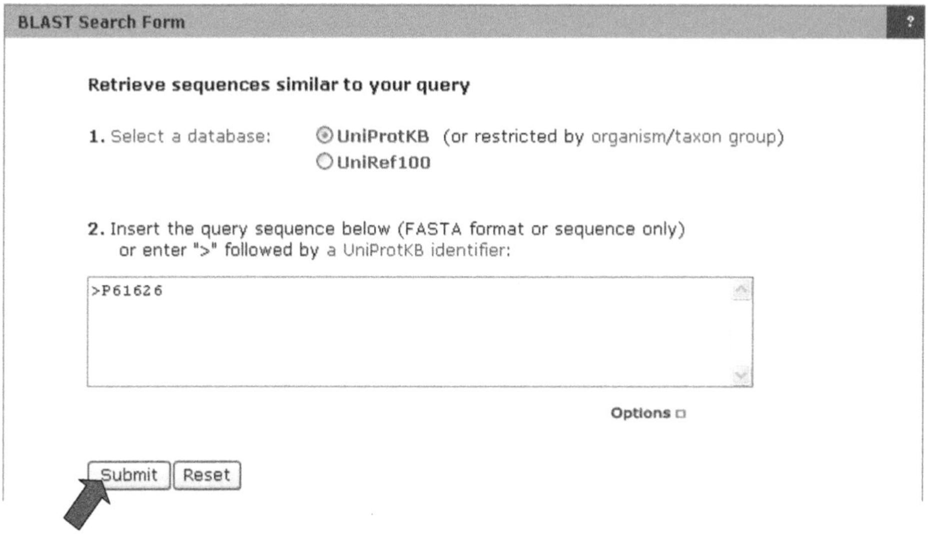

FIGURE **1.11**   Utility of similar search option available in PIR. UniprotKB identifier for human lysozyme, ">P61626" is given as input.

(protein name, taxonomy, gene name, keywords, function, and subunit), cross-references (bibliography, DNA sequence, genome, ontology, function, interaction, structure, and posttranslational modifications), family classification, and feature and sequence display.

It has several features such as similarity search using BLAST and FASTA, peptide match, pattern search, pairwise sequence alignment, and multiple sequence alignment. The similarity search of human lysozyme against UniProtKB (UniProt knowledgebase) using the alignment program BLAST is shown in **Figure 1.11**. It can also be searched using the program FASTA. The partial results obtained with the search option are depicted in **Figure 1.12**. It indicates the sequences and their codes that match the query sequence along with other details, protein name, organism, length, % identity, overlap, e-value, etc (**Figure 1.12a**). Furthermore, it shows the alignment details with other proteins (**Figure 1.12b**). This will be helpful to identify the homologous sequences of any query protein. PIR can also be searched for any specific patterns, for example, alternating hydrophilic and hydrophobic residues as a pattern for β-strands (see **Chapter 2**), and continuous stretches of hydrophobic residues (e.g., AVILLIVWFFGA) in transmembrane helical proteins, etc.

1.4.2   SWISS-PROT and TrEMBL

SWISS-PROT (Bairoch and Apweiler, 1996) is an annotated protein sequence database established in 1986 and maintained collaboratively, since 1987, by the Department of Medical Biochemistry of the University of Geneva and the EMBL Data Library. It is a curated protein sequence database, which strives to provide a high level of annotation (such as the description of the function of a protein, its domain structure, posttranslational modifications and variants), a minimal level of redundancy, and a high level of integration with other databases. TrEMBL is a

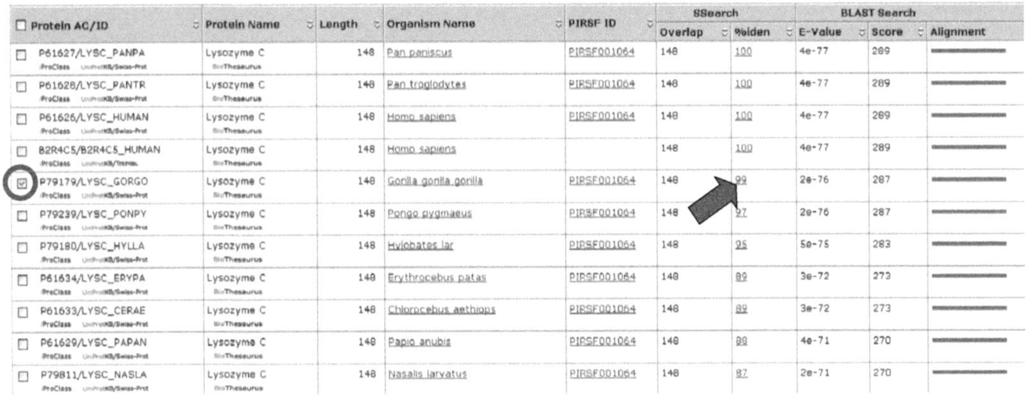

☐ Protein AC/ID	Protein Name	Length	Organism Name	PIRSF ID	SSearch		BLAST Search		
					Overlap	%Iden	E-Value	Score	Alignment
☐ P61627/LYSC_PANPA *ProClass* *UniProtKB/Swiss-Prot*	Lysozyme C *BioThesaurus*	148	Pan paniscus	PIRSF001064	148	100	4e-77	289	———
☐ P61626/LYSC_PANTR *ProClass* *UniProtKB/Swiss-Prot*	Lysozyme C *BioThesaurus*	148	Pan troglodytes	PIRSF001064	148	100	4e-77	289	———
☐ P61626/LYSC_HUMAN *ProClass* *UniProtKB/Swiss-Prot*	Lysozyme C *BioThesaurus*	148	Homo sapiens	PIRSF001064	148	100	4e-77	289	———
☐ B2R4C5/B2R4C5_HUMAN *ProClass* *UniProtKB/TrEMBL*	Lysozyme C *BioThesaurus*	148	Homo sapiens		148	100	4e-77	289	———
☑ P79179/LYSC_GORGO *ProClass* *UniProtKB/Swiss-Prot*	Lysozyme C *BioThesaurus*	148	Gorilla gorilla gorilla	PIRSF001064	148	99	2e-76	287	———
☐ P79239/LYSC_PONPY *ProClass* *UniProtKB/Swiss-Prot*	Lysozyme C *BioThesaurus*	148	Pongo pygmaeus	PIRSF001064	148	97	2e-76	287	———
☐ P79180/LYSC_HYLLA *ProClass* *UniProtKB/Swiss-Prot*	Lysozyme C *BioThesaurus*	148	Hylobates lar	PIRSF001064	148	95	5e-75	283	———
☐ P61634/LYSC_ERYPA *ProClass* *UniProtKB/Swiss-Prot*	Lysozyme C *BioThesaurus*	148	Erythrocebus patas	PIRSF001064	148	89	3e-72	273	———
☐ P61633/LYSC_CERAE *ProClass* *UniProtKB/Swiss-Prot*	Lysozyme C *BioThesaurus*	148	Chlorocebus aethiops	PIRSF001064	148	89	3e-72	273	———
☐ P61629/LYSC_PAPAN *ProClass* *UniProtKB/Swiss-Prot*	Lysozyme C *BioThesaurus*	148	Papio anubis	PIRSF001064	148	88	4e-71	270	———
☐ P79811/LYSC_NASLA *ProClass* *UniProtKB/Swiss-Prot*	Lysozyme C *BioThesaurus*	148	Nasalis larvatus	PIRSF001064	148	87	2e-71	270	———

(a)

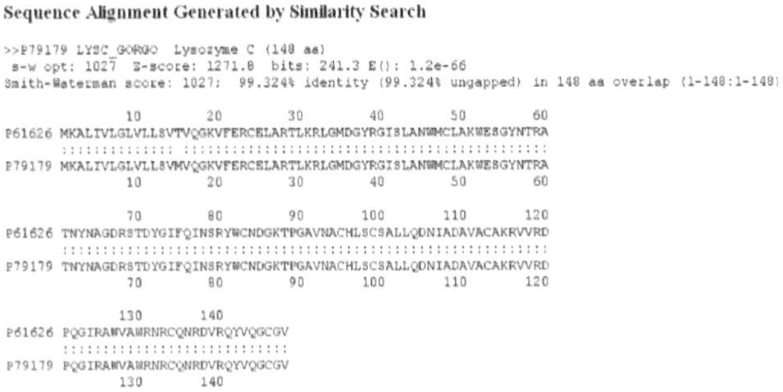

Sequence Alignment Generated by Similarity Search

```
>>P79179 LYSC_GORGO  Lysozyme C (148 aa)
 s-w opt: 1027  Z-score: 1271.8  bits: 241.3 E(): 1.2e-66
Smith-Waterman score: 1027;  99.324% identity (99.324% ungapped)  in 148 aa overlap  (1-148:1-148)

              10        20        30        40        50        60
P61626  MKALIVLGLVLLSVTVQGKVFERCELARTLKRLGMDGYRGISLANWMCLAKWESGYNTRA
        ::::::::::::: ::::::::::::::::::::::::::::::::::::::::::::::::
P79179  MKALIVLGLVLLSVMVQGKVFERCELARTLKRLGMDGYRGISLANWMCLAKWESGYNTRA
              10        20        30        40        50        60

              70        80        90       100       110       120
P61626  TNYNAGDRSTDYGIFQINSRYWCNDGKTPGAVNACHLSCSALLQDNIADAVACAKRVVRD
        ::::::::::::::::::::::::::::::::::::::::::::::::::::::::::::::
P79179  TNYNAGDRSTDYGIFQINSRYWCNDGKTPGAVNACHLSCSALLQDNIADAVACAKRVVRD
              70        80        90       100       110       120

             130       140
P61626  PQGIRAWVAWRNRCQNRDVRQYVQGCGV
        ::::::::::::::::::::::::::::::
P79179  PQGIRAWVAWRNRCQNRDVRQYVQGCGV
             130       140
```

(b)

FIGURE 1.12 Results obtained with the search: (a) details of proteins that have high sequence identity and (b) alignment of residues (see **Chapter 2**) for the two proteins that have high sequence identity.

computer-annotated supplement of SWISS-PROT that contains all the translations of EMBL nucleotide sequence entries, which are not yet integrated in SWISS-PROT. Currently, SWISS-PROT and TrEMBL have 0.5 and 7.6 million sequences, respectively. These databases are freely available at http://www.expasy.org/sprot/ and http://www.ebi.ac.uk/swissprot/.

SWISS-PROT contains the information about the name and origin of the protein, protein attributes, general information, ontologies, sequence annotation, amino acid sequence, bibliographic references, cross-references with sequence, structure and interaction databases, and entry information. An example for human lysozyme is shown in **Figure 1.13**. Furthermore, it has several search options, including sequence retrieval system (SRS), full-text search, advanced search, or by description or identification number.

In bioinformatics, developing a dataset plays a key role for any analysis or prediction. One can easily develop the dataset of amino acid sequences using SWISS-PROT. For example, the procedure for retrieving the data of "transcription factors" is shown in **Figure 1.14a**. Searching in UniProtKB with the keyword "transcription factors" will display all the relevant entries deposited in SWISS-PROT and TrEMBL.

FIGURE 1.13   Sample entry for human lysozyme in UniProtKB/SWISS-PROT.

The retrieval system has the options to restrict the data deposited only in SWISS-PROT and the data at different sequence identities, 90% or 50%. Furthermore, the final data can be downloaded in different file formats, such as Tab delimited, Excel, FASTA, GFF, Flat text, XML, RDF/XML, and list. The result obtained for the sequences of transcriptions factors in FASTA format is shown in **Figure 1.14b**. This set of sequences can be used for the analysis of DNA-binding proteins. In a similar way, sequences for any kind of proteins can be easily obtained with SWISS-PROT.

1.4.3   UniProt: The Universal Protein Resource

Recently, the SWISS-PROT, TrEMBL, and PIR protein database activities have united to form the Universal Protein Knowledgebase (UniProt) consortium. It provides the scientific community with a single, centralized, authoritative resource for protein sequences and functional information (Bairoch et al. 2005). The UniProt produces three layers of protein sequence databases: UniProt Archive, Knowledgebase, and Reference database. The UniProt Knowledgebase is a comprehensive, fully classified, richly and accurately annotated protein sequence knowledgebase with extensive cross-references. It is freely available at: http://www.uniprot.org/

1.4.4   Other protein sequence databases

There are several other protein sequence databases, which aim for specific classes or functions. EXProt (http://www.cmbi.kun.nl/EXProt/) is a nonredundant protein database containing a selection of entries from genome annotation projects and

1 - 25 of 11,439 results for **transcription**⊠ **AND factors**⊠ in **UniProtKB** sorted by **score** descending⊠

.⁂ Browse by taxonomy, keyword, gene ontology, enzyme class or pathway | ⠿ Reduce sequence redundancy to 100%, 90% or 50% | 🖫 Customize display

› Show only reviewed ✭ (UniProtKB/Swiss-Prot) or unreviewed ✩ (UniProtKB/TrEMBL) entries

› Quote terms: "transcription factors"

› Restrict term "transcription" to protein family, gene ontology, keyword, protein name

› Restrict term "factors" to protein family, gene ontology, protein name, web resource

Page 1   of 458 | Next »

None	Accession	Entry name	Status	Protein names	Gene names	Organism	Length
☑	P18850	ATF6A_HUMAN	✩	**Cyclic AMP-dependent transcription factor ATF-6 alpha** (cAMP-dependent transcription factor ATF-6 alpha) (Activating transcription factor 6 alpha) (ATF6-alpha) [Cleaved into: Processed cyclic AMP-dependent transcription factor ATF-6 alpha]	ATF6	Homo sapiens (Human)	670
☑	P0A4H2	BVGA_BORPE	✩	**Virulence factors putative positive transcription regulator bvgA**	bvgA (BP1878)	Bordetella pertussis	209
☑	Q9Y2D1	ATF5_HUMAN	✩	**Cyclic AMP-dependent transcription factor ATF-5** (cAMP-dependent transcription factor ATF-5)	ATF5 (ATFX)	Homo sapiens (Human)	282

(a)

```
>sp|P18850|ATF6A_HUMAN Cyclic AMP-dependent transcription factor ATF-6 alpha OS=Homo s
MGEPAGVAGTMESPFSPGLFHRLDEDWDSALFAELGYFTDTDELQLEAANETYENNFDNL
DFDLDLMPWESDIWDINNQICTVKDIKAEPQPLSPASSSYSVSSPRSVDSYSSTQHVPEE
LDLSSSSQMSPLSLYGENSNSLSSAEPLKEDKPVTGPRNKTENGLTPKKKIQVNSKPSIQ
PKPLLLPAAPKTQTNSSVPAKTIIIQTVPTLMPLAKQQPIISLQPAPTKGQTVLLSQPTV
VQLQAPGVLPSAQPVLAVAGGVTQLPNHVVNVVPAPSANSPVNGKLSVTKPVLQSTMRNV
GSDIAVLRRQQRMIKNRESACQSRKKKKEYMLGLEARLKAALSENEQLKKENGTLKRQLD
EVVSENQRLKVPSPKRRVVCVMIVLAFIILNYGPMSMLEQDSRRMNPSVSPANQRRHLLG
FSAKEAQDTSDGIIQKNSYRYDHSVSNDKALMVLTEEPLLYIPPPPCQPLINTTESLRLN
HELRGWVHRHEVERTKSRRMTNNQQKTRILQGALEQGSNSQLMAVQYTETTSSISRNSGS
ELQVYYASPRSYQDFFEAIRRRGDTFYVVSFRRDHLLLPATTHNKTTRPKMSIVLPAINI
NENVINGQDYEVMMQIDCQVMDTRILHIKSSSVPPYLRDQQRNQTNTFFGSPPAATEATH
VVSTIPESLQ
>sp|P0A4H2|BVGA_BORPE Virulence factors putative positive transcription regulator bvgA
MYNKVLIIDDHPVLRFAVRVLMEKEGFEVIGETDNGIDGLKIAREKIPNLVVLDIGIPKL
DGLEVIARLQSLGLPLRVLVLTGQPPSLFARRCLNSGAAGFVCKHENLHEVINAAKAVMA
GYTYFPSTTLSEMRMGDNAKSDSTLISVLSNRELTVLQLLAQGMSNKDIADSMFLSNKTV
STYKTRLLQKLNATSLVELIDLAKRNNLA
>sp|Q9Y2D1|ATF5_HUMAN Cyclic AMP-dependent transcription factor ATF-5 OS=Homo sapiens
MSLLATLGLELDRALLPASGLGWLVDYGKLPPAPAPLAPYEVLGGALEGGLPVGGEPLAG
DGFSDWMTERVDFTALLPLEPPLPPGTLPQPSPTPPDLEAMASLLKKELEQMEDFFLDAP
LLPPPSPPPLPPPPLPPAPSLPLSLPSFDLPQPPVLDTLDLLAIYCRNEAGQEEVGMPPL
PPPQQPPPPSPPQPSRLAPYPHPATTRGDRKQKKRDQNKSAALRYRQRKRAEGEALEGEC
QGLEARNRELKERAESVEREIQYVKDLLIEVYKARSQRTRSC
```

FIGURE 1.14 Sequence retrieval in UniProtKB/SWISS-PROT: (a) The query and retrieved results and (b) sequences of "transcription factors" in FASTA format (the first line starts with ">" followed by amino acid sequences in single-letter code. Each line has 60 amino acid residues; see **Chapter 2**).

public databases, aiming at including only proteins with an experimentally verified function. The NCBI Entrez Protein database (http://www.ncbi.nlm.nih.gov/entrez) comprises sequences taken from a variety of sources, including SWISS-PROT, PIR, the Protein Research Foundation, the Protein Data Bank, and translations from annotated coding regions in the GenBank and RefSeq databases. Protein sequence records in Entrez have links to precomputed protein BLAST alignments, protein structures, conserved protein domains, nucleotide sequences, genomes, and genes. The Transport Classification Database (TCDB) is a curated, relational database containing sequence, classification, structural, functional, and evolutionary information about transport systems from a variety of living organisms (Busch and Saier, 2002; http://www.tcdb.org/).

Pongor's group (Pongor et al. 1993; Vlahovicek et al. 2005) developed a database of annotated protein domain sequences, SBASE. It facilitates the detection of domain homologies based on direct sequence database search using BLAST or a high-speed Smith Waterman algorithm, and returns a predicted domain architecture of the query sequence. Unlike traditional consensus representations, such as HMM (hidden Markov models), profiles and regular expressions, the SBASE domain library approach gives equal weights to all representatives, and a search against this library will detect both the typical and atypical (rare) representatives. It is to be noted that the domain library approach does not require either multiple alignment or learning algorithms to achieve accuracy. SBASE is freely available at http://www.icgeb.trieste.it/sbase.

1.5   Protein structure databases

Kendrew et al. (1958) solved the first three-dimensional structure of the protein myoglobin using X-ray crystallography. Subsequently, several structures have been determined with this technique. After two decades, NMR spectroscopy has been used to determine the protein structures in solution. Recently, neutron diffraction and electron microscopy have been used to determine protein structures. The number of protein structures has been increased every year, and they are deposited in a database, Protein Data Bank (PDB). On the basis of the structures available in PDB, several other databases have been established for the structural classification of proteins (SCOP), topology and architecture, and amino acid properties, etc.

1.5.1   Protein Data Bank

The PDB was established at Brookhaven National Laboratories, USA, in 1971 as an archive for biological macromolecular crystal structures (Bernstein et al. 1977). Recently, the management of PDB became the responsibility of the Research Collaboratory for Structural Bioinformatics (RCSB), and it has six mirror sites at San Diego Supercomputer Center and Rutgers University in the USA; Cambridge Crystallographic Data Center, UK; National University of Singapore; Osaka University, Japan; and Max Delbruck Center for Molecular Medicine, Germany (Berman et al. 2000). PDB is available at http://www.rcsb.org/.

PDB stores the data in a uniform format atomic coordinates and partial bond connectivities, as derived from crystallographic studies. Text included in each data entry gives pertinent information for the structure at hand (e.g., species from which

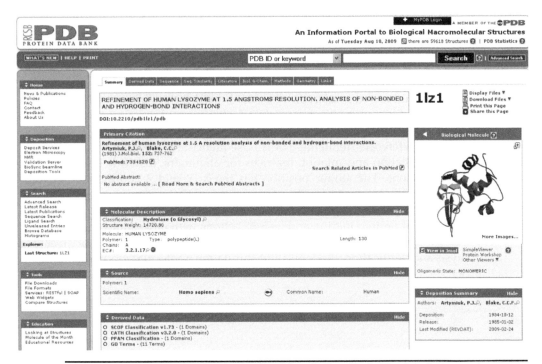

FIGURE **1.15**   Snapshot showing the details of Protein Data Bank. The data for human lysozyme are shown.

the molecule has been obtained, resolution of diffraction data, literature citations and specifications of secondary structure). In addition to atomic coordinates and connectivities, the PDB stores the temperature factor for each atom. The PDB has been widely used in structural analysis on various aspects, such as atomic and residue contacts, amino acid clusters, developing potentials, and amino acid properties.

Currently, PDB has more than 57,000 structures, and 53,000 of them are proteins and their complexes (May 12, 2009). PDB has the search option with its code, authors, or full-text search. It contains the summary information about the name and source of the protein, experimental method, authors, and references. In addition, the details about crystallographic conditions are also provided. An example to human lysozyme is shown in **Figure 1.15**. Furthermore, it has the options to view the structure, display and download the files with/without three-dimensional coordinates, and it provides information about structural neighbors, geometry, sequence details, and other sources.

The sequence provides the detailed information about the number of amino acid residues in the protein, chain information, and amino acid sequence for each chain along with their secondary structures derived using DSSP (Kabsch and Sander, 1983), and number and percentage of each secondary structure through external links. For example, human lysozyme has four α-helices and three β-strands, and the content of α-helices is about 31%. Furthermore, it has the facility to save the sequence in FASTA format, which can be used in other programs.

PDB has the option to see the structural coordinates and view the structure on the screen or to save the structural information in a disk. It also has the possibility to download the data by ftp (ftp://ftp.wwpdb.org/pub/pdb/) or request the structures in CD. The structures can also be visualized using other programs, SWISS-PDB viewer (http://spdbv.vital-it.ch/), RASMOL (Sayle and Milner-White, 1995; http://www.umass.edu/microbio/rasmol/), Jmol (http://jmol.sourceforge.net/), KiNG (http://kinemage.biochem.duke.edu/software/king.php), PyMOL (De-Lano, 2002; http://www.pymol.org), etc. Structural data have the information about the protein, references, refinement details, sequence and secondary structure information, translation and rotation matrices for oligomers, disulfide bonds, and the atomic positions. The atomic position has the X, Y, and Z coordinates and temperature factors (**Figure 1.16**). The atomic coordinates have been extensively used to analyze the principles governing the folding and stability of protein structures, mechanism of protein folding and its interactions with other molecules, etc.

In addition, PDB has plenty of external links to other software for structural analysis and verification, modeling and simulation, molecular graphics, and so on.

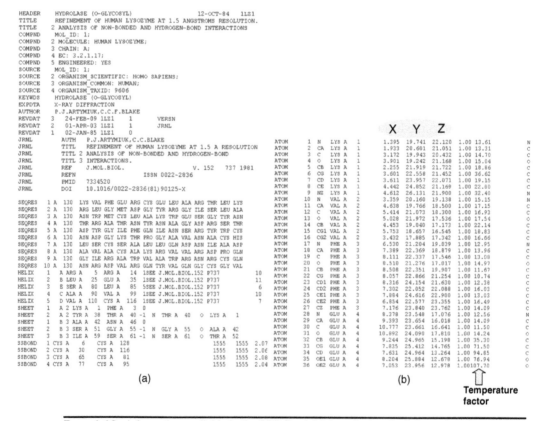

(a)   (b)

FIGURE 1.16 Information available in the PDB file: (a) It has the header information along with compound, source, authors, references, complete amino acid sequence, secondary structural details, and three-dimensional coordinate; (b) coordinate has the information about the residue name and number, atom name and number, chain and X, Y, Z coordinates. It also has the temperature factor for each atom.

1.5.2   Database for nonredundant PDB structures

A dataset with several similar structures influences the characteristic features of such data, which cause a bias on the result. Hence, a dataset of nonredundant proteins is necessary for analyzing the features of protein structures. The level of redundancy may vary depending upon the nature of the problem and the number of proteins in the database. The ASTRAL database provides the PDB codes and amino acid sequences for two cutoff values, i.e., less than 40% and 95% sequence identities. Furthermore, it has the option of selecting the cutoff values from 10% to 100%. The description about the selection of nonredundant dataset of proteins with 25% sequence identity is shown in **Figure 1.17**. The PDB identifiers for this cutoff have been selected to display the results, and it is also possible to get the sequences for these protein structures. Furthermore, ASTRAL has the options to get the sequences with e-values, fold, superfamily, etc.

For a general analysis, one can get the information about the nonredundant structures from ASTRAL. Furthermore, the nonredundant dataset of any

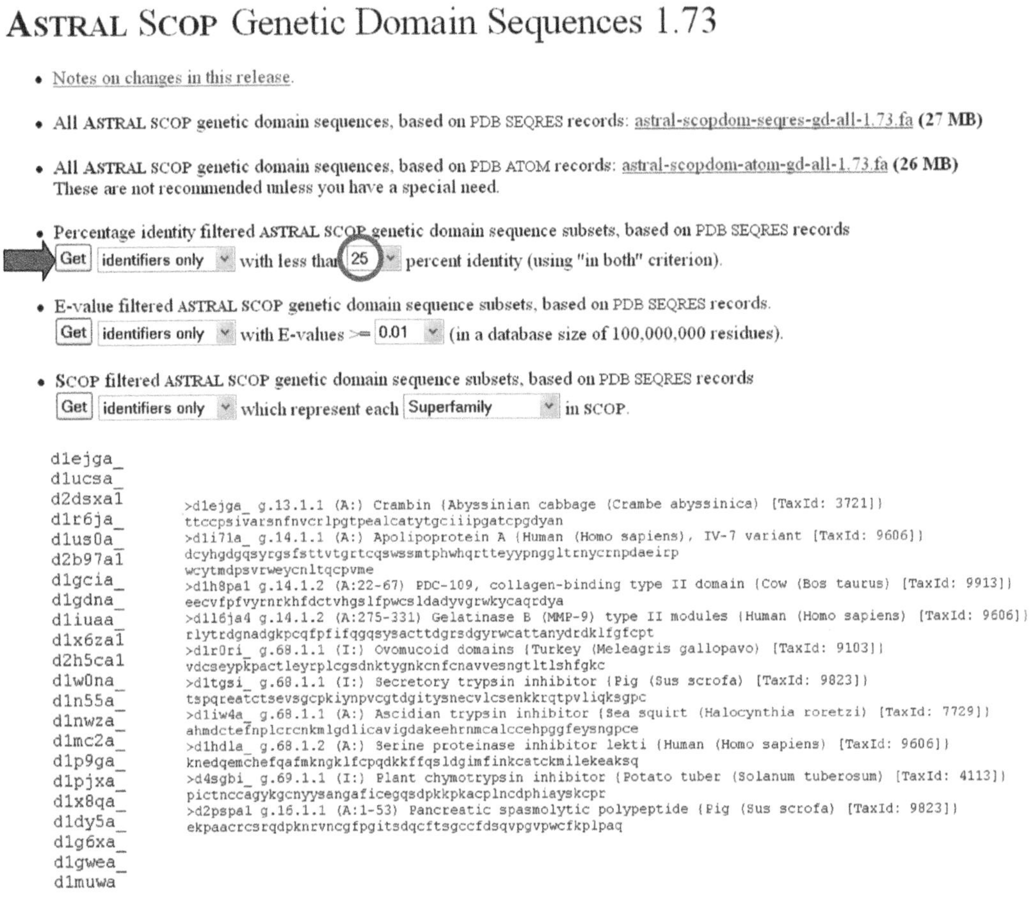

ASTRAL SCOP Genetic Domain Sequences 1.73

- Notes on changes in this release.

- All ASTRAL SCOP genetic domain sequences, based on PDB SEQRES records: astral-scopdom-seqres-gd-all-1.73.fa **(27 MB)**

- All ASTRAL SCOP genetic domain sequences, based on PDB ATOM records: astral-scopdom-atom-gd-all-1.73.fa **(26 MB)**
  These are not recommended unless you have a special need.

- Percentage identity filtered ASTRAL SCOP genetic domain sequence subsets, based on PDB SEQRES records
  Get | identifiers only | with less than 25 | percent identity (using "in both" criterion).

- E-value filtered ASTRAL SCOP genetic domain sequence subsets, based on PDB SEQRES records.
  Get | identifiers only | with E-values >= 0.01 | (in a database size of 100,000,000 residues).

- SCOP filtered ASTRAL SCOP genetic domain sequence subsets, based on PDB SEQRES records
  Get | identifiers only | which represent each | Superfamily | in SCOP.

```
d1ejga_
d1ucsa_
d2dsxa1      >d1ejga_ g.13.1.1 (A:) Crambin (Abyssinian cabbage (Crambe abyssinica) [TaxId: 3721])
d1r6ja_      ttccpsivarsnfnvcrlpgtpealcatytgciiipgatcpgdyan
d1us0a_      >d1i71a_ g.14.1.1 (A:) Apolipoprotein A (Human (Homo sapiens), IV-7 variant [TaxId: 9606])
d2b97a1      dcyhgdgqsyrgsfsttvtgrtcqswssmtphwhqrtteyypnggltrnycrnpdaeirp
d1gcia_      wcytmdpsvrweycnltqcpvme
d1gdna_      >d1h8pa1 g.14.1.2 (A:22-67) PDC-109, collagen-binding type II domain (Cow (Bos taurus) [TaxId: 9913])
d1iuaa_      eecvfpfvyrnrkhfdctvhgslfpwcsldadyvgrwkycaqrdya
d1x6za1      >d116ja4 g.14.1.2 (A:275-331) Gelatinase B (MMP-9) type II modules (Human (Homo sapiens) [TaxId: 9606])
d2h5ca1      rlytrdgnadgkpcqfpfifqgqsysacttdgrsdgyrwcattanydrdklfgfcpt
d1w0na_      >d1r0ri_ g.68.1.1 (I:) Ovomucoid domains (Turkey (Meleagris gallopavo) [TaxId: 9103])
d1n55a_      vdcseypkpactleyrplcgsdnktygnkcnfcnavvesngtltlshfgkc
d1nwza_      >d1tgsi_ g.68.1.1 (I:) Secretory trypsin inhibitor (Pig (Sus scrofa) [TaxId: 9823])
d1mc2a_      tspqreatctsevsgcpkiynpvcgtdgitysnecvlcsenkkrqtpvliqksgpc
d1p9ga_      >d1iw4a_ g.68.1.1 (A:) Ascidian trypsin inhibitor (Sea squirt (Halocynthia roretzi) [TaxId: 7729])
d1pjxa_      ahmdcteinplcrcnkmlgdlicavigdakeehrnmcalccehpggfeysngpce
d1x8qa_      >d1hdla_ g.68.1.2 (A:) Serine proteinase inhibitor lekti (Human (Homo sapiens) [TaxId: 9606])
d1dy5a_      knedqemchefqafmkngklfcpqdkkffqsldgimfinkcatckmilekeaksq
d1g6xa_      >d4sgbi_ g.69.1.1 (I:) Plant chymotrypsin inhibitor (Potato tuber (Solanum tuberosum) [TaxId: 4113])
d1gwea_      pictnccagykgcnyysangaficegqsdpkkpkacplncdphiayskcpr
d1muwa       >d2pspa1 g.16.1.1 (A:1-53) Pancreatic spasmolytic polypeptide (Pig (Sus scrofa) [TaxId: 9823])
             ekpaacrcsrqdpknrvncgfpgitsdqcftsgccfdsqvpgvpwcfkplpaq
```

FIGURE 1.17   Retrieval of nonredundant protein sequences from PDB using ASTRAL. It uses SCOP classification for the domains, and the data obtained with the sequence identity of less than 25% are shown.

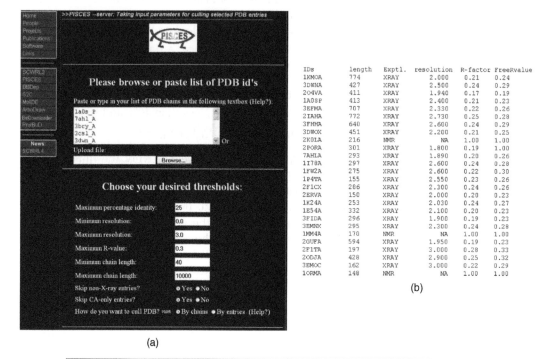

(a)

(b)

FIGURE 1.18   Retrieval of nonredundant protein structures using PISCES. The users have the feasibility of selecting the identity, resolution, etc. and provide own dataset of proteins. (a) The search options and (b) results.

specific class of proteins (e.g., all proteins, membrane proteins, DNA-binding proteins, protein–protein complexes, etc.) can be derived by matching the dataset of interest with ASTRAL nonredundant dataset. ASTRAL is available at http://astral.berkeley.edu/.

Wang and Dunbrack (2005) developed a sequence-culling database server, PISCES, for producing lists of nonredundant proteins from the PDB using entry- and chain-specific criteria and mutual sequence identity. It uses a combination of PSI-BLAST and structure-based alignments to determine sequence identities. An example is shown in **Figure 1.18**. It takes the PDB codes with chain information along with other conditions, such as % sequence identity, resolution, R-factor, inclusion of non–X-ray structures, number of amino acid residues, etc (**Figure 1.18a**), and sends the results via e-mail. The output has four files: (i) original id, (ii) culled id, (iii) FASTA format sequence, and (iv) similarity log. The result obtained for the culled sequences from the list of 34 β-barrel membrane proteins is shown in **Figure 1.18b**. PISCES is available at http://dunbrack.fccc.edu/pisces/.

Noguchi and Akiyama (2003) developed a database of representative chains from PDB. The search options to retrieve nonredundant set of proteins are shown in **Figure 1.19a**. It has several features, including the limits with resolution, R-factor, number of residues, fragments, complex structures, and membrane proteins. Furthermore, the users have the options to use any cutoff for sequence identity and root-mean-square deviation between the sequences and structures, respectively as shown in **Figure 1.19b**. The results obtained with the search are displayed in

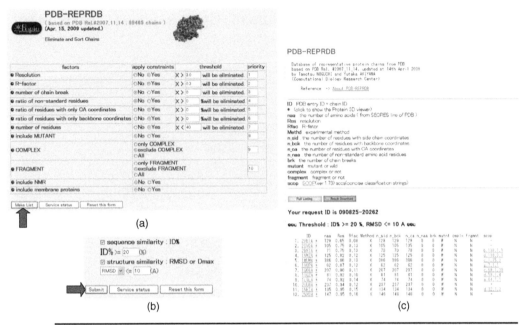

FIGURE 1.19 Retrieval of nonredundant proteins with PDB-REPRDB. This server will have several options to retrieve protein sequences, including the sequence identity and RMSD between two structures. (a) The initial parameters, (b) similarities, and (c) results are shown in the figure.

Figure 1.19c. It provides several details, including the links to SCOP database (see Section 1.5.4). It is available at http://www.cbrc.jp/pdbreprdb/.

1.5.3 PDB-related databases

On the basis of PDB, several other databases have been developed for different sets of proteins, such as membrane proteins, protein–protein complexes, protein–nucleic acid complexes, and ligand binding proteins. Tusnady et al. (2005) developed a database of transmembrane proteins, PDBTM (http://pdbtm.enzim.hu/), which includes the sequences and structures of redundant and nonredundant α-helical and β-barrel membrane proteins along with their membrane-spanning segments. Jayasinghe et al. (2001) compiled the structures of known membrane proteins, MPtopo (http://blanco.biomol.uci.edu/mptopo), and classified them into several groups, such as monotopic, GPCRs, rhodopsins, and β-barrel membrane proteins. They have included the PDB codes, structures, and their respective references. Ikeda et al. (2003) developed a database of transmembrane protein topologies, TMPDB (http://bioinfo.si.hirosaki-u.ac.jp/~TMPDB/), which is based on the experimental evidences from X-ray crystallography, NMR spectroscopy, etc. Sarai and colleagues (An et al. 1998) developed a database for protein–nucleic acid complex structures, and these structures have been classified into different groups based on the recognition motif of proteins and DNA involved in the complex. Puvanendrampillai and Mitchell (2003) developed the Protein Ligand Database (PLD), which has the PDB codes for protein-ligand complexes along with binding information.

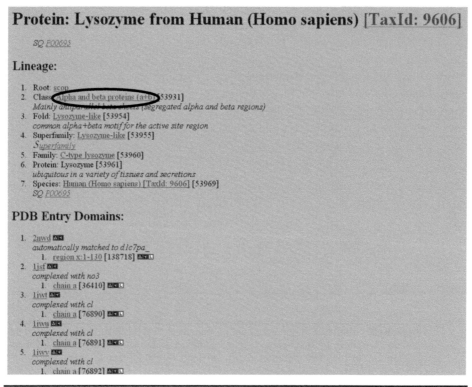

Protein: Lysozyme from Human (Homo sapiens) [TaxId: 9606]

SQ P00695

Lineage:

1. Root: scop
2. Class: Alpha and beta proteins (a+b) [53931]
   Mainly antiparallel beta sheets (segregated alpha and beta regions)
3. Fold: Lysozyme-like [53954]
   common alpha+beta motif for the active site region
4. Superfamily: Lysozyme-like [53955]
   Superfamily
5. Family: C-type lysozyme [53960]
6. Protein: Lysozyme [53961]
   ubiquitous in a variety of tissues and secretions
7. Species: Human (Homo sapiens) [TaxId: 9606] [53969]
   SQ P00695

PDB Entry Domains:

1. 2nwd
   *automatically matched to d1c7pa_*
   1. region x:1-130 [138718]
2. 1jsf
   complexed with no3
   1. chain a [36410]
3. 1iwt
   complexed with cl
   1. chain a [76890]
4. 1iwu
   complexed with cl
   1. chain a [76891]
5. 1iwv
   complexed with cl
   1. chain a [76892]

FIGURE 1.20 Sample entry for human lysozyme in SCOP database. It is classified as $\alpha + \beta$ protein. Furthermore, links are available for the different folds and families within this class.

1.5.4 Databases for structural classification of proteins

Murzin et al. (1995) constructed the SCOP database, which provides a detailed and comprehensive description of the structural and evolutionary relationships of the proteins of known structures. For each protein, the classification has the hierarchical levels, family, superfamily, fold, and structural class. An example for human lysozyme is shown in **Figure 1.20**. It belongs to the family of C-type lysozyme and the fold of lysozyme-like under $\alpha + \beta$ structural class. The structure can be identified with a six-letter code (1lz1__). The first four letters of the code (1lz1) represent the PDB code followed by the chain name (_; it indicates there are no multiple chains in this protein) and domain name (_). SCOP database has been linked from the PDB to obtain the structural class information of each protein directly. On the other hand, one can search the SCOP database for obtaining the structural class, fold, and domain information.

Orengo et al. (1997) developed a semiautomatic procedure for deriving a novel hierarchical classification of protein domain structures (CATH) and created a database in providing the class information for all the structures in PDB. The four main levels of CATH classification are protein class (C), architecture (A), topology (T), and homologous superfamily (H). Class is the simplest level, and it essentially describes the secondary structure composition of each domain. Architecture summarizes the shape revealed by the orientations of the secondary structure units,

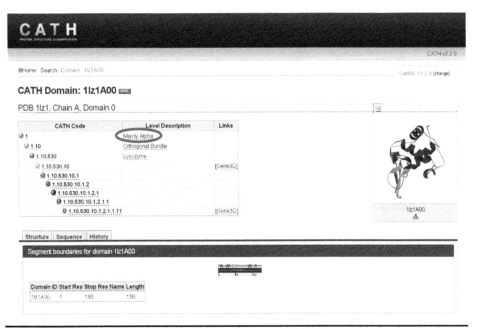

FIGURE 1.21   Structural classification of human lysozyme with CATH.

such as barrels and sandwiches. At the topology level, sequential connectivity is considered, such that members of the same architecture might have quite different topologies. The homologous superfamilies contain proteins with highly similar structures and functions. The CATH classification for human lysozyme is shown in **Figure 1.21**. It has the architecture of an orthogonal bundle, and it is designated an all-α class proteins in CATH. It may be possible to have different assignments in SCOP and CATH for some proteins, especially if the percentage of helical content is high and strand content is less or vice versa. In human lysozyme, the content of α-helical structures is 31% and that of β-strands is 8%. On the basis of the high content of α-helix it was classified as all-α proteins in CATH, and due to the presence of α-helices and β-strands SCOP classified it as α + β protein.

1.6   Literature databases

Literature databases play a vital role to understand the current status of research on various fields of interest. At present, several databases provide the information with/without restrictions. These databases include PubMed, Google Scholar, Scopus, Science Citation Index, Chemical abstracts, and so on. The PubMed literature database covers most of the protein-related works published in international journals, and it has different features to search the database and display the results. It can be searched with author's name, year of publication, name of the journal, and key words. As an example, a search for the articles published in *Nature* about protein structure is shown in **Figure 1.22**. The journal name is specified with *Nature*. The search retrieved 3240 records, including 45 reviews. **Figure 1.22** shows the summary of results, and it is possible to display abstract, abstract plus with

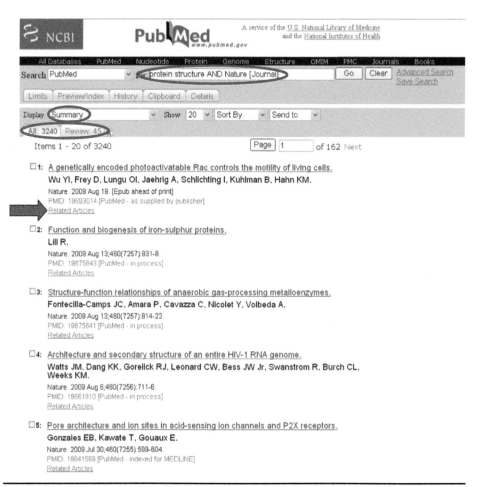

FIGURE 1.22 Retrieval of results using PubMed literature database. The search with keywords and journal name, display options, and number of retrieved records are indicated. The link to related articles is shown with an arrow.

key words, etc. It is also possible to see the related articles for all the references listed in the result page. Each hit is linked with its respective journal site, and one can get the complete article if the journal has open access options or the users have the license to access the full text of the journal. PubMed can be freely accessed at http://www.ncbi.nlm.nih.gov/pubmed/.

1.7 Exercises

1. Retrieve the sequence of E. coli Triose phosphate isomerase in FASTA format. *Hint:* Use the text search in iProClass, and search with Triose phosphate isomerase and E. coli; select the correct one among the displayed results; "show" the selected one; and click on "save the result" as FASTA.

2. Obtain the amino acid sequences of membrane proteins located in outer membrane using SWISS-PROT/UniProtKB database.
Hint: use the keyword, "membrane protein" and location, "outer membrane." Click on Download and select the format. Number of hits is about 500.

3. Find the three-dimensional coordinates of porin from Rhodobactor capsulatus and store the data.
Hint: Search at the top of the page of PDB using the key words. Select the best match (e.g., 3POR or 2POR). Click on Display files and select PDB text. Click on Download files to save the coordinates.

4. Get the list of PDB codes for RNA-binding proteins.
Hint: Search with RNA-binding proteins. In the result page, click on "Results ID List" to get the list of codes.

5. Check the interacting partners (protein and DNA) in 1RXW.
Hint: Use Jmol to view the protein–DNA complex, and find the chain information.

6. Obtain the PDB codes that have ligands and the sequence identity is less than 30%.
Hint: Go to Advanced Search and search with "ligand" in Structure title. Select the sequence identity as 30%.

7. How many helices are there in 2LZM?
Hint: Check the PDB file of 2LZM and the presence of helices and strands.

8. What is the average fluctuation for the residue Phe at position 4 in 2LZM?
Hint: Identify the temperature factors of all atoms in Phe4, and get the average. The information is shown in **Figure 1.16**.

9. List the major transport systems in Transport classification database?
Hint: Check the TCDB database and click on TC system

10. What is the major class of transporters?
Hint: Find the number of proteins in each transporter, and find the one with the maximum number of proteins.

11. Get the dataset of protein structures with less than 40% sequence identity.
Hint: Click on ASTRAL 1.73 with less than 40% sequence identity.

12. Get the dataset of protein structures with less than 20% sequence identity.
Hint: Select the sequences and follow **Figure 1.17**.

13. Get the PDBs with less than 20% sequence identity using the IDs obtained for RNA-binding proteins (question 4).
Hint: Follow **Figure 1.18** using the list of PDB codes obtained as the answer to question 4.

14. Obtain the list of nonmembrane proteins without mutation and not complexed with other molecules and the sequence identity of less than 25%.
Hint: Use PDB-REPRDB and provide the necessary conditions.

15. Get the nonredundant structures of β-barrel membrane proteins.
Hint: Go to PDBTM and click on download. The codes are listed with the link at the end of the page.

16. Retrieve the proteins with seven α-helical segments in PDBTM.
Hint: Search with type of transmembrane protein, and select the number of segments.

17. Compare the folding types and number of proteins belonging to different structural classes of proteins.
   Hint: SCOP has the information.

18. What are the major folding types in four structural classes?
   Hint: Compare the folding types with number of entries.

19. Compare the heme-binding proteins (family: globin) and contents in SCOP and CATH databases.
   Hint: SCOP: All-Globin like-Globin; CATH: 1.10.490.10.

20. Find the domain information for outer membrane protein TolC.
   Hint: search CATH database with FASTA sequence, which can be found from UniProtKB/SWISS-PROT (P02930) or PDB.

21. Find the articles published about "protein interactions" and "shape complimentarity."
   Hint: Search PubMed with relevant keywords.

22. Identify the papers published in the journal *Cell* about mitochondrial β-barrel membrane proteins.
   Hint: Search with mitochondrial β-barrel membrane proteins AND *Cell* [Journal].

References

An J, Nakama T, Kubota Y, Sarai A. 3DinSight: an integrated relational database and search tool for the structure, function and properties of biomolecules. Bioinformatics. 1998;14(2):188–195.

Anfinsen CB. Principles that govern the folding of protein chains. Science. 1973;181(96): 223–230.

Bairoch A, Apweiler R, Wu CH, Barker WC, Boeckmann B, Ferro S, Gasteiger E, Huang H, Lopez R, Magrane M, Martin MJ, Natale DA, O'Donovan C, Redaschi N, Yeh LS. The Universal Protein Resource (UniProt). Nucleic Acids Res. 2005;33(Database issue):D154–159.

Bairoch A, Apweiler R. (1996) The SWISS-PROT protein sequence data bank and its new supplement TREMBL. Nucleic Acids Res. 1996;24(1):21–25.

Barker WC, Garavelli JS, Huang H, McGarvey PB, Orcutt BC, Srinivasarao GY, Xiao C, Yeh LS, Ledley RS, Janda JF, Pfeiffer F, Mewes HW, Tsugita A, Wu C. The protein information resource (PIR). Nucleic Acids Res. 2000;28(1):41–44.

Berman HM, Westbrook JZ, Feng G, Gilliland TN, Bhat H, Weissig IN, Shindyalov, Bourne PE. The Protein Data Bank. Nucleic Acids Res. 2000;28:235–242.

Bernstein FC, Koetzle TF, Williams GJ, Meyer EF Jr, Brice MD, Rodgers JR, Kennard O, Shimanouchi T, Tasumi M. The Protein Data Bank: a computer-based archival file for macromolecular structures. J Mol Biol. 1977;112(3):535–542.

Busch W, Saier MH Jr. The transporter classification (TC) system, 2002. Crit Rev Biochem Mol Biol. 2002;37:287–337.

DeLano WL. The PyMOL Molecular Graphics System. San Carlos, CA: DeLano Scientific; 2002. Available online at http://www.pymol.org.

Dayhoff MO, Eck RV, Chang MA, Sochard MR. Atlas of Protein Sequence and Structure, Vol. 1. Silver Spring, MD: National Biomedical Research Foundation; 1965.

Gromiha MM, Selvaraj S. Inter-residue interactions in protein folding and stability. Prog Biophys Mol Biol. 2004;86(2):235–277.

Ikeda M, Arai M, Okuno T, Shimizu T. TMPDB: a database of experimentally-characterized transmembrane topologies. Nucleic Acids Res. 2003;31(1):406–409.

Jayasinghe S, Hristova K, White SH. MPtopo: A database of membrane protein topology. Protein Sci. 2001;10(2):455–458.

Kabsch W, Sander C, Dictionary of protein secondary structure: pattern recognition of hydrogen-bonded and geometrical features. Biopolymers. 1983;22:2577–2637.

Kendrew JC, Bodo G, Dintzis HM, Parrish RG, Wyckoff H, Phillips DC. A three-dimensional model of the myoglobin molecule obtained by x-ray analysis. Nature. 1958;181(4610):662–666.

Levitt M, Chothia C. Structural patterns in globular proteins. Nature. 1976;261(5561): 552–558.

Murzin AG, Brenner SE, Hubbard T, Chothia C. SCOP: a structural classification of proteins database for the investigation of sequences and structures. J Mol Biol. 1995;247(4):536–540.

Nelson DL, Cox MM. Lehninger Principles of Biochemistry. New York: W.H. Freeman and Company; 2005.

Noguchi T, Akiyama Y. PDB-REPRDB: a database of representative protein chains from the Protein Data Bank (PDB) in 2003. Nucleic Acids Res. 2003;31(1):492–493.

Orengo CA, Michie AD, Jones S, Jones DT, Swindells MB, Thornton JM. CATH: a hierarchic classification of protein domain structures. Structure. 1997;5(8):1093–1108.

Pauling L, Corey RB, Branson HR. The structure of proteins: two hydrogen-bonded helical configurations of the polypeptide chain. Proc Natl Acad Sci U S A. 1951;37(4):205–211.

Pongor S, Skerl V, Cserzö M, Hátsági Z, Simon G, Bevilacqua V. The SBASE domain library: a collection of annotated protein segments. Protein Eng. 1993;6(4):391–395.

Puvanendrampillai D, Mitchell JB. Protein Ligand Database (PLD): additional understanding of the nature and specificity of protein–ligand complexes. Bioinformatics. 2003;19(14):1856–1857.

Ramachandran GN, Ramakrishnan C, Sasisekharan V. Stereochemistry of polypeptide chain configurations. J Mol Biol. 1963;7:95–99.

Sayle RA, Milner-White EJ. RASMOL: biomolecular graphics for all. Trends Biochem Sci. 1995;20(9):374.

Schulz GE. Transmembrane beta-barrel proteins. Adv Protein Chem. 2003;63:47–70.

Tusnady GE, Dosztanyi Z, Simon I. PDB_TM: selection and membrane localization of transmembrane proteins in the protein data bank. Nucleic Acids Res. 2005;33(Database issue):D275–278.

Vlahovicek K, Kaján L, Agoston V, Pongor S. The SBASE domain sequence resource, release 12: prediction of protein domain-architecture using support vector machines. Nucleic Acids Res. 2005;33(Database issue):D223–D225.

Wang G, Dunbrack RL Jr. PISCES: a protein sequence culling server. Bioinformatics. 2003;19:1589–1591.

White SH, Wimley WC. Membrane protein folding and stability: physical principles. Annu Rev Biophys Biomol Struct. 1999;28:319–365.

CHAPTER 2

Protein Sequence Analysis

The analysis of protein sequences provides the information about the preference of amino acid residues and their distribution along the sequences for understanding the secondary and tertiary structures of proteins and their functions. The identification of similar motifs in protein sequences would help to predict the structurally or functionally important regions. The profiles obtained with the single amino acid properties based on amino acid sequence would reveal the clustering of amino acids with similar property. Furthermore, the comparison of different amino acid sequences using alignment methods would enhance our knowledge about the availability of similar sequences, and these sequences could be used as a template for protein three-dimensional structure prediction.

2.1   Sequence alignment

The comparison of two proteins is mainly carried out by aligning the sequences or structures. In this method, a one-to-one correspondence is set up between the residues of the two proteins. The simplest observation is the global alignment of two sequences, in which the two proteins have maintained a correspondence over the entire length. An alternative is the local alignment in which the alignment is made only with the most similar part of the proteins.

   An alignment of two sequences A and B must obey the following conditions: (i) All residues should be used in the alignment and all should be in the same order, (ii) align one residue from A with another from B, (iii) a residue can be aligned with a blank (-), and (iv) two blanks cannot be aligned. The different ways of aligning two sequences, VEITGEIST and PRETERIT, are shown in **Figure 2.1**. From these alignments, one could estimate the score for each aligned positions and hence the total score. The scoring scheme will be as follows: (i) Score = 1, if both the residues in the same positions of the sequences A and B are the same (e.g., in Alignment 1 [**Figure 2.1**], both the sequences A and B at position 3 are E, and hence it will have the score of 1), (ii) if the residues are different, score = 0 (e.g., position 1, the residues are V and P, respectively in sequences A and B), and (iii) score = −1 if there is a gap in the alignment (e.g., positions 2 and 4 in Alignment 1). The added score for all the residues gives the net score for the aligned sequences. In alignments, the positioning of residues with similar properties (e.g., Val and Ile are hydrophobic,

Position

Alignment 1

```
          1  2  3  4  5  6  7  8  9 10
Seq A: V  -  E  I  T  G  E  I  S  T
Seq B: P  R  E  -  T  E  R  I  -  T
Score: 0 -1  1 -1  1  0  0  1 -1  1
Total: 1
```

Alignment 2

```
Seq A: V  E  I  T  G  E  I  S  T
Seq B: P  R  E  T  -  E  R  I  T
Score: 0  0  0  1 -1  1  0  0  1
Total: 2
```

Alignment 3

```
Seq A: -  V  E  I  T  G  E  -  I  S  T
Seq B: P  R  E  -  T  -  E  R  I  -  T
Score:-1  0  1 -1  1 -1  1 -1  1 -1  1
Total: 0
```

FIGURE 2.1   Sequence alignment and scoring schemes for two typical sequences: score = 1 for same residue (shown in boxes); score = 0 for different residues and score = −1 for gap.

Glu and Asp are negatively charged, etc.) is used to find *similar* sequences (Eidhammer et al. 2004).

2.2   Programs for aligning sequences

Several computer programs have been developed for estimating the similarity score of two sequences and for finding similar sequences from available databases using pairwise and multiple alignments.

2.2.1   Basic Local Alignment Search Tool (BLAST)

Altschul et al. (1990) developed an approach for a rapid sequence comparison, basic local alignment search tool (BLAST), which directly approximates alignments that optimize a measure of local similarity and the maximal segment pair score. This algorithm has been applied in a variety of contexts, including straightforward DNA and protein sequence database searches, motif searches, gene identification searches, and in the analysis of multiple regions of similarity in long DNA sequences. In this method, the query protein sequence can be searched with several databases, including the nonredundant structures available in PDB, protein sequences at SWISS-PROT, etc. Furthermore, BLAST has several features such as (i) identifying protein sequences similar to the query, (ii) finding members of a protein family or building a custom position-specific scoring matrix, (iii) finding proteins similar to the query around a given pattern, (iv) finding conserved domains in the query, and (v) searching for peptide motifs. BLAST is available at http://www.ncbi.nlm.nih.gov/BLAST/. An example to identify protein sequences similar to the query is shown in **Figure 2.2**. BLAST has several options for querying a sequence:

(i) Accepts the sequence with accession number, gi, and FASTA format. The input data can be given by copying and pasting the details directly on the Web or by uploading a file from a local computer. Accession number is the number allotted in UniProt for each sequence (e.g., P61626); gi is a bar-separated NCBI sequence identifier (e.g., gi|48428995). A sequence in FASTA format begins with a single-line description, followed by lines of sequence data. The description line is distinguished from the sequence data by a greater than (">") symbol at the beginning. An example sequence in FASTA format is given below:

```
>gi|48428995|sp|P61626.1|LYSC_HUMAN RecName: Full=Lysozyme C
MKALIVLGLVLLSVTVQGKVFERCELARTLKRLGMDGYRGISLANWMCLAKWESGYNTRA
TNYNAGDRSTDYGIFQINSRYWCNDGKTPGAVNACHLSCSALLQDNIADAVACAKRVVRD
PQGIRAWVAWRNRCQNRDVRQYVQGCGV
```

The complete amino acid sequence in FASTA format has been provided in **Figure 2.2a**. It is also possible to specify a fragment of the sequence by providing a sub-range of the query sequence.

(ii) Allows selecting from a database to search against the input sequence. The nonredundant protein sequences (nr) have been selected as the database in **Figure 2.2a**.

(iii) The algorithm of the program can be selected and, for finding similar sequences, BLASTP is used.

(iv) It is possible to adjust several parameters: (a) displaying the maximum number of aligned sequences, expect threshold, and word size. Expect threshold (e-value) is the expected number of chance matches in a random model, and it is set at 10 as the default value. Word size is the length of the seed that initiates an alignment. In addition, scoring parameters can be selected for matrix, gap cost, and compositional adjustments. The substitution matrix is a key element in evaluating the quality of a pairwise sequence alignment, which assigns a score for aligning any possible pair of residues. Generally BLOSUM62 is used as the substitution matrix, which is a 20×20 matrix obtained for all possible substitutions of 20 amino acid residues (**Table 2.1**). It is based on a likelihood method by estimating the occurrence of each possible pairwise substitution using the biochemical character of amino acid residues (aliphatic, aromatic, positive charged, negative charged, polar, sulfur containing, etc., see **Figure 1.2**), and the development of BLOSUM62 has been described in Eddy (2004). The gap cost is a cost to create and extend a gap in an alignment. Furthermore, options are available to filter the low-complexity regions and mask query and lowercase letters in the sequence.

The output shows the closest sequence in the database with the alignment score and e-value for statistical significance (**Figure 2.2b**). The low e-value indicates that the alignment is statistically significant, and it is not obtained for random. For each sequence hit with BLAST, one can get the details about the alignment of residues with the query sequence, and an example is shown in **Figure 2.2c**. In this figure,

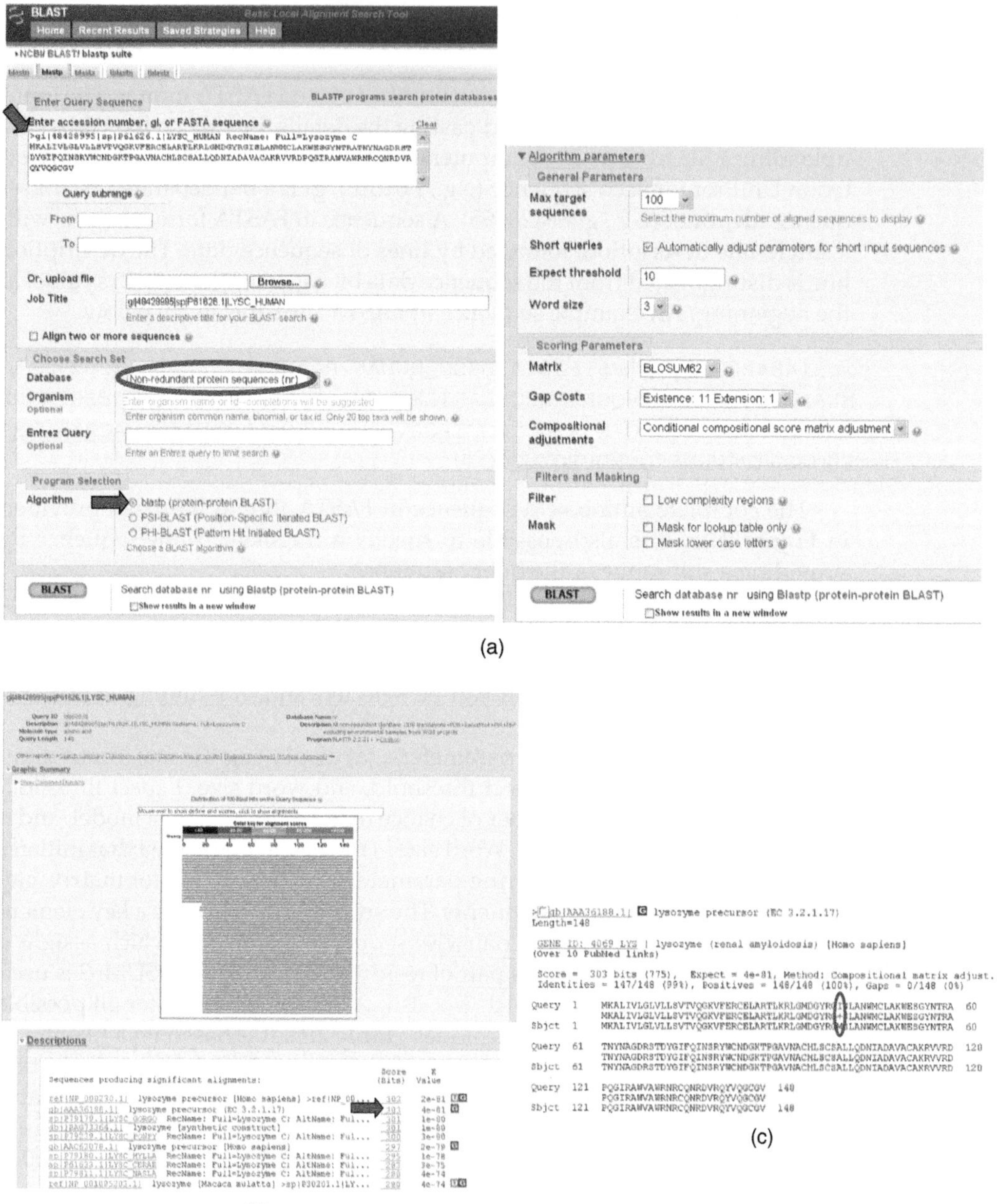

(a)

(b)

(c)

FIGURE 2.2 Retrieval of similar sequences using BLAST: (a) the input page showing the query sequence and other options, (b) the sequences that are showing high sequence identity with the query sequence, and (c) the sequence alignment of the two homologous sequences.

TABLE **2.1**   Blosum62 matrix

	A	R	N	D	C	Q	E	G	H	I	L	K	M	F	P	S	T	W	Y	V
A	4	-1	-2	-2	0	-1	-1	0	-2	-1	-1	-1	-1	-2	-1	1	0	-3	-2	0
R	-1	5	0	-2	-3	1	0	-2	0	-3	-2	2	-1	-3	-2	-1	-1	-3	-2	-3
N	-2	0	6	1	-3	0	0	0	1	-3	-3	0	-2	-3	-2	1	0	-4	-2	-3
D	-2	-2	1	6	-3	0	2	-1	-1	-3	-4	-1	-3	-3	-1	0	-1	-4	-3	-3
C	0	-3	-3	-3	9	-3	-4	-3	-3	-1	-1	-3	-1	-2	-3	-1	-1	-2	-2	-1
Q	-1	1	0	0	-3	5	2	-2	0	-3	-2	1	0	-3	-1	0	-1	-2	-1	-2
E	-1	0	0	2	-4	2	5	-2	0	-3	-3	1	-2	-3	-1	0	-1	-3	-2	-2
G	0	-2	0	-1	-3	-2	-2	6	-2	-4	-4	-2	-3	-3	-2	0	-2	-2	-3	-3
H	-2	0	1	-1	-3	0	0	-2	8	-3	-3	-1	-2	-1	-2	-1	-2	-2	2	-3
I	-1	-3	-3	-3	-1	-3	-3	-4	-3	4	2	-3	1	0	-3	-2	-1	-3	-1	3
L	-1	-2	-3	-4	-1	-2	-3	-4	-3	2	4	-2	2	0	-3	-2	-1	-2	-1	1
K	-1	2	0	-1	-3	1	1	-2	-1	-3	-2	5	-1	-3	-1	0	-1	-3	-2	-2
M	-1	-1	-2	-3	-1	0	-2	-3	-2	1	2	-1	5	0	-2	-1	-1	-1	-1	1
F	-2	-3	-3	-3	-2	-3	-3	-3	-1	0	0	-3	0	6	-4	-2	-2	1	3	-1
P	-1	-2	-2	-1	-3	-1	-1	-2	-2	-3	-3	-1	-2	-4	7	-1	-1	-4	-3	-2
S	1	-1	1	0	-1	0	0	0	-1	-2	-2	0	-1	-2	-1	4	1	-3	-2	-2
T	0	-1	0	-1	-1	-1	-1	-2	-2	-1	-1	-1	-1	-2	-1	1	5	-2	-2	0
W	-3	-3	-4	-4	-2	-2	-3	-2	-2	-3	-2	-3	-1	1	-4	-3	-2	11	2	-3
Y	-2	-2	-2	-3	-2	-1	-2	-3	2	-1	-1	-2	-1	3	-3	-2	-2	2	7	-1
V	0	-3	-3	-3	-1	-2	-2	-3	-3	3	1	-2	1	-1	-2	-2	0	-3	-1	4

147 out of 148 residues were exact matches with each other, and hence the sequence identity is 99%.

BLAST can also be used to understand the similarity between two sequences (Tatiana and Madden, 1999). An example is shown in **Figure 2.3**. In this figure, the sequence similarity between α-lactalbumin (1ALC) and hen egg lysozyme (4LYZ) has been computed using the BLASTP option (**Figure 2.3a**). It showed that 115 residues have been aligned, and the sequence identity is 38% (**Figure 2.3b**). On the other hand, when similar residues are grouped together (e.g., hydrophobic, polar, etc.), 58% of residues are identified as similar residues.

Furthermore, BLAST has the option to download the programs and databases and is available at http://www.ncbi.nlm.nih.gov/BLAST/download.shtml. Once the software is downloaded, it is convenient to use the program locally. It can be done with the commands, *formatdb –i inputfile –n databasefile* (to create a database to search against a query sequence; inputfile is the sequences in the database, and databasefile is the created database) and *blastpgp –i queryfile –d databasefile –o output-file* (queryfile is the query sequence, and outputfile is the result file). The commands can be executed with different options, including specific e-value, tabular form, and the details are available at the homepage (tutorial) of BLAST.

2.2.2 FASTA

Pearson and Lipman (1988) developed the program, FASTA, for comparing protein and DNA sequences. This program was developed with optimized searches for local alignments using substitution matrices, and it has high level of sensitivity

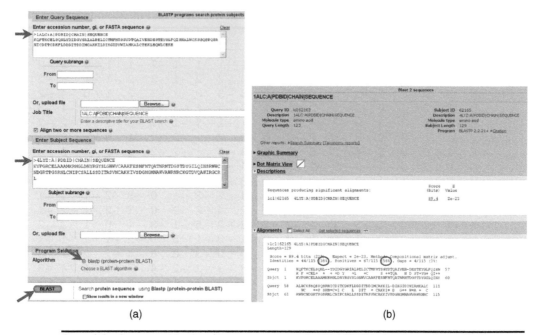

Figure 2.3 Sequence identity of two sequences using BLAST. This has been done with the option, BLAST2 sequences: (a) snapshot showing the two sequences to be aligned to obtain the sequence identity and (b) the alignment details and score between the two query sequences. Here the sequence identity is 38%. Positives show the amino acids that are similar based on chemical behavior (e.g., positively charged, hydrophobic, etc.)

for similarity searching. It has several features, including the similarity search with several sequence and structure databases, genomics, and proteomics. An example to obtain the related sequences of T4 lysozyme is shown in **Figure 2.4**. The related sequences with the details of UniProt ID, length, percentage of identity and similarity, aligned number of residues, and *p*-values are displayed in the output. In addition, it has several features to get the pairwise alignment details, sequences of aligned proteins, visual representation, and so on. The executable files and databases for the sequence alignment program, FASTA, can be obtained from ftp://ftp.ebi.ac.uk/pub/software/unix/fasta/.

2.2.3 ClustalW

ClustalW is a general purpose, multiple sequence alignment program for biomolecules. In multiple sequence alignment, the tables contain patterns of amino acid conservation, from which distant relationships may be reliably detected. Essentially, multiple sequence alignment should have a distribution of closely and distantly related sequences. If all sequences are very closed related, the information content is largely redundant, and few inferences can be drawn. If all sequences are distantly related, it will be difficult to construct an accurate alignment; and in such cases, the quality of the results is questionable.

ClustalW produces biologically meaningful multiple sequence alignments of divergent sequences. It calculates the best match for the selected sequences and lines them up so that the identities, similarities, and differences can be easily seen.

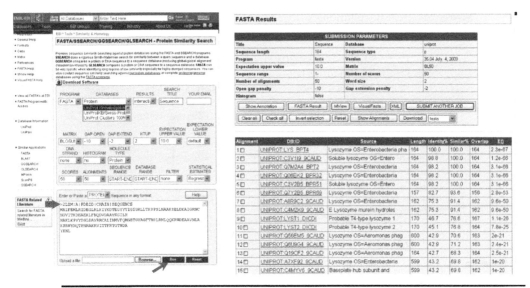

FIGURE 2.4   Sequence similarity search using FASTA.

It is available at http://www.ebi.ac.uk/clustalw/. It is also possible to download the software and run locally. An example to get the multiple sequence alignment of few proteins is shown in **Figure 2.5**. **Figure 2.5a** shows the input page showing the conditions (here default values have been used). The multiple sequence alignment obtained with ClustalW is shown in **Figure 2.5b**. **Figure 2.5c** shows the score for each pair of sequences with the details about the name and number of residues in each pair. The multiple sequence alignments have a wide range of applications, including the secondary structure prediction of proteins, identification of functionally important residues, motifs, structurally similar regions, etc (Pickett et al. 1992; Cuff and Barton, 2000; Kunin et al. 2001; Garg et al. 2005; Cheng and Baldi, 2006; Muramatsu and Suwa, 2006; Chu et al. 2006; Dor and Zhou, 2007; López et al. 2007; Kumar et al. 2008; Cole et al. 2008; Shen and Chou, 2008; Liu et al. 2008; Dou et al. 2009; Sankararaman et al. 2009).

Yamada et al. (2006) proposed a group-to-group sequence alignment algorithm using the concept of piecewise linear gap cost. Utilizing the algorithm, a program PRIME (Profile-based Randomized Iteration Method) has been developed to optimize the well-defined sum-of-pairs score, and it can construct accurate alignments without employing pairwise alignment information. PRIME is available at http://prime.cbrc.jp/.

Pei and Grishin (2007) developed a multiple sequence alignment method, PROMALS, for protein homologs with sequence identity below 10%. The PROMALS algorithm is mainly based on the following features: (i) sequence database searches to retrieve additional homologs, (ii) an accurate secondary structure prediction, (iii) a Hidden Markov Model (HMM) that uses a novel combined scoring of amino acids and secondary structures, and (iv) probabilistic consistency-based scoring applied to progressive alignment of profiles. It is available at http://prodata.swmed.edu/promals/. Recently, Kemena and Notredame (2009) analyzed the upcoming challenges for multiple sequence alignment methods in the high-throughput era.

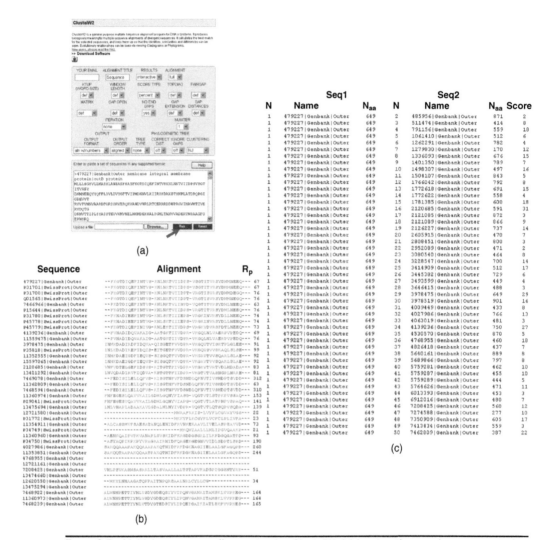

(a)

(b)

(c)

FIGURE 2.5 Multiple sequence alignment using ClustalW: (a) input options and sequences to be aligned, (b) the alignment of sequences, and (c) the score between two aligned sequences and the same for all possible combinations. R_p: residue position; N: sequence number; N_{aa}: number of residues.

2.2.4 PSI-BLAST

PSI-BLAST (Position-Specific Iterative BLAST) is a program that searches a database of sequences similar to a query sequence. PSI-BLAST begins with search results obtained with BLAST and derives pattern information from a multiple sequence alignment of the initial hits. It then repeats the process and fine-tunes the pattern in successive cycles. It is available at http://www.ebi.ac.uk/Tools/psiblast/. In BLAST, the PSSM profiles can be obtained with the option –Q imposed in the program (e.g., ./blastpgp -d [database] -j 2 -i [input file name] -Q [output file name]) along with other options. The tutorial about the usage of PSI-BLAST is available at http://www.ncbi.nlm.nih.gov/Education/BLASTinfo/psi1.html.

2.2.5   Position-specific scoring matrices (profiles)

Position-Specific Scoring Matrices (PSSM) or profiles express the patterns inherent in a multiple sequence alignment of a set of homologous sequences. The basic idea to use profiles is to match the query sequences from the database against the sequences in the alignment table, giving higher weight to positions that are conserved than to those that are variable. These profiles are obtained with a set of probability scores for each amino acid (or gap) at each position of the alignment. Profiles have several applications such as (i) they permit greater accuracy in alignments of distantly related sequences, (ii) the conservation patterns facilitate the identification of other homologous sequences, (iii) patterns from the sequences are useful in classifying subfamilies within a set of homologues, (iv) most structure prediction methods are reliable if based on a multiple sequence alignment rather than on a single sequence, etc. Recently, PSSM profiles have been successfully used for discriminating proteins of different folding types, identification of binding residues, functional residues, etc (Kelley et al. 2000; Reche et al. 2002; Ahmad and Sarai, 2005; Sim et al. 2005; Su et al. 2006; Hwang et al. 2007; Ou et al. 2008, 2009; Kumar et al. 2007; 2008).

2.2.6   Algorithm to develop PSSM-400 for residue pairs

Kumar et al. (2007) reported an algorithm for deriving PSSM for the 400 residue pairs and applied it to identify DNA-binding proteins/domains from the amino acid sequence. The development of PSSM-400 is illustrated in **Figure 2.6**. **Step 1**: Generate PSSM for the query sequence using PSI-BLAST search against "nr" database with an e-value cut off of 0.001. The PSSM contains probability of occurrence of each type of amino acid at each residue position of the protein sequence. **Step 2**: Normalize the PSSM in the range of 0 to 1 using the formula: (X-min)/max-min, where X is the data in PSSM. **Step 3**: Combine the pairs of amino acids in the PSSM and get the average value for each residue pair. The examples for LA and EC are shown in **Figure 2.6**. The PSSM obtained for all the residue pairs (20×20) is termed as PSSM-400.

2.2.7   Hidden Markov Models

The HMM is a computational structure for describing the subtle patterns that define families of homologous sequences. HMMs are powerful tools for detecting distant relatives and for predicting protein folding patterns. It has been widely used for discriminating β-barrel membrane proteins, recognizing protein folds, etc. (Bienkowska et al. 2000; Martelli et al. 2002; Alexandrov and Gerstein, 2004; Scheeff and Bourne, 2006; Bigelow and Rost, 2006). HMMs are the only methods based entirely on sequences that are competitive with PSI-BLAST for identifying distant homologues.

HMMs are usually represented as procedures for generating sequences. They include the possibility of introducing gaps into the generated sequence with position-dependent gap penalties. HMMs have the ability to carry out both the alignment and the assignment of probabilities together. Software for applying HMMs to biological sequence analysis can achieve training, detection of distant homologues, and alignment of additional sequences (Lesk, 2002).

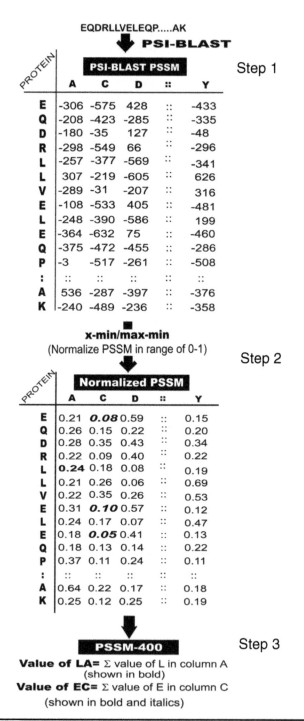

EQDRLLVELEQP.....AK

PSI-BLAST

PSI-BLAST PSSM   Step 1

PROTEIN	A	C	D	::	Y
E	-306	-575	428	::	-433
Q	-208	-423	-285	::	-335
D	-180	-35	127	::	-48
R	-298	-549	66	::	-296
L	-257	-377	-569	::	-341
L	307	-219	-605	::	626
V	-289	-31	-207	::	316
E	-108	-533	405	::	-481
L	-248	-390	-586	::	199
E	-364	-632	75	::	-460
Q	-375	-472	-455	::	-286
P	-3	-517	-261	::	-508
:	::	::	::	::	::
A	536	-287	-397	::	-376
K	-240	-489	-236	::	-358

■
x-min/max-min
(Normalize PSSM in range of 0-1)

Step 2

Normalized PSSM

PROTEIN	A	C	D	::	Y
E	0.21	*0.08*	0.59	::	0.15
Q	0.26	0.15	0.22	::	0.20
D	0.28	0.35	0.43	::	0.34
R	0.22	0.09	0.40	::	0.22
L	**0.24**	0.18	0.08	::	0.19
L	0.21	0.26	0.06	::	0.69
V	0.22	0.35	0.26	::	0.53
E	0.31	*0.10*	0.57	::	0.12
L	0.24	0.17	0.07	::	0.47
E	0.18	*0.05*	0.41	::	0.13
Q	0.18	0.13	0.14	::	0.22
P	0.37	0.11	0.24	::	0.11
:	::	::	::	::	::
A	0.64	0.22	0.17	::	0.18
K	0.25	0.12	0.25	::	0.19

PSSM-400   Step 3

Value of LA= Σ value of L in column A
(shown in bold)
Value of EC= Σ value of E in column C
(shown in bold and italics)

FIGURE **2.6**   Steps to develop PSSM-400 for a query sequence (Kumar et al. 2007).

2.3   Amino acid properties

Amino acid sequences have a lot of hidden information, which can be used for developing sequence-based prediction methods. Several amino acid properties can be derived from the knowledge of protein sequences. Recently, it has been reported that the composition of amino acid residues plays a vital role to discriminate proteins belonging to different structural classes, folding types, and that perform different functions. These studies include the prediction of protein structural classes (Chou and Zhang, 1995), discriminating DNA-binding proteins (Ahmad et al. 2004; Yu et al. 2006), RNA-binding proteins (Yu et al. 2006), α-helical membrane proteins (Chou and Elrod, 1999; Qiu et al. 2009), β-barrel membrane proteins (Gromiha and Suwa, 2005, 2006), proteins belonging to different folds (Chou, 1995), functional classification of membrane proteins (Gromiha and Yabuki, 2008), folding rates (Gromiha, 2005a; Gromiha et al. 2006; Huang and Gromiha, 2007), and secretary proteins (Zuo and Li, 2009). In addition, the preference of amino acid residue pairs along the sequence, motifs specific to different folding types of proteins has been determined.

2.3.1   Amino acid occurrence

Amino acid occurrence is the number of amino acids of each type present in a protein. For example, the T4 lysozyme has 164 residues, and the amino acid occurrence is the information about each of the 20 amino acid residues in this protein, i.e., Ala: 15, Asp: 10, Cys: 2, etc.

2.3.2   Amino acid composition

The amino acid composition is the number of amino acids of each type normalized with the total number of residues. It is defined as

$$\text{Comp}(i) = \sum n_i {}^* 100/N, \qquad (2.1)$$

where i stands for the 20 amino acid residues; n_i is the number of residues of each type, and N is the total number of residues. The summation is through all the residues in the considered protein.

For example, the compositions of Ala, Asp, and Cys in T4 lysozyme are 9.15%, 6.10%, and 1.22%, respectively. The computed amino acid composition for a set of globular and transmembrane β-barrel proteins (TMBs) is presented in **Table 2.2**. Several residues showed a significant difference between the compositions in globular and β-barrel membrane proteins (**Figure 2.7**). This result reveals the importance of specific residues in these classes of proteins. The polar residues, especially Ser, Asn, and Gln, have higher occurrence in TMBs, which are important for the folding, stability, and function of such class of proteins (Pautsch and Schulz, 2000; Vandeputte-Rutten et al. 2001; Chimento et al. 2003a,b; Yue et al. 2003; Zeth et al. 2000).

2.3.3   Total and average amino acid property

The total amino acid property for each residue type in a protein was computed using the following formula:

$$P_{\text{total}}(i) = P(i){}^* \sum n_i, \qquad (2.2)$$

TABLE **2.2**   Amino acid composition for the 20
amino acid residues in globular and TMBs

Residue	Composition (%)	
	Globular	**TMB**
Ala	8.47	8.95
Asp	5.97	5.91
Cys	**1.39**	**0.47**
Glu	**6.32**	**4.78**
Phe	3.91	3.68
Gly	7.82	8.54
His	**2.26**	**1.25**
Ile	**5.71**	**4.77**
Lys	5.76	4.93
Leu	8.48	8.78
Met	2.21	1.56
Asn	**4.54**	**5.74**
Pro	4.63	3.74
Gln	**3.82**	**4.75**
Arg	4.93	5.24
Ser	**5.94**	**8.05**
Thr	5.79	6.54
Val	7.02	6.76
Trp	1.44	1.24
Tyr	3.58	4.13

The amino acid residues that have large difference
(>0.9) between globular proteins and TMBs are
highlighted in bold. Data were taken from Gromiha
and Suwa (2005).

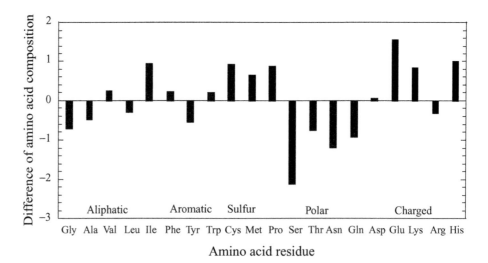

FIGURE **2.7**   Amino acid compositional difference between globular and β-barrel membrane
proteins. Figure was adapted from Gromiha and Suwa (2005).

where $P(i)$ and n_i are the property value and number of residues of ith-type residue. The summation is over N, the total number of residues in a protein. The amino acid properties for the 20 amino acid residues have been discussed in **Section 2.5** and **Chapter 3**. The average amino acid property of a protein can be obtained by dividing the total amino acid property with the total number of residues in the protein.

2.3.4 Residue pair preference

The composition of dipeptides is a measure to quantify the preference of amino acid residue pairs in a sequence. This has been computed using the following expression (Gromiha et al. 2005):

$$\text{Dipep}(i,j) = \sum N_{ij} \, {}^{*}100 / \left(\sum N_i + \sum N_j \right),\tag{2.3}$$

where i, j stands for the distribution of 20 amino acid residues at positions i and $i+1$. $N_{i,j}$ is the number of residues of type i followed by the residue j. ΣN_i and ΣN_j are the total number of residues of type i and j, respectively.

The dipeptide composition for all possible 400 pairs in globular and TMBs has been computed using Equation 2.3, and the differences between them are presented in **Table 2.3a**. In this table, positive values indicate higher occurrence in TMBs than globular proteins, and negative values show the preference in globular proteins. The occurrence of dipeptides, AQ, GY, LG, LS, SA, SL, SV, and VS, is significantly higher in TMBs than in globular proteins. A majority of these dipeptides involve the residue Ser, which is one of the most favored residues in TMBs. On the other hand, the dipeptides, EE, HH, HP, RE, and YP, have higher occurrence in globular proteins than TMBs. It is noteworthy that most of these dipeptides involve the charged residues, Glu and His, which have significantly higher occurrence in globular proteins compared with TMBs.

The distribution of amino acid residues in transmembrane helical (TMH) proteins is entirely different due to the presence of the stretch of hydrophobic residues in this class of proteins. The difference between the dipeptide composition of α-helical and TMBs is presented in **Table 2.3b**. As expected α-helical membrane proteins have a higher occurrence of hydrophobic neighbors (AI, IF, IL, LI, LL, LV, etc.) than do TMBs.

2.3.5 Motifs

The concept of motifs provides the information about the preference of residue pairs with a gap (any residue between the pair of residues). This has been computed by using the same expression that was used for dipeptide composition (Eqn. 2.3). The main difference is that the residues i and j are the distribution of 20 amino acid residues at positions i and $i+1$ for AB, i and $i+2$ for A*B, i and $i+3$ for A**B, and so on (Gromiha, 2005b).

The difference of the A*B motif composition between globular and TMBs for all possible 400 pairs is presented in **Table 2.4**. This table shows that the occurrence of dipeptide motifs, S*S, V*S, R*D, Q*Q, Q*N, S*N, R*N, Y*F, T*N and G*S, is significantly higher in TMBs than in globular proteins. These dipeptide motifs mainly contain the residues Ser, Asn, and Gln, which play an important role to the structure and stability of TMBs. On the other hand, the dipeptide motifs, E*L, A*K,

TABLE 2.3a Difference of dipeptide composition in globular and TMBs

Residue	Ala	Asp	Cys	Glu	Phe	Gly	His	Ile	Lys	Leu	Met	Asn	Pro	Gln	Arg	Ser	Thr	Val	Trp	Tyr
Ala	0.16	0.18	−0.72	−0.18	−0.05	0.75	−0.69	−0.77	−0.50	0.41	−0.75	0.16	0.08	**1.29**	0.18	0.82	0.37	−0.40	−0.15	−0.01
Asp	−0.11	−0.03	−0.77	−0.33	−0.27	−0.44	−0.55	0.35	−0.01	0.11	−0.22	0.84	−0.90	0.13	0.03	−0.07	0.51	0.27	0.02	0.46
Cys	−0.55	−0.43	−0.58	−0.52	−0.56	−0.85	−0.84	−0.73	−0.47	−0.68	−0.38	−0.76	−0.87	−0.79	−0.93	−0.81	−0.65	−0.56	−0.17	−0.69
Glu	−0.57	−0.71	−0.40	**−1.54**	−0.31	−0.44	−0.11	−0.20	−0.82	−0.52	−0.38	0.12	−0.44	0.42	0.06	0.10	−0.44	−0.40	−0.26	−0.32
Phe	−0.03	0.33	−0.42	−0.41	−0.19	0.43	−0.62	−0.42	−0.14	−0.83	−0.54	0.59	−0.60	−0.22	0.11	0.48	−0.32	−0.38	−0.42	−0.06
Gly	0.49	−0.10	−0.89	0.05	0.40	0.31	−1.01	−0.35	−0.54	0.42	−0.70	0.98	−0.67	−0.17	0.09	0.57	0.09	0.43	0.26	**1.20**
His	−0.63	−0.32	−0.65	−0.12	−0.85	−0.89	−1.06	−0.49	−0.20	−0.92	−0.45	−0.23	−0.10	−0.17	−0.45	−0.27	−0.55	−0.86	−0.85	−0.25
Ile	−0.45	−0.39	−0.62	0.25	−0.50	−0.08	−0.96	−0.77	−0.11	−0.76	−0.65	0.34	0.17	−0.10	0.16	0.10	0.03	−0.70	−0.20	−0.43
Lys	−0.64	−0.14	−0.59	−0.25	−0.39	−0.18	−0.36	−0.31	−0.12	−0.60	−0.53	0.10	−0.29	0.21	−0.26	0.09	0.46	−0.60	0.06	0.35
Leu	−0.03	0.23	−0.69	−0.70	0.58	1.23	−0.78	−0.24	−0.83	−0.07	−0.37	0.72	0.08	0.10	−0.38	**1.32**	0.20	−0.37	−0.15	−0.07
Met	−0.39	−0.74	−0.48	−0.59	−0.28	−0.58	−0.51	−0.27	0.12	−0.38	−0.32	−0.38	−0.12	−0.22	−0.39	−0.08	−0.33	−0.47	−0.20	−0.37
Asn	0.58	0.33	−0.76	−0.26	0.20	0.17	−0.40	0.48	0.81	0.54	−0.20	0.87	−0.67	0.56	0.31	0.65	0.50	0.49	0.04	0.52
Pro	−0.15	−0.62	−0.05	−0.83	0.12	−0.67	−0.72	−0.38	−0.21	0.14	−0.19	−0.73	−0.39	0.04	0.05	−0.73	−0.05	0.10	−0.08	−0.21
Gln	0.85	0.30	−0.56	−0.20	0.00	0.05	−0.36	0.04	−0.29	0.35	−0.40	0.13	0.55	0.63	0.64	0.77	0.59	0.45	0.14	0.54
Arg	0.36	−0.12	−0.73	**−1.05**	0.17	0.11	0.22	−0.09	−0.23	0.29	−0.14	0.66	0.17	0.46	−0.17	0.34	0.41	0.04	−0.36	0.41
Ser	**1.10**	0.13	−0.86	−0.32	0.14	0.59	−0.73	0.26	0.32	**1.48**	−0.41	0.35	−0.38	0.25	0.44	0.79	0.55	**1.32**	−0.16	0.86
Thr	0.49	0.59	−0.90	−0.22	−0.49	−0.09	−0.47	−0.19	0.14	0.84	−0.33	0.56	0.25	0.46	0.21	0.52	0.17	0.34	−0.38	0.16
Val	−0.16	−0.31	−0.77	−0.18	−0.49	0.50	−0.77	−0.68	−0.58	−0.21	−0.54	0.14	0.16	0.39	0.33	**1.10**	0.31	0.03	−0.37	−0.05
Trp	−0.23	−0.02	0.06	−0.18	−0.33	−0.01	−0.13	−0.59	−0.17	−0.13	−0.11	−0.18	−0.34	0.05	0.02	0.20	−0.25	−0.56	0.40	−0.05
Tyr	0.49	0.97	−0.72	0.04	0.06	0.76	−0.15	−0.05	0.02	−0.03	−0.47	0.32	**−1.13**	0.84	0.72	0.56	0.45	−0.36	−0.41	−0.18

The dipeptides that have large difference (>1.0) between TMB and globular proteins are highlighted in bold. Data were taken from Gromiha et al. (2005).

TABLE 2.3b Difference of dipeptide composition in TMH and TMBs

Residue	Ala	Asp	Cys	Glu	Phe	Gly	His	Ile	Lys	Leu	Met	Asn	Pro	Gln	Arg	Ser	Thr	Val	Trp	Tyr
Ala	-0.33	1.19	-0.46	0.80	-1.02	0.00	-0.02	-2.11	0.52	-1.54	-1.37	0.82	0.09	1.47	0.52	0.16	0.15	-1.20	-0.59	0.65
Asp	1.26	0.37	-0.01	0.30	0.72	1.36	-0.33	1.32	0.50	1.56	-0.06	1.09	-0.31	0.13	-0.36	0.69	0.90	1.51	-0.03	0.98
Cys	-0.23	0.16	0.10	-0.02	-0.52	-0.40	-0.13	-0.58	-0.18	-0.28	-0.32	-0.15	0.23	0.06	-0.27	-0.14	-0.03	-0.40	-0.32	-0.50
Glu	0.47	-0.24	0.11	-0.64	0.71	0.78	-0.30	0.59	0.01	1.11	-0.44	0.76	-0.34	-0.50	-0.11	-0.06	0.47	0.59	-0.08	0.24
Phe	-1.55	1.28	-0.52	0.63	-1.94	-1.00	-0.19	-1.54	0.67	-1.63	-1.02	0.79	-0.44	0.46	0.22	-0.13	-0.49	-1.40	1.71	-0.11
Gly	-0.55	0.93	-0.35	0.44	-0.71	0.17	-0.11	-1.50	0.72	-0.74	-1.59	1.32	-0.36	0.44	0.23	0.94	0.47	0.03	-0.38	1.63
His	-0.15	-0.16	-0.04	-0.49	-0.46	-0.43	-0.94	0.07	-0.03	-0.35	-0.10	0.18	-0.43	0.47	-0.47	-0.15	-0.20	-0.09	-0.53	-0.69
Ile	-1.85	0.90	-0.47	0.79	-2.13	-1.84	-0.08	-1.73	1.09	-2.15	-1.40	0.98	0.13	0.81	0.85	-0.58	0.07	-1.58	-0.62	-0.09
Lys	0.62	0.41	-0.09	0.17	0.66	0.62	-0.30	1.06	0.29	0.90	-0.22	0.69	0.11	0.14	-0.15	0.81	0.99	0.81	0.43	0.72
Leu	-1.64	1.82	-0.49	0.69	-1.54	-0.36	-0.28	-2.00	0.84	-2.70	-1.22	1.42	0.08	0.71	0.31	0.66	0.10	-2.01	-0.88	-0.03
Met	-1.44	-0.56	-0.16	-0.63	-1.14	-1.43	-0.63	-1.03	0.74	-1.22	-0.99	-0.70	-0.62	-0.30	-0.61	-0.83	-1.12	-1.27	-0.59	-0.37
Asn	0.85	0.60	-0.08	0.37	0.49	1.21	-0.17	1.03	1.32	1.37	-0.14	1.29	-0.43	0.58	0.51	1.25	1.06	1.21	-0.03	1.02
Pro	0.02	0.01	0.49	-0.66	-0.48	0.00	-0.67	-0.40	0.06	-0.18	-0.62	-0.16	-0.40	0.10	0.13	0.16	0.29	-0.31	-0.25	-0.01
Gln	1.21	0.74	0.17	0.16	0.73	0.23	-0.60	0.49	-0.03	0.73	-0.49	0.02	0.37	-0.53	0.33	0.84	0.49	0.77	0.36	0.63
Arg	0.81	0.03	-0.09	-0.89	0.76	0.54	0.10	0.31	-1.00	0.70	-0.46	0.68	-0.12	0.13	-1.42	0.50	0.32	0.65	-0.67	0.87
Ser	0.65	0.82	-0.01	-0.02	-0.17	0.85	-0.32	-0.06	0.93	0.58	-0.92	0.84	-0.10	0.54	0.14	1.03	0.58	0.93	-0.33	1.23
Thr	0.16	0.99	-0.15	0.23	-0.18	0.38	-0.35	-0.24	1.00	0.18	-1.04	1.24	0.54	0.23	0.44	0.98	0.52	0.60	-0.19	0.62
Val	-0.89	0.70	-0.54	1.05	-1.05	-0.27	-0.05	-1.81	0.87	-1.71	-1.24	1.20	-0.17	0.58	0.86	0.80	0.43	-1.21	-0.71	0.14
Trp	-0.46	0.54	0.01	0.29	-1.10	-0.29	-0.34	-0.84	-0.19	-1.23	-0.49	-0.19	-0.78	-0.34	-0.15	-0.08	-0.27	-0.95	-0.45	-0.20
Tyr	0.71	1.83	-0.22	0.83	-0.15	0.80	-0.05	0.03	1.03	-0.10	-0.59	1.14	-0.39	0.94	0.99	0.68	0.80	0.12	-0.54	0.32

The dipeptides that have large difference (>2.0) between TMB and α-helical membrane proteins are highlighted in bold. Data were taken from Gromiha et al. (2005).

TABLE 2.4 Difference of dipeptide composition of A*B motif in globular and TMBs

Residue	Ala	Asp	Cys	Glu	Phe	Gly	His	Ile	Lys	Leu	Met	Asn	Pro	Gln	Arg	Ser	Thr	Val	Trp	Tyr
Ala	0.92	−0.27	−0.76	−0.71	0.06	−0.11	−0.89	−0.05	−1.24	0.71	−0.73	0.13	0.73	0.23	−0.15	0.81	0.57	0.47	−0.01	0.65
Asp	−0.06	0.26	−0.64	−0.52	−0.54	0.53	−0.12	−0.86	0.24	−0.54	0.08	0.37	−0.23	0.74	0.84	0.37	0.13	−0.50	−0.05	−0.17
Cys	−0.60	−1.00	0.75	−0.82	−0.51	−1.03	−0.84	−0.47	−0.33	−0.74	−0.34	−0.75	−0.42	−0.78	−0.84	−0.78	−0.43	−0.69	−0.51	−0.61
Glu	−0.66	−0.16	−0.51	−0.03	−0.82	−0.36	−0.46	−0.95	−0.13	−1.56	−0.86	0.28	−0.43	0.45	0.16	0.55	0.12	−1.06	−0.69	−0.19
Phe	0.14	−0.50	−0.70	−0.75	0.25	0.44	−0.78	−0.07	−0.57	0.49	−0.49	−0.42	−0.24	−0.31	−0.25	−0.20	0.07	0.06	0.24	0.21
Gly	0.31	−0.39	−1.02	0.17	0.29	0.49	−0.64	−0.29	−0.05	0.47	−0.24	0.90	−1.03	0.23	0.32	1.03	0.32	−0.30	0.08	0.38
His	−0.80	−0.55	−0.60	−0.36	0.01	−0.99	−1.04	−0.70	−0.40	−0.58	−0.78	−0.23	−0.80	0.08	−0.43	−0.26	−0.62	−0.94	−0.21	−0.80
Ile	−0.44	0.10	−0.58	−0.94	−0.23	−0.26	−0.81	−0.13	−0.73	−0.01	−0.42	−0.45	0.40	0.27	−0.75	−0.03	−0.30	0.06	0.04	0.16
Lys	−0.20	0.26	−0.51	−0.32	−0.85	−0.40	−0.18	−0.88	0.27	−0.91	−0.58	0.46	−0.86	0.23	0.38	0.52	0.49	−0.70	−0.18	−0.12
Leu	0.84	−0.11	−0.69	−1.23	0.12	0.67	−0.73	0.13	−1.15	0.93	−0.40	−0.20	−0.15	0.02	0.06	0.35	−0.22	0.94	−0.01	0.79
Met	−0.41	−0.44	−0.50	−0.81	−0.25	−0.30	−0.59	−0.48	−0.14	−0.22	−0.03	−0.29	−0.38	−0.27	−0.27	−0.45	−0.17	−0.44	−0.16	−0.12
Asn	0.47	0.19	−0.77	0.38	0.23	0.72	−0.35	−0.38	0.08	0.44	−0.29	0.86	−0.26	0.68	0.36	1.40	0.50	0.31	−0.15	0.14
Pro	−0.43	−0.58	−0.61	−0.54	−0.15	−0.70	−0.52	0.55	0.01	−0.05	−0.34	0.15	−0.28	−0.42	−0.02	−0.01	−0.53	−0.20	−0.35	−0.67
Gln	0.04	0.69	−0.61	0.27	0.15	0.33	−0.29	0.05	0.49	0.08	−0.54	1.11	−0.05	1.14	0.21	0.62	0.80	0.14	−0.42	0.01
Arg	0.14	1.21	−0.85	0.58	−0.27	0.51	−0.65	−0.61	−0.03	−0.79	−0.50	1.05	−0.35	0.40	0.38	0.68	0.28	−0.44	−0.60	0.14
Ser	0.55	0.67	−0.68	−0.08	−0.10	0.95	−0.51	−0.20	0.14	0.44	−0.39	1.06	0.21	0.90	0.66	1.65	0.57	0.34	−0.20	0.30
Thr	0.18	0.12	−0.70	−0.54	−0.02	0.58	−0.78	0.00	0.06	0.45	−0.18	1.04	−0.70	0.30	0.32	0.40	0.69	0.32	−0.19	0.43
Val	0.34	−0.38	−0.85	−0.63	−0.26	0.10	−0.71	0.09	−0.52	0.73	−0.69	−0.28	0.14	0.26	−0.54	−0.15	0.05	0.48	0.24	0.78
Trp	0.05	−0.14	−0.03	−0.48	0.20	−0.06	−0.09	−0.06	−0.47	0.20	−0.28	−0.27	−0.34	−0.13	−0.04	−0.25	−0.65	0.06	−0.26	−0.06
Tyr	−0.02	0.22	−0.43	−0.07	1.05	0.03	0.17	0.24	0.48	0.29	−0.40	0.32	−0.83	0.59	0.13	0.21	0.54	0.15	0.03	0.40

The dipeptides that have large difference (>1.0) between TMB and globular proteins are highlighted in bold. Data were taken from Gromiha (2005b).

L*E, L*K, H*H, C*G, and G*C, have a higher occurrence in globular proteins than do TMBs. It is noteworthy that most of these dipeptide motifs involve the charge residues, Lys, Glu, and His, which have significantly higher occurrence in globular proteins compared with TMBs.

Kleiger et al. (2002) showed that GxxxG and AxxxA are common α-helical interaction motifs in proteins, and particularly in extremophiles. Schneider and Engelman (2004) carried out statistical searches for specific motifs that mediate transmembrane helix–helix interactions and showed that two glycine residues separated by three intervening residues (GxxxG) provide a framework for specific interactions. Johnson et al. (2006) reported that the position of the Gly-xxx-Gly motif in transmembrane segments modulates dimer affinity. Furthermore, other motifs of small residues can mediate the interaction of transmembrane domains, so that the AxxxA-motif could also drive strong interactions of α-helices in soluble proteins. Recently, the motif Po.GHy.Hy.Hy (Po, polar residue; G, glycine; Hy, large hydrophobic residue), which occurs near the end of the most C-terminal β-strand, has been identified as a sorting signal to be specific for mitochondrial β-barrel outer membrane proteins (Kutik et al. 2008; Imai et al. 2008).

The motifs in protein sequences can be searched from UniProt using the option "pattern search." The rules followed to perform a pattern search are given below: (i) Use capital letters for amino acid residues, (ii) use "[...]" for a choice of multiple amino acids in a particular position. [LIVM] means that L, I, V, or M can be in the first position, (iii) use "{...}" to exclude amino acids. {CF} means C and F should not be in that particular position, (iv) use "x" or "X" for a position that can be any amino acid, and (v) use "(n)," where n is a number, for multiple positions; x(3) is the same as "xxx." As an example, consider the pattern, [LIVM]-[VIC]-x(2)-G-[DENQTA]-x-[GAC]-x(2)-[LIVMFY](4)-x (2)-G. This denotes a 17 amino acid peptide that has a L, I, V, or M at position 1; a V, I, or, C at position 2; any residue at positions 3 and 4; a G at position 5; and so on. One can obtain any type of patterns from UniProt, and **Figure 2.8** illustrates the steps to obtain such patterns. **Figure 2.8a** shows the desired input pattern, and **Figure 2.8b** is the output obtained from sequence databases. In addition, the matching region of the search pattern is also shown in **Figure 2.8c**.

2.3.6 Conservation score

The conservation score for all the residues in a protein can be obtained by comparing the sequence of a PDB chain with the proteins deposited in Swiss-Prot (see **Section 1.4.2**) and finds the ones that are homologous to the PDB sequence. The number of PSI-BLAST iterations and the e-value cutoff used in all similarity searches were 1 and 0.001, respectively. The protein sequence alignments are used to classify the residues in the protein into nine categories: from very variable (score = 1) to highly conserved (score = 9). Glaser et al. (2003) developed a server for calculating the conservation score, and it is available at http://consurf.tau.ac.il/. An example is shown in **Figure 2.9**. It takes the PDB ID (4LYZ) and chain information (A) as inputs (**Figure 2.9a**) and displays the output with several options: amino acid conservation scores, view the scores in Jmol and Chimera, sequences obtained with PSI-BLAST and used to compute the conservation score, and so on. The Jmol view of the results and detailed conservation scores are shown in **Figures 2.9b** and **c**,

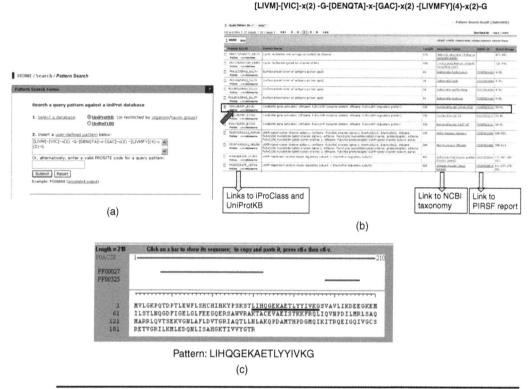

FIGURE 2.8   Searching for patterns in protein sequences: (a) the input showing desired pattern, (b) the sequences in databases with the same pattern, and (c) the position of the pattern in protein sequence.

respectively. It provides the details about the sequence in one- and three-letter codes, the normalized score, the conservation details (conserved: 9 and the variable: 1), the number of aligned sequences, and the residues presented in the alignment at each position.

Pei and Grishin (2001) developed a program based on amino acid frequencies at each position in a multiple sequence alignment. This program takes the output obtained from a multiple alignment program (ClustalW in **section 2.2.3**) and displays the conservation score in the output. The executable file and online server are available at http://prodata.swmed.edu/al2co/al2co.php. **Figure 2.10** shows the input parameters and output results obtained with AL2CO server. The high positive values indicate the conserved residues.

2.3.7   Nonredundant protein sequences

Li and Godzik (2006) developed a program, CD-HIT, for obtaining the nonredundant sequences at different cutoff of sequence identities. It uses the clustering algorithm and eliminates the redundant sequences. The main advantages of this program are as follow: (i) It can handle huge datasets, (ii) it is easy to download, and (iii) the results can be obtained quickly. CD-HIT is freely available at http://

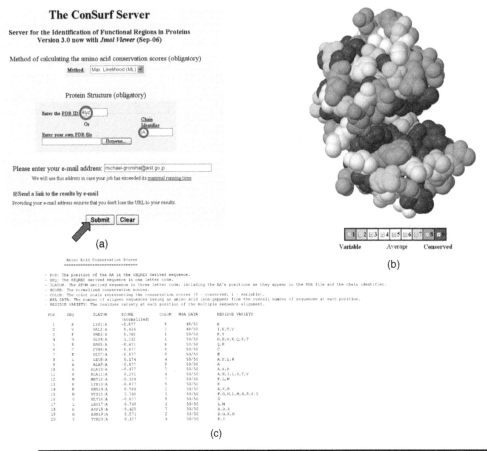

Figure 2.9 Computation of conservation score using ConSurf server: (a) input parameters, (b) Jmol view of conservation score, and (c) detailed results.

cd-hit.org/. CD-HIT can be used to create the nonredundant dataset of less than 40% sequence identity.

Another popular program to create nonredundant sets of protein sequences is blastclust. It is a program within the standalone BLAST package that is used to cluster either protein or nucleotide sequences. The program begins with pairwise matches and places a sequence in a cluster if the sequence matches at least one sequence already in the cluster. In the case of proteins, the BLASTP algorithm is used to compute the pairwise matches. The general command to create a set of nonredundant protein sequences is *blastclust -i infile -o outfile -p T -L .9 -b T -S 95*, where <u>infile</u> and <u>outfile</u> are input and output files, resepectively. T stands for protein; the coverage of the length and sequence identity cutoff are 90% (-L .9) and 95% (-S 95), respectively.

The PISCES server (http://dunbrack.fccc.edu/pisces/), which is used to obtain nonredundant structures, is also used to reduce the redundancy in protein sequences. It takes the amino acid sequence in FASTA format and sends the list of nonredundant protein sequences by e-mail. An example is shown in **Figure 2.11**. It

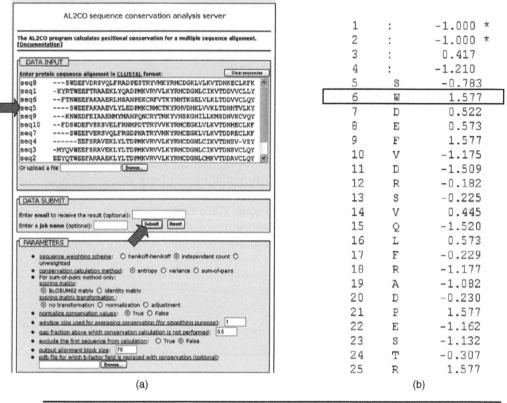

FIGURE 2.10   Utility of AL2CO server for obtaining conservation score from ClustalW aligned sequences: (a) input options and (b) conservation score. The residue W has high positive score (1.577) indicating the conservation.

shows the protein sequences in the FASTA format given as the input (**Figure 2.11a**) and the output (**Figure 2.11b**) files.

2.4   Amphipathic character of α-helices and β-strands

The amino acid sequence information has also been used to detect the amphipathic character of α-helices and β-strands, which are used for secondary structure prediction (Cid et al. 1992; Gromiha and Ponnuswamy, 1995).

2.4.1   α-helices

The amphipathic character of α-helices was determined by the following procedure (Muthusamy and Ponnuswamy, 1990; Ponnuswamy and Gromiha, 1993): In this procedure, the residues of an α-helical segment are considered on four adjacent edges along the direction of the helical axis. The average hydrophobicity (numerical values for the 20 amino acid residues; see AAindex in **Chapter 3**) of the residues constituting the edge i ($i = 1, 4$) is given by

$$\alpha_i = \left(\sum h_{i+j} \right) / n, \tag{2.4}$$

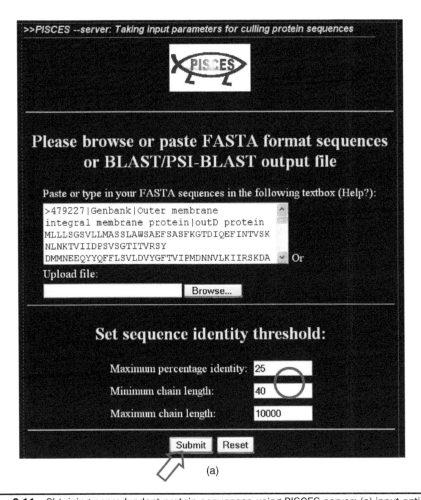

(a)

FIGURE 2.11   Obtaining nonredundant protein sequences using PISCES server: (a) input options in the server and (b) output files, sequence id and FASTA sequences.

where n is the total number of residues in the edge, j increases at an interval of 4 from 0 to m, m being the number of residues in the helix; h is the hydrophobic index of the residue. The pictorial representation for the amphipathic character of α-helices is shown in **Figure 2.12a**. The average hydrophobicity is given by $<H> = (\Sigma h_i)/N$, where N is the number of residues in a protein.

The power of amphipathicity of a helix is taken to be

$$A_\alpha = |(\alpha_1 + \alpha_2) - (\alpha_3 + \alpha_4)| \text{ or } |(\alpha_1 + \alpha_4) - (\alpha_2 + \alpha_3)|. \qquad (2.5)$$

Note that only the above two combinations can constitute the opposing faces for a helix (**Figure 2.12a**). It has been reported that 75% of the helical segments in known structures are amphipathic in nature.

2.4.2   β-strands

The amphipathic character of a β-strand has been determined as follows (Muthusamy and Ponnuswamy, 1990; Gromiha and Ponnuswamy, 1993): In this

```
> 7467903|Genbank|Outer membrane integral membrane
MSKFTITIFITTLLFTGSVIALDLEQALTEGYKNNEELKAAQIKFLNAIE
QFPQAFSGFMPNVGLQINRQNSKTKYNKKYVNRLGITPRETASTQGILTI
EQSLFNGGASIAALKAAQSGFRASRSEYYAGEQKVLLNLITAYLDCVESK
EKYDISESRVRTNIQQVKTVEEKLRLGEATAIDIAAARAGLAAAETNKLA
AYADFQGKKANFIKVFGIEANDITMPDLPDRLPISLDEFTRKAAKFNPDI
NSARHNVTVTKALEMVQKGKLLPQVSVKLLSGGTNYNPQEPVIQNINNRI
YTTTLSVNIFIYPEGGAQYSRIRSAKNQTRNSVVQLDSAIKQIKAGVVSV
WEGFETAKSRIVAANQGVEAAQISYNGIVQEEIVGSKTILDVLDAEQKLY
EAKITRVDAYKNSVLASYQMKLLTGELTAKSLKLKVKYFSPEEEFNNLKK
KMFIGF
> 11559475|Genbank|Outer membrane integral membrane
MTRNRFVMRRIATTLLVAGIIVSQAAYAQVTLNFVNADIDQVAKAIGAAT
GKTIIVDPRVKGQLNLVAERPVPEDQALKTLQSALRMQGFALVQDHGVLK
VVPEADAKLQGVPTYIGNAFQARGDQVITQVFELHNESANNLLPVLRPLI
SFNNTVTAYPANNTIVVTDYADNVRRIAQIISGVDSAAGAQVQVVPLRNA
NAIDLAAQLQKMLDPGAIGNSDATLKVSVTADPRTNALLLRASNASRLAA
AKRLVQQLDAPSAVPGNMHVVPLRNADAVKLAKTLRGMLGKGGNDSGSSA
SSNDANSFNQNGGSSASGNFSTGTSGTPPLPSGGLGGSSSSSYGGSGGSS
GGGLGTGGLLGGDKDKSGDDNQPGGMIQADSATNSLIITASDPVYRNLRS
VIDQLDARRAQVYIEALIVELNSTTQGNLGIQWQVASGQFLGGTNLAPTA
GNGLGNSIINLTAGGLTNAAGGITGGGLASNLGQLSQGLNIGWLHNMFGV
QGLGALLQYFAGVSDANVLSTPNLITLDNEEAKIVVGQNVPIATGSYSNL
TSGTTSNAFNTYDRRDVGLTLHVKPQITDGGILKLQLYTEDSAVVNGTTN
SQTGPTFTKRSIQSTILADNGEIIVLGGLMQDNYQVSNSKVPLLGDIPWI
GQLFRSESKVRAKTNLMVFLRPVIISDRSTAQEVTSNRYDYIQGVTGAYK
SDNNVIRDKDDFVVPFMPLGPSQGGTAAGNLFDLDKMRRQQLQRQVVPVP
AQPLPEATPAQPQGVPLQAVPQQPLTTAPGASQ
> 7469324|Genbank|Outer membrane integral membrane
MRSNSVKNFRFWLTTEIATCCLLALAPAQAETVSQSNTLDGDLRTAIAGD
SSRDWLQFRKSLEQSLKQKEETDSWKPSLELMQAKSLVKPGQKLTNIELL
VQELEALSDFLALNFFEPNQTSVAQMAPPSRPMPPPPAGSGQVMFPNPEI
IIQQQGGVPQRGASPQVGNPSILSPAVPVAPVRSRAVPPPVGDLAISNIN
ASFDMIDLGQRGQVNVPSLVLREAPAREVLAVLTRYAGMNLIFTDNQNNE
GTPTPGTPPGGQVAPPQAQSTITLDIQNESVQDVFNYVLMASGLKASRRG
NTIFAGANLLPSARNIITRTIRLNQASAESVASTLASQGAEVNILFEGQE
DVQLAENAPFRVIKQFPTLVPLTVQKFANDSSVLILEGLVVSTDPRLNTV
TLVGEFRNVELASSMITQMDARRRQVAVNVKIIDINLNNIQDYDSSFSFG
IGDSFFVQDSGSAVMRFGDTAPVQEIDINNNLGRITNPPAIVNPFQDGEI
FFDLNRITNIEVPLGPGTIFINFFTSGSGAVSNNPLFNGVTEFPIVEVDE
QGLLTITQPEFGLPSFYQYPKKFQAQIDAQIRSGNAKILTDPTLIVQEGE
AAQVKLTESVIASVDTQVDTQGDTAVRTITPVLEDVGLTLNVIVDRIDDN
GFITLRVNPIVASPAGTQVFDSGAGAINEITLINKRELTSGVVRLRDDQT
FILSGIISELQRSTTSKVPILGDLPVIGALFRQSTDTTDRSEVIILMTPK
IIHDSTEAQFGFRYNPDAATAEFLRQKGFPVQAQP
```

Sequence id

IDs	length
7467903|Genbank|Outer	456
11559475|Genbank|Outer	783
7469324|Genbank|Outer	785
15596906|Genbank|Outer	295
P26466|SwissProt|Outer	452
15597487|Genbank|Outer	452
5640161|Genbank|Outer	889
P13949|SwissProt|Outer	201
P10170|SwissProt|Outer	260
P16945|SwissProt|Outer	292
7208425|Genbank|Outer	560
15598604|Genbank|Outer	891
13470835|Genbank|Outer	794
P19196|SwissProt|Outer	835
12620518|Genbank|Outer	230
P31600|SwissProt|Outer	990
15596468|Genbank|Outer	616
P15727|SwissProt|Outer	482
7520765|Genbank|Outer	778
P16466|SwissProt|Outer	1577
3228547|Genbank|Outer	700
P06970|SwissProt|Outer	812
P22340|SwissProt|Outer	505
P44601|SwissProt|Outer	565
P35077|SwissProt|Outer	584
12721580|Genbank|Outer	444
7470479|Genbank|Outer	654
P13794|SwissProt|Outer	350
P06111|SwissProt|Outer	257
P24126|SwissProt|Outer	530
P29041|SwissProt|Outer	759
5759281|Genbank|Outer	462

(b)

FIGURE 2.11   (Continued)

procedure, a β-strand segment is considered to have two faces, and the average hydrophobicity of residues constituting the face i ($i = 1, 2$) is given by

$$\beta_i = \left(\sum h_{i+j}\right)/n, \tag{2.6}$$

where n is the total number of residues in the face, j increases at an interval of 2 from 0 to m, m being the number of residues in the strand. The pictorial representation for the amphipathic character of β-strands is shown in **Figure 2.12b**.

The amphipathicity index of a strand is computed using the equation,

$$A_\beta = |\beta_1 - \beta_2|. \tag{2.7}$$

The structural analysis showed that about 65% of the β-strands possess amphipathic character.

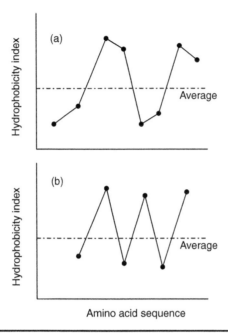

FIGURE **2.12**   Amphipathic character of (a) α-helices and (b) β-strands.

2.4.3   Hydrophobic moment for measuring amphipathicity

Eisenberg et al. (1984) proposed the concept of hydrophobic moment, which measures the amphipathic character of amino acid residues in a protein segment. In this criterion, the periodicities in the polar/apolar character of the amino acid sequence of a protein have been examined by assigning to each residue a numerical hydrophobicity and searching for periodicity in the resulting one-dimensional function. The hydrophobic moment is the strength of each periodic component.

If the three-dimensional structure of the protein is known, the hydrophobic moment can be calculated from the relationship,

$$\mu_s = \sum H_n s_n, \tag{2.8}$$

where H_n is the numerical hydrophobicity of the nth residue and s_n is a unit vector in the direction from the nucleus of the α carbon toward the geometric center of the side chain; n varies from 1 to N, N being the number of amino acid residues in a segment.

The hydrophobic moment of a segment of protein can also be estimated from the amino acid sequence, provided the segment is periodic and the period is known. Let the periodic structure be specified by m, the number of residues per turn, or alternatively, $\delta = 2\pi/m$, in which δ is the angle in radians at which successive side chains emerge from the backbone, when the periodic segment is viewed down its axis. Thus, for an α helix, δ is 100° ($m = 3.6$); and for a strand of β structure, δ is expected to be in the range of 160° ($m = 2.3$) to 180° ($m = 2.0$). A periodic structure that is amphipathic will yield a large value for μ, given by

$$\mu = \left\{ \left[\sum H_n \sin(\delta n) \right]^2 + \left[\sum H_n \cos(\delta n) \right]^2 \right\}^{1/2}, \tag{2.9}$$

in which δ is measured in radians and the length of the hydrophobic moment, μ, is given by the sum of the components of the hydrophobicity vectors.

2.4.4   Detecting amphipathic structures in proteins

Cornette et al. (1987) defined the amphipathic index of α-helices as a measure of the Fourier transform power spectrum concentrated around $95°$ compared to the total area under the spectrum,

$$\frac{1/25 \int_{85}^{110} P(\omega)d\omega}{1/180 \int_{0}^{180} P(\omega)d\omega}, \tag{2.10}$$

where $P(\omega)$ is the power spectrum of frequency ω used to detect periodic variation in a sequence: $P(\omega) = [\Sigma H_n \sin(\delta n)]^2 + [\Sigma H_n \cos(\delta n)]^2$, where H_n is the hydrophobicity values of the residues along the sequence and n is the number of residues in a sequence.

In a similar manner, the amphipathic index for the power spectrum of a collection of strands is given by (Cornette et al. 1987):

$$\frac{1/20 \int_{160}^{180} P(\omega)d\omega}{1/180 \int_{0}^{180} P(\omega)d\omega}. \tag{2.11}$$

This is also referred as β-amphipathic index.

2.5   Amino acid properties for sequence analysis

Several tools are available on the Internet to compute different parameters from amino acid sequences. For example, the computation of amino acid composition of a protein is available at http://www.expasy.ch/tools/protparam.html, http://pir.georgetown.edu/pirwww/search/comp_mw.shtml. This is illustrated in **Figure 2.13**. These servers take the amino acid sequence as the input (**Figure 2.13a**) and display the amino acid composition and molecular weight in the output (**Figure 2.13b**).

An et al. (1998) developed a program, 3Dinsight, to analyze several amino acid properties, (http://gibk26.bse.kyutech.ac.jp/jouhou/3dinsight/3dinsight_main.html). An example for the analysis of T4 lysozyme (PDB code: 2LZM) is shown in **Figure 2.14a**. The property, surrounding hydrophobicity, has been selected to understand the hydrophobic characteristics of residues along the chain. The Web server has several options to obtain the values for each residue in tabular form

ProtParam (References / Documentation) is a tool which allows the computation of various physical and chemical parameters for a given pr stored in Swiss-Prot or TrEMBL or for a user entered sequence. The computed parameters include the molecular weight, theoretical pI, am composition, atomic composition, extinction coefficient, estimated half-life, instability index, aliphatic index and grand average of hydropathici (GRAVY) (Disclaimer).

www.expasy.ch

Please note that you may only fill out **one** of the following fields at a time.

Enter a Swiss-Prot/TrEMBL accession number (AC) (for example **P05130**) or a sequence identifier (ID) (for example **KPC1_DROME**):

Or you can paste your own sequence in the box below:

```
KVFGRCELAAAMKRHGLDNYRGYSLGNWVCAAKFESNFNTQATNRNTDGSTDYGILQIN
SRWWCNDGRTPGSRNLCNIPC
SALLSSDITASVNCAKKIVSDGNGMNAWVAWRNRCKGTDVQAWIRGCRL
```

RESET   Compute parameters

■ HOME / Search / *Composition/Molecular Weight Calculation*

Composition/Molecular Weight Calculation Form   **?**

pir.georgetown.edu

Enter any UniProtKB identifiers:
(separated by a space)

and/or
Insert your sequences below using the single letter amino acid code:
(separate sequences by an empty line)

```
>4LYZ:A|PDBID|CHAIN|SEQUENCE
KVFGRCELAAAMKRHGLDNYRGYSLGNWVCAAKFESNFNTQATNRNTDGSTDYGILQINS
RWWCNDGRTPGSRNLCNIPC
SALLSSDITASVNCAKKIVSDGNGMNAWVAWRNRCKGTDVQAWIRGCRL
```

Submit   Reset

Example: P53039 (sample output/annotated output)

(a)

FIGURE **2.13** Calculation of amino acid composition using two different servers: (a) input, showing the amino acid sequence of the query protein and (b) output, showing the amino acid composition of the protein.

along with average and sum of the property values, and graphical output. The graphical representation of the variation of hydrophobicity along the sequence of T4 lysozyme is shown in **Figure 2.14b**.

In addition, Gromiha et al. (1999a) created a set of 49 properties for amino acids and utilized them for the analysis and prediction of protein folding rates and protein mutant stability (Gromiha et al. 1999b; 2006). The same set of properties has been widely used by several researchers on various applications of protein

ProtParam

www.expasy.ch

User-provided sequence:

```
         10         20         30         40         50         60
KVFGRCELAA AMKRHGLDNY RGYSLGNWVC AAKFESNFNT QATNRNTDGS TDYGILQINS
         70         80         90        100        110        120
RWWCNDGRTP GSRNLCNIPC SALLSSDITA SVNCAKKIVS DGNGMNAWVA WRNRCKGTDV
QAWIRGCRL
```

References and documentation are available.

• Please note the modified algorithm for extinction coefficient.

Number of amino acids: 129

Molecular weight: 14313.1

Theoretical pI: 9.32

Amino acid composition: [CSV format]

Ala (A)	12	9.3%
Arg (R)	11	8.5%
Asn (N)	14	10.9%
Asp (D)	7	5.4%
Cys (C)	8	6.2%
Gln (Q)	3	2.3%
Glu (E)	2	1.6%
Gly (G)	12	9.3%
His (H)	1	0.8%
Ile (I)	6	4.7%
Leu (L)	8	6.2%
Lys (K)	6	4.7%
Met (M)	2	1.6%
Phe (F)	3	2.3%
Pro (P)	2	1.6%
Ser (S)	10	7.8%
Thr (T)	7	5.4%
Trp (W)	6	4.7%
Tyr (Y)	3	2.3%
Val (V)	6	4.7%

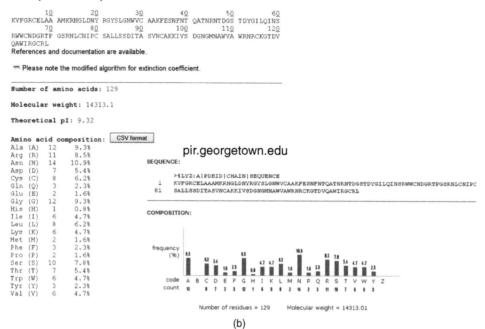

pir.georgetown.edu

SEQUENCE:

```
        >4LYZ:A|PDBID|CHAIN|SEQUENCE
      1 KVFGRCELAAAMKRHGLDNYRGYSLGNWVCAAKFESNFNTQATNRNTDGSTDYGILQINSRWWCNDGRTPGSRNLCNIPC
     81 SALLSSDITASVNCAKKIVSDGNGMNAWVAWRNRCKGTDVQAWIRGCRL
```

COMPOSITION:

Number of residues = 129   Molecular weight = 14313.01

(b)

FIGURE 2.13   (Continued)

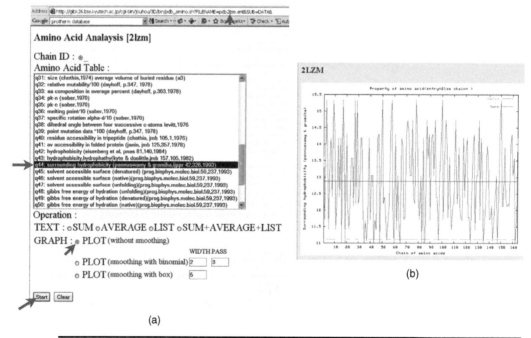

(a)

(b)

FIGURE 2.14   Amino acid analysis for different properties: (a) options to select the property (here surrounding hydrophobicity is selected) and results (plot without smoothing is selected) and (b) the hydrophobicity plot for 2LZM.

Numerical values for amino acid properties

Property	Ala	Asp	Cys	Glu	Phe	Gly	His	Ile	Lys	Leu	Met	Asn	Pro	Gln	Arg	Ser	Thr	Val	Trp	Tyr
1 KO	-25.50	-33.12	-32.82	-36.17	-34.54	-27.00	-31.84	-31.78	-32.40	-31.78	-31.18	-30.90	-23.25	-32.60	-26.62	-29.88	-31.23	-30.62	-30.24	-35.01
2 Ht	0.87	0.66	1.52	0.67	2.87	0.10	0.87	3.15	1.64	2.17	1.67	0.09	2.77	0.00	0.85	0.07	0.07	1.87	3.77	2.67
3 Hp	13.05	11.10	14.30	11.41	13.89	12.20	12.42	15.34	11.01	14.19	13.62	11.72	11.06	11.78	12.40	11.68	12.12	14.73	13.96	13.57
4 P	0.00	49.70	1.48	49.90	0.35	0.00	51.60	0.10	49.50	0.13	1.43	3.38	1.58	3.53	52.00	1.67	1.66	0.13	2.10	1.61
5 pH1	6.00	2.77	5.05	5.22	5.48	5.97	7.59	6.02	9.74	5.98	5.74	5.41	6.30	5.65	10.76	5.68	5.66	5.96	5.89	5.66
6 pK'	2.34	2.01	1.65	2.19	1.89	2.34	1.82	1.36	2.10	2.36	2.28	2.02	1.99	2.17	1.81	2.21	2.10	2.32	2.38	2.20
7 Mw	89.00	133.00	121.00	147.00	165.00	75.00	155.00	131.00	146.00	131.00	149.00	132.00	115.00	146.00	174.00	105.00	119.00	117.00	204.00	181.00
8 Bl	11.50	11.68	13.46	13.57	19.80	3.40	13.67	21.40	15.71	21.40	16.25	12.82	17.43	14.45	14.28	9.47	15.77	21.57	21.61	18.03
9 Rf	9.90	2.80	2.80	3.20	18.80	5.60	8.20	17.10	3.50	17.60	14.70	5.40	14.80	9.00	4.60	6.90	9.50	14.30	17.00	15.00
10 Mu	14.34	12.00	35.77	17.26	29.40	0.00	21.81	19.06	21.29	18.78	21.64	13.28	10.93	17.56	26.66	6.35	11.01	13.92	42.53	31.55
11 Hnc	0.62	0.90	0.29	-0.74	1.19	0.48	-0.40	1.38	-1.50	1.06	0.64	-0.78	0.12	-0.85	-2.53	-0.18	-0.05	1.08	0.81	0.26
12 Esm	1.40	1.16	1.37	1.16	1.14	1.36	1.22	1.19	1.07	1.32	1.30	1.18	1.24	1.12	0.92	1.30	1.25	1.25	1.03	1.03
13 El	0.49	0.35	0.67	0.37	0.72	0.53	0.54	0.76	0.30	0.65	0.65	0.38	0.46	0.40	0.55	0.45	0.52	0.73	0.83	0.65
14 Et	1.90	1.52	2.04	1.54	1.86	1.90	1.76	1.95	1.37	1.97	1.96	1.56	1.70	1.52	1.48	1.75	1.77	1.98	1.87	1.69
15 Pa	1.42	1.01	0.70	1.51	1.13	0.57	1.00	1.08	1.16	1.21	1.45	0.67	0.57	1.11	0.98	0.77	0.83	1.06	1.08	0.69
16 Pb	0.83	0.54	1.19	0.37	1.38	0.75	0.87	1.60	0.74	1.30	1.05	0.89	0.55	1.10	0.93	0.75	1.19	1.70	1.37	1.47
17 Pt	0.66	1.46	1.19	0.74	0.60	1.56	0.95	0.47	1.01	0.59	0.60	1.56	1.52	0.98	0.95	1.43	0.96	0.50	0.96	1.14
18 Pc	0.71	1.21	1.19	0.84	0.71	1.52	1.07	0.66	0.99	0.69	0.59	1.37	1.61	0.87	1.07	1.34	1.08	0.63	0.76	1.07
19 Ca	20.00	26.00	25.00	33.00	46.00	13.00	37.00	39.00	46.00	35.00	43.00	28.00	22.00	36.00	55.00	20.00	28.00	33.00	61.00	46.00
20 F	0.96	1.14	0.87	1.07	0.69	1.16	0.80	0.76	1.14	0.79	0.78	1.04	1.16	1.07	1.05	1.13	0.96	0.79	0.77	1.01
21 Br	0.38	0.14	0.57	0.09	0.51	0.38	0.31	0.56	0.04	0.50	0.42	0.15	0.18	0.11	0.07	0.23	0.23	0.48	0.40	0.26
22 Ra	3.70	2.60	3.03	3.30	6.60	3.13	3.57	7.69	1.79	5.88	5.21	2.12	2.12	2.70	2.53	2.43	2.60	7.14	6.25	3.03
23 Ns	6.05	4.95	7.86	5.10	6.62	6.16	5.80	7.51	4.88	7.37	6.39	5.04	5.65	5.45	5.70	5.53	5.81	7.62	6.98	6.73
24 aN	1.59	0.53	0.33	1.45	1.14	0.53	0.89	1.22	1.13	1.91	1.25	0.53	0.00	0.98	0.67	0.70	0.75	1.42	1.33	0.58
25 aC	1.44	2.13	0.76	2.01	1.01	0.62	0.56	0.68	0.59	0.58	0.73	0.93	2.19	1.20	0.39	0.81	1.25	0.63	1.40	0.72
26 aM	1.22	0.56	1.53	1.28	1.13	0.40	2.23	0.77	1.65	1.05	1.47	0.93	0.00	1.63	1.59	0.87	0.46	1.20	0.46	0.52
27 VO	60.46	73.83	67.70	85.88	121.48	43.25	98.79	107.72	108.50	107.75	105.35	78.01	82.83	93.90	127.34	60.62	76.83	90.78	143.91	123.60
28 Nm	2.11	1.80	1.88	2.09	1.98	1.53	1.98	1.77	1.96	2.19	2.27	1.84	1.32	2.03	1.94	1.57	1.57	1.63	1.90	1.67
29 Nl	3.92	2.85	5.55	2.72	4.53	4.31	3.77	5.58	2.79	4.59	4.14	3.64	3.57	3.06	3.78	3.75	4.09	5.43	4.83	4.93
30 Hgm	13.85	11.61	15.37	11.38	13.93	13.34	13.82	15.28	11.58	14.13	13.86	13.02	12.35	12.61	13.10	13.39	12.70	14.56	15.48	13.88
31 ASAD	104.00	132.20	132.50	161.90	182.00	73.40	165.80	171.50	165.20	161.40	189.00	134.90	135.10	164.90	210.20	111.40	130.40	143.90	208.80	196.40
32 ASAN	23.20	62.40	17.90	81.00	33.10	29.20	57.70	28.30	107.50	31.10	41.30	60.50	60.70	71.50	94.50	48.70	52.00	28.10	39.50	50.40
33 dASA	70.90	69.60	114.30	80.50	148.40	44.00	107.90	142.70	87.50	129.80	147.90	74.00	73.50	93.30	116.00	62.80	78.00	115.60	167.80	145.90
34 dGh	-0.54	-2.97	-1.64	-3.71	-1.06	-0.59	-3.38	0.32	-2.19	0.27	-0.60	-3.55	0.32	-3.92	-5.96	-3.82	-1.97	0.13	-3.80	-5.64
35 dGhD	-0.58	-6.10	-1.91	-7.37	-1.35	-0.82	-5.57	0.40	-5.97	0.35	-0.71	-6.63	0.56	-7.12	-12.76	-6.18	-3.66	0.18	-4.71	-8.45
36 GhN	-0.06	-3.11	-0.27	-3.62	-0.28	-0.23	-2.18	0.07	-1.70	0.07	-0.10	-3.03	0.23	-3.15	-6.85	-2.36	-1.69	0.04	-0.98	-2.82
37 dHh	-2.24	-4.54	-3.43	-5.63	-5.11	-1.46	-6.83	-3.84	-5.02	-3.32	-4.16	-5.68	-1.95	-6.23	-10.43	-5.94	-4.39	-3.15	-8.99	-10.67
38 -Tdsh	1.70	1.57	1.79	1.92	4.05	0.87	3.45	4.16	2.83	3.79	3.56	2.13	2.27	2.31	4.47	2.12	2.42	3.28	5.19	5.03
39 dcph	14.22	2.73	9.41	3.17	39.06	4.88	20.05	41.98	17.68	38.26	31.67	3.91	3.80	3.74	16.66	6.14	16.11	32.58	37.69	30.54
40 dGc	0.51	2.89	2.71	3.59	3.22	0.68	3.95	-0.40	1.87	-0.35	1.13	3.26	-0.39	3.69	5.25	3.42	1.74	-0.19	5.59	6.56
41 dHc	2.77	4.72	8.64	5.69	11.93	1.23	7.64	4.03	3.57	3.69	7.06	3.64	1.97	4.47	6.03	5.80	4.42	3.45	13.46	14.41
42 -Tdsc	-2.25	-1.83	-5.92	-2.11	-8.71	-0.55	-3.69	-4.42	-1.70	-4.04	-5.93	-0.39	-2.36	-0.78	-0.78	-2.38	-2.68	-3.64	-7.87	-7.95
43 dG	-0.02	-0.06	1.08	-0.13	2.16	0.09	0.56	-0.08	-0.32	-0.08	0.53	-0.30	-0.06	-0.23	-0.71	-0.40	-0.06	1.78	0.91	
44 dH	0.51	0.18	5.21	0.05	6.82	-0.23	0.79	0.19	-1.45	0.17	2.89	-2.03	0.02	-1.76	-4.40	-0.16	0.04	0.30	4.47	3.73
45 -Tds	-0.54	-0.26	-4.14	-0.19	-4.66	0.31	-0.23	-0.27	1.13	-0.24	-2.36	1.74	-0.08	1.53	3.69	-0.24	-0.28	-0.36	-2.69	-2.82
46 v	1.00	4.00	2.00	5.00	7.00	0.00	6.00	4.00	5.00	4.00	4.00	4.00	3.00	5.00	7.00	2.00	3.00	3.00	10.00	8.00
47 s	0.00	2.00	0.00	3.00	2.00	0.00	2.00	1.00	0.00	2.00	0.00	2.00	0.00	3.00	5.00	0.00	1.00	1.00	2.00	2.00
48 f	0.00	2.00	1.00	3.00	2.00	0.00	2.00	2.00	4.00	2.00	3.00	2.00	0.00	3.00	5.00	1.00	1.00	1.00	2.00	2.00
49 Pf-s	0.65	2.60	0.95	0.80	0.55	3.55	2.80	0.05	1.70	0.35	0.55	5.25	0.00	1.40	1.45	1.10	0.05	0.10	0.65	0.95

(a)

FIGURE 2.15 Parameters for 49 amino acid properties: (a) numerical values and (b) normalized ones. Data were taken as a screenshot from the Web site http://www.cbrc.jp/~gromiha/fold_rate/property.html

folding and stability (Caballero et al. 2006). The numerical and normalized values for the 49 properties for the amino acids are presented in **Figures 2.15a** and **b**, and brief descriptions about the properties along with data are available at http://www.cbrc.jp/~gromiha/fold_rate/property.html. The amino acid properties were normalized between 0 and 1 using the expression,

$$P_{norm}(i) = [P(i) - P_{min}]/[P_{max} - P_{min}], \tag{2.12}$$

where $P(i)$, $P norm(i)$ are, respectively, the original and normalized values of amino acid i for a particular property, and $P min$ and $P max$ are, respectively, the minimum and maximum values. Further details on amino acid properties are described in **Chapter 3**.

Normalized values for amino acid properties

Property	Ala	Asp	Cys	Glu	Phe	Gly	His	Ile	Lys	Leu	Met	Asn	Pro	Gln	Arg	Ser	Thr	Val	Trp	Tyr
1 KO	0.83	0.24	0.26	0.00	0.13	0.71	0.34	0.34	0.29	0.34	0.39	0.41	1.00	0.28	0.74	0.49	0.38	0.43	0.46	0.09
2 Ht	0.23	0.18	0.40	0.18	0.76	0.03	0.23	0.84	0.44	0.58	0.44	0.02	0.73	0.00	0.23	0.02	0.02	0.50	1.00	0.71
3 Hp	0.47	0.02	0.76	0.09	0.67	0.27	0.33	1.00	0.00	0.73	0.60	0.16	0.01	0.18	0.32	0.15	0.26	0.86	0.68	0.59
4 P	0.00	0.96	0.03	0.96	0.01	0.00	0.99	0.00	0.95	0.00	0.03	0.07	0.03	0.07	1.00	0.03	0.03	0.00	0.04	0.03
5 pHi	0.40	0.00	0.29	0.31	0.34	0.40	0.60	0.41	0.87	0.40	0.37	0.33	0.44	0.36	1.00	0.36	0.36	0.40	0.39	0.36
6 pK'	0.96	0.64	0.28	0.81	0.52	0.96	0.45	0.00	0.80	0.98	0.90	0.65	0.62	0.79	0.44	0.83	0.73	0.94	1.00	0.82
7 Mw	0.11	0.45	0.36	0.56	0.70	0.00	0.62	0.43	0.55	0.43	0.57	0.44	0.31	0.55	0.77	0.23	0.34	0.33	1.00	0.82
8 Bl	0.44	0.45	0.55	0.56	0.90	0.00	0.56	0.99	0.68	0.99	0.71	0.52	0.77	0.61	0.60	0.33	0.68	1.00	1.00	0.80
9 Rf	0.44	0.00	0.00	0.03	1.00	0.18	0.34	0.89	0.04	0.93	0.74	0.16	0.75	0.39	0.11	0.26	0.42	0.72	0.89	0.76
10 Mu	0.34	0.28	0.84	0.41	0.69	0.00	0.51	0.45	0.50	0.44	0.51	0.31	0.26	0.41	0.63	0.15	0.26	0.33	1.00	0.74
11 Hnc	0.81	0.88	0.72	0.46	0.95	0.77	0.54	1.00	0.26	0.92	0.81	0.45	0.68	0.43	0.00	0.60	0.63	0.92	0.85	0.71
12 Esm	1.00	0.50	0.94	0.50	0.46	0.92	0.63	0.56	0.31	0.83	0.79	0.54	0.67	0.42	0.00	0.79	0.69	0.69	0.23	0.23
13 El	0.36	0.09	0.70	0.13	0.79	0.43	0.45	0.87	0.00	0.66	0.66	0.15	0.30	0.19	0.47	0.28	0.42	0.81	1.00	0.66
14 Et	0.79	0.22	1.00	0.25	0.73	0.79	0.58	0.87	0.00	0.90	0.88	0.28	0.49	0.22	0.16	0.57	0.60	0.91	0.75	0.48
15 Pa	0.90	0.47	0.14	1.00	0.60	0.00	0.46	0.54	0.63	0.68	0.94	0.11	0.00	0.57	0.44	0.21	0.28	0.52	0.54	0.13
16 Pb	0.35	0.13	0.62	0.00	0.76	0.29	0.38	0.92	0.28	0.70	0.51	0.39	0.14	0.55	0.42	0.29	0.62	1.00	0.75	0.83
17 Pt	0.17	0.91	0.66	0.25	0.12	1.00	0.44	0.00	0.50	0.11	0.12	1.00	0.96	0.47	0.44	0.88	0.45	0.03	0.45	0.61
18 Pc	0.12	0.61	0.59	0.25	0.12	0.91	0.47	0.07	0.39	0.10	0.00	0.76	1.00	0.27	0.47	0.74	0.48	0.04	0.17	0.47
19 Ca	0.15	0.27	0.25	0.42	0.69	0.00	0.50	0.54	0.69	0.46	0.62	0.31	0.19	0.48	0.88	0.15	0.31	0.42	1.00	0.69
20 F	0.57	0.96	0.38	0.81	0.00	0.23	0.15	0.96	0.21	0.19	0.74	1.00	0.81	0.77	0.94	0.57	0.21	0.17	0.68	
21 Br	0.64	0.19	1.00	0.09	0.89	0.64	0.51	0.98	0.00	0.87	0.72	0.21	0.26	0.13	0.06	0.36	0.36	0.83	0.68	0.42
22 Ra	0.32	0.14	0.21	0.26	0.82	0.23	0.30	1.00	0.00	0.69	0.58	0.06	0.06	0.15	0.13	0.11	0.14	0.91	0.76	0.21
23 Nm	0.39	0.02	1.00	0.07	0.58	0.43	0.31	0.88	0.00	0.84	0.51	0.05	0.26	0.19	0.28	0.22	0.31	0.92	0.70	0.62
24 aN	0.83	0.28	0.17	0.76	0.60	0.28	0.47	0.64	0.59	1.00	0.65	0.28	0.00	0.51	0.35	0.37	0.39	0.74	0.70	0.30
25 aC	0.58	0.97	0.21	0.90	0.34	0.13	0.09	0.16	0.11	0.11	0.19	0.30	1.00	0.45	0.00	0.23	0.48	0.13	0.56	0.18
26 aM	0.55	0.25	0.69	0.57	0.51	0.18	1.00	0.35	0.74	0.47	0.66	0.42	0.00	0.73	0.71	0.39	0.21	0.54	0.21	0.23
27 V0	0.17	0.30	0.24	0.42	0.78	0.00	0.55	0.64	0.65	0.64	0.62	0.35	0.39	0.50	0.84	0.17	0.33	0.47	1.00	0.80
28 Nm	0.83	0.51	0.59	0.81	0.69	0.22	0.69	0.47	0.67	0.92	1.00	0.55	0.00	0.75	0.65	0.26	0.26	0.33	0.61	0.37
29 Nl	0.42	0.05	0.99	0.00	0.63	0.56	0.37	1.00	0.02	0.65	0.50	0.32	0.30	0.12	0.37	0.36	0.48	0.95	0.74	0.77
30 Hgm	0.60	0.06	0.97	0.00	0.62	0.48	0.60	0.95	0.05	0.67	0.60	0.40	0.24	0.30	0.42	0.49	0.32	0.78	1.00	0.61
31 ASAD	0.22	0.43	0.43	0.65	0.79	0.00	0.68	0.72	0.89	0.64	0.85	0.45	0.45	0.67	1.00	0.28	0.42	0.52	0.99	0.90
32 ASAN	0.17	0.50	0.00	0.70	0.17	0.13	0.44	0.12	1.00	0.15	0.26	0.48	0.48	0.60	0.85	0.34	0.38	0.11	0.24	0.36
33 dASA	0.22	0.21	0.57	0.29	0.84	0.00	0.52	0.00	0.35	0.69	0.84	0.24	0.24	0.40	0.58	0.15	0.27	0.58	1.00	0.82
34 dGh	0.86	0.48	0.69	0.36	0.78	0.86	0.41	1.00	0.60	0.99	0.85	0.38	1.00	0.32	0.00	0.34	0.64	0.97	0.34	0.05
35 GhD	0.91	0.50	0.81	0.41	0.86	0.90	0.54	0.99	0.51	0.98	0.90	0.46	1.00	0.42	0.00	0.49	0.68	0.97	0.60	0.32
36 GhN	0.96	0.53	0.93	0.46	0.93	0.94	0.66	0.98	0.73	0.98	0.95	0.54	1.00	0.52	0.00	0.63	0.73	0.97	0.84	0.57
37 dHh	0.92	0.67	0.79	0.55	0.60	1.00	0.42	0.74	0.61	0.78	0.71	0.54	0.95	0.48	0.03	0.51	0.68	0.82	0.18	0.00
38 -TdSh	0.19	0.16	0.21	0.24	0.74	0.00	0.60	0.76	0.45	0.68	0.62	0.29	0.32	0.33	0.83	0.29	0.36	0.56	1.00	0.96
39 dCph	0.29	0.00	0.17	0.01	0.93	0.05	0.44	1.00	0.38	0.91	0.74	0.03	0.53	0.03	0.35	0.09	0.34	0.76	0.89	0.71
40 dGc	0.13	0.47	0.45	0.57	0.52	0.16	0.62	0.00	0.33	0.01	0.22	0.53	0.00	0.59	0.81	0.55	0.31	0.03	0.86	1.00
41 dHc	0.12	0.26	0.56	0.34	0.81	0.00	0.49	0.21	0.18	0.19	0.44	0.18	0.06	0.25	0.36	0.35	0.24	0.17	0.93	1.00
42 -TdSc	0.78	0.83	0.34	0.79	0.00	0.98	0.60	0.52	0.84	0.56	0.33	1.00	0.76	0.95	0.95	0.76	0.72	0.61	0.10	0.09
43 dG	0.24	0.22	0.62	0.20	1.00	0.28	0.44	0.22	0.14	0.22	0.43	0.14	0.23	0.17	0.00	0.11	0.16	0.23	0.07	0.56
44 dH	0.44	0.41	0.86	0.40	1.00	0.37	0.46	0.41	0.26	0.41	0.65	0.21	0.39	0.24	0.00	0.38	0.40	0.42	0.79	0.72
45 -TdS	0.49	0.53	0.06	0.54	0.00	0.60	0.53	0.53	0.69	0.53	0.28	0.77	0.55	0.74	1.00	0.53	0.52	0.51	0.24	0.22
46 v	0.10	0.40	0.20	0.50	0.70	0.00	0.60	0.40	0.50	0.40	0.40	0.40	0.30	0.50	0.70	0.20	0.30	0.30	1.00	0.80
47 s	0.00	0.40	0.00	0.60	0.40	0.00	0.40	0.20	0.00	0.40	0.00	0.40	0.00	0.60	1.00	0.00	0.20	0.20	0.40	0.40
48 f	0.00	0.40	0.20	0.60	0.40	0.00	0.40	0.40	0.80	0.40	0.60	0.40	0.00	0.60	1.00	0.20	0.20	0.20	0.40	0.40
49 Pf-s	0.12	0.50	0.18	0.15	0.10	0.68	0.53	0.01	0.32	0.07	0.10	1.00	0.00	0.27	0.28	0.21	0.01	0.02	0.12	0.18

(b)

FIGURE 2.15   (Continued)

2.6   Exercises

1. Analyze the occurrence of similar proteins in "nr" and SWISS-PROT database for the sequence given below:

```
>1336093|Genbank|Outer membrane integral membrane protein|HrcC
MVEKRELRCRLLGALLMLCATLPAGAQTPADWKEQSYAYSADRTPLSTVLQDFADGHSVD
LHLGNVEDTEVTAKIRAENASAFLDRLALEHHFQWFVYNNTLYVSPQDEQSSERLEISPD
AAPDIKQALSGIGLLDPRFGWGELPDDGVVLVTGPPQYLELVKRFSEQREKKEDRRKVMT
FPLRYASVADRTIHYRDQTVVIPGVATMLNELMNGKRAAPASASGIDSTPGGPDTNSMMQ
NTQTLLSRLSSRNKTSNRAGGRDNEIEDVSGRISADVRNNALLIRDDDKRHDEYSQLIAK
IDVPQNLVEIDAVILDIDRTALNRLEANWQATLGGVTGGSSLMSGSGTLFVSDFKRFFAD
IQALEGEGTASIVANPSVLTLENQPAVIDFSQTAYITATGERVADIQPVTAGTSLQVTPR
AVGNEGHSSIQLMIDIEDGHVQTNGDGQATGVKRGTVSTQALISENRALVLGGFHVEESA
DRDRRIPLLGDIPWLGQLFSSKRHEISQRQRLFILTPRLIGDQTDPTRYVTADNRQQLSD
```

```
AMGRVERRHSSVNQHDVVENALRDLAEGQSPAGFQPQTSGTRLSEVCRSTPALLFESTRG
QWYSSSTNGVQLSVGVVRNTSSKPLRFDEANCASKRTLAVAVWPHSALAPGESAEVYLAM
DPSRVLHASRESLLNR
```

Hint: Go to BLAST page and input the sequence. Select nr or SWISS-PROT in "database" option.

2. What is the sequence identity of the query sequence with gb|AAS45460.1|?
   Hint: Search the result page to find the alignment details.

3. Identify the similar sequences with different e-values 10 and e-50.
   Hint: Change the expected threshold.

4. Compare the sequences with PDB ids 1TIM and 2BTM.
   Hint: Obtain the sequences in FASTA format from the PDB site. Input in BLAST with the option, alignment of two sequences and analyze the alignment.

5. Comment on the alignment with BLAST and FASTA.
   Hint: Use the two sequences given in previous question and repeat with the FASTA program.

6. Carry out the multiple sequence alignment for TIM barrel proteins.
   Hint: Obtain the sequences of TIM barrel proteins from CATH or SCOP database and carry out multiple sequence alignment.

7. Comment on the conservation of residues for the first sequence.
   Hint: Input the multiple sequence alignment in the AL2CO server and discuss the conservation score.

8. Discuss the conservation of residues in 1TIM.
   Hint: Find the chain information from the PDB and input in the Consurf server.

9. Obtain the nonredundant sequences of TIM barrel proteins at less than 40% sequence identity.
   Hint: Get the sequences from CATH or SCOP. Use CD-HIT and blastclust to reduce the redundancy.

10. Obtain the nonredundant sequences of TIM barrel proteins at less than 40% sequence identity using the PISCES server.
    Hint: Follow the instructions given in **Figure 2.11**.

11. Compare the nonredundant sequences of TIM barrel proteins obtained with CD-HIT, PISCES, and blastclust algorithms.

12. Compute the amino acid composition of proteins belonging to different structural classes and folding types of proteins, and discuss their similarities and differences.
    Hint: Obtain the sequences from CATH or SCOP databases, and compute the composition.

13. How far do the amino acid residue pair preferences vary in different folding types of globular and membrane proteins?
    Hint: Use the given type of proteins and compute the residue pair preference.

14. What is the average hydrophobicity of the protein 1PRC?
    Hint: Obtain the sequence from the PDB. Get the hydrophobicity value of each amino acid residue (**Section 2.5** and/or **Chapter 3**), and compute the average value.

15. Compare the amphipathic character of α-helices in membrane proteins and long helices in globular proteins.

Hint: Prepare a dataset of long helices in globular and membrane proteins, and use equations 2.4 and 2.5.

16. Discuss the amphipathicity of β-strands in β-barrel membrane proteins and long β-strands in globular proteins.

Hint: Prepare a dataset of long strands in globular and membrane proteins, and use equations 2.6 and 2.7.

17. Analyze the variation of amino acid properties at different secondary structures of T4 lysozyme (2LZM), bacteriorhodopsin (2BRD), and porin (2POR).

Hint: Use different properties and analyze the results with 3Dinsight.

18. Compute the average Hp, P, P_α and P_β for the PDB codes 2POR, L chain of 1PRC, and 4LYZ using the numerical values given in **Figure 2.15a**.

Hint: Get the sequence from PDB. Compute the amino acid occurrence and total as well as average property value using Equation. 2.2.

References

Ahmad S, Gromiha MM, Sarai A. Analysis and prediction of DNA-binding proteins and their binding residues based on composition, sequence and structural information. Bioinformatics. 2004;20(4):477–486.

Ahmad S, Sarai A. PSSM-based prediction of DNA binding sites in proteins. BMC Bioinformatics. 2005;6:33.

Alexandrov V, Gerstein M. Using 3D Hidden Markov Models that explicitly represent spatial coordinates to model and compare protein structures. BMC Bioinformatics. 2004;5:2.

Altschul SF, Gish W, Miller W, Myers EW, Lipman DJ. Basic local alignment search tool. J. Mol. Biol. 1990;215:403–410.

An J, Nakama T, Kubota T, Sarai A. 3DinSight: an integrated relational database and search tool for structure, function and property of biomolecules. Bioinformatics. 1998;14:188–195.

Bienkowska JR, Yu L, Zarakhovich S, Rogers RG Jr, Smith TF. Protein fold recognition by total alignment probability. Proteins. 2000;40(3):451–462.

Bigelow H, Rost B. PROFtmb: a web server for predicting bacterial transmembrane beta barrel proteins. Nucleic Acids Res. 2006;34(Web Server issue):W186–W188.

Caballero J, Fernández L, Abreu JI, Fernández M. Amino acid sequence autocorrelation vectors and ensembles of Bayesian-regularized genetic neural networks for prediction of conformational stability of human lysozyme mutants. J Chem Inf Model. 2006;46(3):1255–1268.

Cheng J, Baldi P. A machine learning information retrieval approach to protein fold recognition. Bioinformatics. 2006;22(12):1456–1463.

Chimento DP, Mohanty AK, Kadner RJ, Wiener MC. Substrate-induced transmembrane signaling in the cobalamin transporter BtuB. Nat Struct Biol. 2003a;10:394–401.

Chimento DP, Kadner RJ, Wiener MC. The Escherichia coli outer membrane cobalamin transporter BtuB: structural analysis of calcium and substrate binding, and identification of orthologous transporters by sequence/structure conservation. J Mol Biol. 2003b;332:999–1014.

Chou KC, Elrod DW. Prediction of membrane protein types and subcellular locations. Proteins. 1999;34(1):137–153.

Chou KC, Zhang CT. Prediction of protein structural classes. Crit Rev Biochem Mol Biol. 1995;30(4):275–349.

Chou KC. Does the folding type of a protein depend on its amino acid composition? FEBS Lett. 1995;363(1–2):127–131.

Chu W, Ghahramani Z, Podtelezhnikov A, Wild DL. Bayesian segmental models with multiple sequence alignment profiles for protein secondary structure and contact map prediction. IEEE/ACM Trans Comput Biol Bioinform. 2006;3(2):98–113.

Cid H, Bunster M, Canales M, Gazitua F. Hydrophobicity and structural classes in proteins. Protein Eng. 1992;5(5):373–375.

Cole C, Barber JD, Barton GJ. The Jpred 3 secondary structure prediction server. Nucleic Acids Res. 2008;36(Web Server issue):W197–W201.

Cornette JL, Cease KB, Margalit H, Spouge JL, Berzofsky JA, DeLisi C. Hydrophobicity scales and computational techniques for detecting amphipathic structures in proteins. J Mol Biol. 1987;195(3):659–685.

Cuff JA, Barton GJ. Application of multiple sequence alignment profiles to improve protein secondary structure prediction. Proteins. 2000;40(3):502–511.

Dor O, Zhou Y. Achieving 80% ten-fold cross-validated accuracy for secondary structure prediction by large-scale training. Proteins. 2007;66(4):838–845.

Dou Y, Zheng X, Wang J. Prediction of catalytic residues using the variation of stereochemical properties. Protein J. 2009;28(1):29–33.

Eddy SR. Where did the BLOSUM62 alignment score matrix come from? Nat Biotechnol. 2004;22(8):1035–1036.

Eidhammer I, Jonassen I, Taylor WR. Protein Bioinformatics: An Algorithmic Approach to Sequence and Structure Analysis. West Sussex, UK: John Wiley and Sons Ltd; 2004.

Eisenberg D, Weiss RM, Terwilliger TC. The hydrophobic moment detects periodicity in protein hydrophobicity. Proc Natl Acad Sci U S A. 1984;81(1):140–144.

Garg A, Kaur H, Raghava GP. Real value prediction of solvent accessibility in proteins using multiple sequence alignment and secondary structure. Proteins. 2005;61(2):318–324.

Glaser F, Pupko T, Paz I, Bell RE, Bechor D, Martz E, Ben-Tal N. ConSurf: identification of functional regions in proteins by surface-mapping of phylogenetic information. Bioinformatics. 2003;19:163–164.

Gromiha MM. A statistical model for predicting protein folding rates from amino acid sequence with structural class information. J Chem Inf Model. 2005a;45(2):494–501.

Gromiha MM. Motifs in outer membrane protein sequences: applications for discrimination. Biophys Chem. 2005b;117(1):65–71.

Gromiha MM, Ponnuswamy PK. Prediction of transmembrane beta-strands from hydrophobic characteristics of proteins. Int J Pept Protein Res. 1993;42(5):420–431.

Gromiha MM, Ponnuswamy PK. Prediction of protein secondary structures from their hydrophobic characteristics. Int J Pept Protein Res. 1995;45(3):225–240.

Gromiha MM, Suwa M. A simple statistical method for discriminating outer membrane proteins with better accuracy. Bioinformatics. 2005;21(7):961–968.

Gromiha MM, Suwa M. Discrimination of outer membrane proteins using machine learning algorithms. Proteins. 2006;63(4):1031–1037.

Gromiha MM, Yabuki Y. Functional discrimination of membrane proteins using machine learning techniques. BMC Bioinformatics. 2008;9:135.

Gromiha MM, Oobatake M, Sarai A. Important amino acid properties for enhanced thermostability from mesophilic to thermophilic proteins. Biophys Chem. 1999a;82(1): 51–67.

Gromiha MM, Oobatake M, Kono H, Uedaira H, Sarai A. Role of structural and sequence information in the prediction of protein stability changes: comparison between buried and partially buried mutations. Protein Eng. 1999b;12(7):549–555.

Gromiha MM, Ahmad S, Suwa M. Application of residue distribution along the sequence for discriminating outer membrane proteins. Comput Biol Chem. 2005;29(2): 135–142.

Gromiha MM, Thangakani AM, Selvaraj S. FOLD-RATE: prediction of protein folding rates from amino acid sequence. Nucleic Acids Res. 2006;34(Web Server issue):W70–W74.

Huang LT, Gromiha MM. Analysis and prediction of protein folding rates using quadratic response surface models. J Comput Chem. 2008;29(10):1675–1683.

Hwang S, Gou Z, Kuznetsov IB. DP-Bind: a web server for sequence-based prediction of DNA-binding residues in DNA-binding proteins. Bioinformatics. 2007;23(5):634–636.

Imai K, Gromiha MM, Horton P. Mitochondrial beta-barrel proteins, an exclusive club? Cell. 2008;135(7):1158–1159.

Johnson RM, Rath A, Deber CM. The position of the Gly-xxx-Gly motif in transmembrane segments modulates dimer affinity. Biochem Cell Biol. 2006;84(6):1006–1012.

Kelley LA, MacCallum RM, Sternberg MJ. Enhanced genome annotation using structural profiles in the program 3D-PSSM. J Mol Biol. 2000;299(2):499–520.

Kemena C, Notredame C. Upcoming challenges for multiple sequence alignment methods in the high-throughput era. Bioinformatics. 2009;25(19):2455–2465.

Kleiger G, Grothe R, Mallick P, Eisenberg D. GXXXG and AXXXA: common alpha-helical interaction motifs in proteins, particularly in extremophiles. Biochemistry. 2002;41(19):5990–5997.

Kumar M, Gromiha MM, Raghava GP. Identification of DNA-binding proteins using support vector machines and evolutionary profiles. BMC Bioinformatics. 2007;8:463.

Kumar M, Gromiha MM, Raghava GP. Prediction of RNA binding sites in a protein using SVM and PSSM profile. Proteins. 2008;71(1):189–194.

Kunin V, Chan B, Sitbon E, Lithwick G, Pietrokovski S. Consistency analysis of similarity between multiple alignments: prediction of protein function and fold structure from analysis of local sequence motifs. J Mol Biol. 2001;307(3):939–949.

Kutik S, Stojanovski D, Becker L, Becker T, Meinecke M, Krüger V, Prinz C, Meisinger C, Guiard B, Wagner R, Pfanner N, Wiedemann N. Dissecting membrane insertion of mitochondrial beta-barrel proteins. Cell. 2008;132(6):1011–1024.

Lesk AM. Introduction to Bioinformatics. New York: Oxford University Press; 2002.

Li W, Godzik A. Cd-hit: a fast program for clustering and comparing large sets of protein or nucleotide sequences. Bioinformatics. 2006;22(13):1658–1659.

Liu B, Wang X, Lin L, Dong Q, Wang X. A discriminative method for protein remote homology detection and fold recognition combining Top-n-grams and latent semantic analysis. BMC Bioinformatics. 2008;9:510.

López G, Valencia A, Tress ML. Firestar–prediction of functionally important residues using structural templates and alignment reliability. Nucleic Acids Res. 2007;35(Web Server issue):W573–W577.

Martelli PL, Fariselli P, Krogh A, Casadio R. A sequence-profile-based HMM for predicting and discriminating beta barrel membrane proteins. Bioinformatics. 2002;18(suppl 1):S46–S53.

Muramatsu T, Suwa M. Statistical analysis and prediction of functional residues effective for GPCR-G-protein coupling selectivity. Protein Eng Des Sel. 2006;19(6):277–283.

Muthusamy R, Ponnuswamy PK. Variation of amino acid properties in protein secondary structures, alpha-helices and beta-strands. Int J Pept Protein Res. 1990;35(5): 378–395.

Ou YY, Chen SA, Gromiha MM. Prediction of membrane spanning segments and topology in beta-barrel membrane proteins at better accuracy. J Comput Chem. 2009 (in press) DOI:10.1002/jcc.21281.

Ou YY, Gromiha MM, Chen SA, Suwa M. TMBETADISC-RBF: discrimination of beta-barrel membrane proteins using RBF networks and PSSM profiles. Comput Biol Chem. 2008;32(3):227–231.

Pautsch A, Schulz GE. High-resolution structure of the OmpA membrane domain. J Mol Biol. 2000;298:273–282.

Pearson WR, Lipman DJ. Improved tools for biological sequence comparison. Proc Natl Acad Sci U S A. 1988;85(8):2444–2448.

Pei J, Grishin NV. AL2CO: calculation of positional conservation in a protein sequence alignment. Bioinformatics. 2001;17(8):700–712.

Pei J, Grishin NV. PROMALS: towards accurate multiple sequence alignments of distantly related proteins. Bioinformatics. 2007;23(7):802–808.

Pickett SD, Saqi MA, Sternberg MJ. Evaluation of the sequence template method for protein structure prediction. Discrimination of the (beta/alpha)8-barrel fold. J Mol Biol. 1992;228(1):170–187.

Ponnuswamy PK, Gromiha MM. Prediction of transmembrane helices from hydrophobic characteristics of proteins. Int J Pept Protein Res. 1993;42(4):326–341.

Qiu JD, Huang JH, Liang RP, Lu XQ. Prediction of G-protein-coupled receptor classes based on the concept of Chou's pseudo amino acid composition: an approach from discrete wavelet transform. Anal Biochem. 2009;390(1):68–73.

Reche PA, Glutting JP, Reinherz EL. Prediction of MHC class I binding peptides using profile motifs. Hum Immunol. 2002;63(9):701–709.

Sankararaman S, Kolaczkowski B, Sjölander K. INTREPID: a web server for prediction of functionally important residues by evolutionary analysis. Nucleic Acids Res. 2009;37(Web Server issue):W390–W395.

Scheeff ED, Bourne PE. Application of protein structure alignments to iterated hidden Markov model protocols for structure prediction. BMC Bioinformatics. 2006;7:410.

Schneider D, Engelman DM. Motifs of two small residues can assist but are not sufficient to mediate transmembrane helix interactions. J Mol Biol. 2004;343(4):799–804.

Shen H, Chou JJ. MemBrain: improving the accuracy of predicting transmembrane helices. PLoS ONE. 2008;3(6):e2399.

Sim J, Kim SY, Lee J. PPRODO: prediction of protein domain boundaries using neural networks. Proteins. 2005;59(3):627–632.

Su CT, Chen CY, Ou YY. Protein disorder prediction by condensed PSSM considering propensity for order or disorder. BMC Bioinformatics. 2006;7:319.

Tatiana AT, Madden TL. Blast 2 sequences: a new tool for comparing protein and nucleotide sequences. FEMS Microbiol Lett. 1999;174:247–250.

Vandeputte-Rutten L, Kramer RA, Kroon J, Dekker N, Egmond MR, Gros P. Crystal structure of the outer membrane protease OmpT from Escherichia coli suggests a novel catalytic site. EMBO J. 2001;20:5033–5039.

Yamada S, Gotoh O, Yamana H. Improvement in accuracy of multiple sequence alignment using novel group-to-group sequence alignment algorithm with piecewise linear gap cost. BMC Bioinformatics. 2006;7:524.

Yu X, Cao J, Cai Y, Shi T, Li Y. Predicting rRNA-, RNA-, and DNA-binding proteins from primary structure with support vector machines. J Theor Biol. 2006;240(2):175–184.

Yue WW, Grizot S, Buchanan SK. Structural evidence for iron-free citrate and ferric citrate binding to the TonB-dependent outer membrane transporter FecA. J Mol Biol. 2003;332:353–368.

Zeth K, Diederichs K, Welte W, Engelhardt H. Crystal structure of Omp32, the anion-selective porin from Comamonas acidovorans, in complex with a periplasmic peptide at 2.1 A resolution. Structure. 2000;8:981–992.

Zuo YC, Li QZ. Using K-minimum increment of diversity to predict secretory proteins of malaria parasite based on groupings of amino acids. Amino Acids. 2009 (in press). DOI:10.1007/s00726-009-0292-1.

CHAPTER 3

Protein Structure Analysis

The analysis on protein structures provides plenty of information about the factors governing the folding and stability of proteins, the nature of interactions between amino acid residues and with the surrounding medium, the preferred amino acid residues in protein environment, the location of residues in the interior/surface of a protein, amino acid clusters, etc. The information obtained from the analysis will be useful for predicting the secondary and tertiary structures of proteins. The assignment of secondary structures based on a hydrogen-bonding pattern will be helpful for predicting the secondary structures from amino acid sequence and evaluating the performance of different methods. In addition, secondary structure information has been used for predicting three-dimensional structures of proteins, protein stability upon mutations, binding site residues in protein-protein, protein-DNA, protein-RNA complexes, etc. The concept of solvent accessibility is widely used to understand the location of amino acid residues in protein structures and their contribution to the stability of proteins. The potentials derived from protein structures and other parameters are helpful for understanding the factors influencing the stability of proteins and the process of protein folding. The details are presented in the following sections.

3.1   Assignment of secondary structures

Protein Data Bank (PDB) records contain the information about the secondary structures, helix, strand, coil, and turn. However, these data are incomplete, and several structures do not have the secondary structure information. In 1983, Kabsch and Sander developed the program, DSSP (Dictionary of Secondary Structures in Proteins), based on the pattern recognition of hydrogen bonding and geometrical features. It assigns eight secondary structures: helix (H), isolated beta bridge (B), extended beta strand (E), 3_{10} helix (G), π helix (I), turn (T), bend (S), and irregular (loop). An example is shown in **Figure 3.1a**. The output shows the number of residues, total accessible surface area, number of hydrogen bonds for the protein, and so on, along with the structural information (secondary structure, solvent accessibility, hydrogen bonding partners, dihedral angles, etc.) for each residue. It has the numbering system of the PDB file as well as the continuous numbers starting

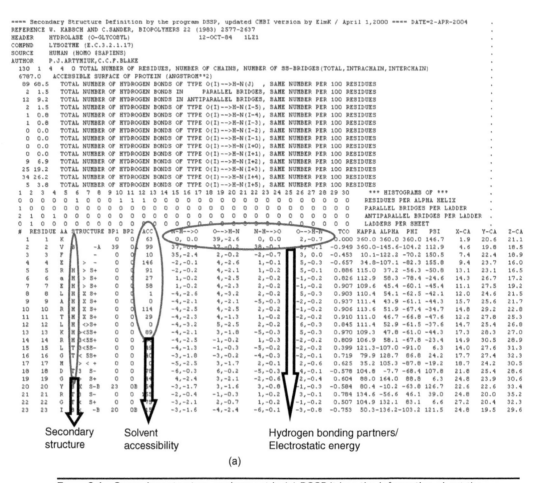

Secondary structure | Solvent accessibility | Hydrogen bonding partners/ Electrostatic energy

(a)

Figure 3.1 Secondary structure assignment in (a) DSSP. It has the information about the secondary structure and the solvent accessibility of human lysozyme. The secondary structure information is given under "Structure (column 4)" and solvent accessibility is given as "ACC (column 7)" and (b) the PDB. The assigned helical and strand regions are indicated.

from one. The residue names and secondary structural assignments are given after the residue number. Furthermore, the solvent accessibility of each residue and the hydrogen-bonding pattern along with electrostatic energy is given in the DSSP file (Kabsch and Sander, 1983). The DSSP output for all the PDB structures can be downloaded using the ftp from the Web site of the developers (ftp://ftp.cmbi.kun.nl/pub/molbio/data/dssp/). Furthermore, one can get the executable file from the developers and run it locally. The secondary structural assignment for each residue in a protein has also been obtained from the PDB. **Figure 3.1b** shows the helical and strand regions in human lysozyme (1LZ1).

3.2   Computation of solvent accessibility

The solvent accessible surface area (ASA) is defined as the locus of the center of the solvent molecule as it rolls over the van der Waals surface of the protein, and it is illustrated in **Figure 3.2a** (Gromiha and Ahmad, 2005). Generally, a sphere of water

```
SEQRES    1 A  130    LYS VAL PHE GLU ARG CYS GLU LEU ALA ARG THR LEU LYS
SEQRES    2 A  130    ARG LEU GLY MET ASP GLY TYR ARG GLY ILE SER LEU ALA
SEQRES    3 A  130    ASN TRP MET CYS LEU ALA LYS TRP GLU SER GLY TYR ASN
SEQRES    4 A  130    THR ARG ALA THR ASN TYR ASN ALA GLY ASP ARG SER THR
SEQRES    5 A  130    ASP TYR GLY ILE PHE GLN ILE ASN SER ARG TYR TRP CYS
SEQRES    6 A  130    ASN ASP GLY LYS THR PRO GLY ALA VAL ASN ALA CYS HIS
SEQRES    7 A  130    LEU SER CYS SER ALA LEU LEU GLN ASP ASN ILE ALA ASP
SEQRES    8 A  130    ALA VAL ALA CYS ALA LYS ARG VAL VAL ARG ASP PRO GLN
SEQRES    9 A  130    GLY ILE ARG ALA TRP VAL ALA TRP ARG ASN ARG CYS GLN
SEQRES   10 A  130    ASN ARG ASP VAL ARG GLN TYR VAL GLN GLY CYS GLY VAL
```

```
HELIX     1   A ARG A    5    ARG A   14  1SEE J.MOL.BIOL.152 P737            10
HELIX     2   B LEU A   25    GLU A   35  1SEE J.MOL.BIOL.152 P737            11
HELIX     3   E SER A   80    LEU A   85  5SEE J.MOL.BIOL.152 P737             6
HELIX     4   C ALA A   90    VAL A   99  1SEE J.MOL.BIOL.152 P737            10
HELIX     5   D VAL A  110    CYS A  116  1SEE J.MOL.BIOL.152 P737             7
SHEET     1   A 2 LYS A    1    PHE A    3  0
SHEET     2   A 2 TYR A   38    THR A   40 -1  N  THR A   40   O  LYS A    1
SHEET     1   B 3 ALA A   42    ASN A   46  0
SHEET     2   B 3 SER A   51    GLY A   55 -1  N  GLY A   55   O  ALA A   42
SHEET     3   B 3 ILE A   59    SER A   61 -1  N  SER A   61   O  THR A   52
SSBOND    1 CYS A    6    CYS A  128                            1555   1555   2.07
SSBOND    2 CYS A   30    CYS A  116                            1555   1555   2.06
SSBOND    3 CYS A   65    CYS A   81                            1555   1555   2.08
SSBOND    4 CYS A   77    CYS A   95                            1555   1555   2.04
```

(b)

FIGURE 3.1   (Continued)

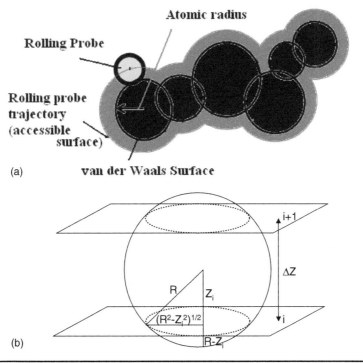

(a)

(b)

FIGURE 3.2   (a) Definition of solvent accessibility. The probe radius, van der Waals surface, and accessible surface are indicated. Figure was adapted from Gromiha and Ahmad (2005); (b) a two-dimensional representation of an isolated van der Waals' sphere cut out by two planes. The parameters involved in the computation of solvent accessible areas are marked.

is assumed to be the solvent molecule with a radius 1.4 Å. The solute molecule is represented by a set of interlocking spheres of appropriate van der Waals radii assigned to each atom, and the solvent molecule is rolled along the envelope of the van der Waals surface at planes conveniently sectioned. The ASA of an atom of radius r is then the area on the surface of the sphere of radius $R = r + r_{solv}$ on each point of which the center of the solvent molecule can be placed in contact with this atom without penetrating any other atoms of the solute molecule. The solvent ASA is calculated using the formula (Lee and Richards, 1971):

$$ASA = \sum \left[R/(R^2 - Z_i^2)^{1/2} \right] L_i \cdot D; D = \Delta Z/2 + \Delta' Z, \tag{3.1}$$

where L_i is the length of the arc computed on a given section i, Z_i is the perpendicular distance from the center of the sphere to the section i, ΔZ is the spacing between the sections, and $\Delta' Z$ is $\Delta Z/2$ or $R - Z_i$, whichever is smaller. Summation is over all of the arcs drawn for the given atom. **Figure 3.2b** shows the relevant parameters for an isolated atom cut by the planes i and $i + 1$.

Several programs have been developed for computing the ASA of atoms and residues from three-dimensional structures of proteins and complexes. The details of selected methods are described in this section.

3.2.1 ACCESS

Richmond and Richards (1978) introduced the program, ACCESS, for computing ASA. This program takes the three-dimensional structure of a protein/nucleic acid and computes the accessible and contact surface area of each atom. It accepts different types of input formats with the default of the PDB. ACCESS uses a standard probe radius of 1.4 Å, and the van der Waals radii to atoms are assigned depending on the atom name and the residue name (Lee and Richards, 1971). ACCESS has different options to (i) calculate the area of an atom, (ii) ignore an atom during accessibility calculation, and (iii) an atom is considered to be a part of protein environment and no calculations will be done for this atom. The final output of ACCESS gives the ASA of each atom, the total ASA of the main chain atoms of each residue, the side chain atoms of each residue, and the total ASA of a protein. ACCESS is available at http://www.csb.yale.edu/.

3.2.2 Naccess

Hubbard and Thornton (1993) developed a program, Naccess, for calculating the atomic accessible area. The program uses the method of Lee and Richards (1971), whereby a probe of a given radius is rolled around the surface of the molecule, and the path traced out by its center is the accessible surface. Naccess is a stand-alone program that calculates the accessible area of a molecule from a PDB format file. It calculates the atomic and residue accessibilities for both proteins and nucleic acids and is freely available for researchers. The main features of this method are that it can be run with user-defined probe size, residues (amino acid, nucleic acid, and hetero atoms), atomic radii, standard state residue accessibilities and with/without hetero, hydrogen, and waters. It will produce the output of atomic ASAs, an absolute and relative residue ASA, and a calculation log file. Naccess is available at http://wolf.bms.umist.ac.uk/naccess/.

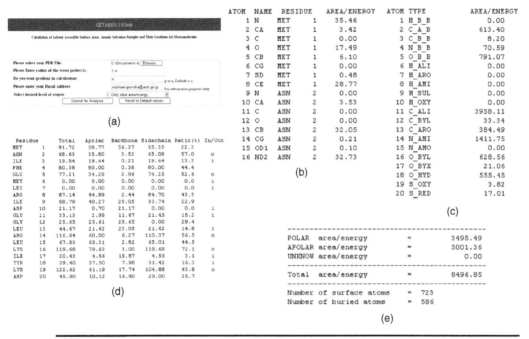

FIGURE 3.3 Input and output options available in GETAREA: (a) input parameters, (b) ASA per atom, (c) ASA per atom type, (d) ASA per residue, and (e) total ASA.

3.2.3 GETAREA

Fraczkiewicz and Braun (1998) developed the program, GETAREA, for calculating the ASA of biomacromolecules on the Web. It is an efficient program, which allows users to submit Cartesian coordinates of atoms in a molecule, to store in a PDB format, and to get back the ASA or solvation energy (depending on parameter setup) in a variety of formats. By default, the submission form is set up to calculate the ASA of nonhydrogen atoms in proteins, but an appropriate change of input parameters will allow to compute any quantity proportional to the ASA for any kind of molecule. It has the options to compute and display the ASA for each atom, atom type, residue, and total area. **Figure 3.3** illustrates the input options **(Figure 3.3a)** and output results obtained for different categories. The output of this program gives the ASA for each atom **(Figure 3.3b)**, atom types **(Figure 3.3c)**, residues **(Figure 3.3d)**, and the whole protein **(Figure 3.3e)**. GETAREA is available at http://www.scsb.utmb.edu/getarea/.

3.2.4 DSSP

Kabsch and Sander (1983) developed a method, DSSP, to compute the solvent accessibility of residues **(Figure 3.1a)** and set up a database of the ASA for most of the proteins deposited in the PDB (Berman et al. 2000). DSSP is one of the commonly used programs/databases for generating the ASA values for prediction algorithms (Rost and Sander, 1994; Pascarella et al. 1998; Mucchielli-Giorgi et al. 1999; Ahmad et al. 2003; Gianese et al. 2003; Qin et al. 2005; Momen-Roknabadi et al. 2008). DSSP is available at http://www.cmbi.kun.nl/gv/dssp/.

TABLE 3.1   Characteristic features of the most common methods for calculating ASA

Comparison index	ACCESS	DSSP	NACCESS	ASC	GETAREA	POPS
Standalone executable availability	Yes	Yes	Licensed	Yes	No	No
Online calculations/ database	No	Yes	No	Yes	Yes	Yes
Polar and nonpolar area	No	No	Yes	No	Yes	Yes
Atom-wise surface area	Yes	No	Yes	Yes	Yes	Yes
Source code availability	No	Yes	No	Yes	No	No
Choice of probe radius	Yes	No	Yes	Yes	No	No
Choice of van der Waals and other parameters	Yes	No	Yes	Yes	By manual editing	No
Secondary structure	No	Yes	No	No	No	No
Reference	Richmond and Richards, 1978	Kabsch and Sander, 1983	Hubbard and Thornton, 1993	Eisenhaber and Argos, 1993	Fraczkiewicz and Braun, 1998	Cavallo et al. 2003

Data were taken from Gromiha and Ahmad (2005).

3.2.5 ASC

Eisenhaber and Argos (1993) developed a program package, analytic surface calculation (ASC), for computing the ASA of each atom and residue in a protein/nucleic acid. ASC has the following options in the program: (i) calculation of the surface of a set of intersecting spheres via the new analytical method (the van der Waals surface or the solvent-accessible surface with any probe radius), (ii) numerical determination of the surface via the double cube lattice method as well as the calculation of volume and of dot surfaces, and (iii) computation of surface energies and the hydrophilic/hydrophobic surface for ensembles of molecules and their constituents. The program is written in C and is available to download for researchers. ASC provides the output in different formats and the ASA for each atom and residue. ASC is available at http://mendel.imp.univie.ac.at/mendeljsp/studies/asc.jsp.

3.2.6 POPS

Cavallo et al. (2003) developed a method, POPS (Parameter Optimzed Surfaces), to calculate atomic and residue level solvent ASAs, which is based on an empirically parameterizable analytical formula. The parameterization has been derived from a selected dataset of proteins with different sizes and topologies. The program POPS is available at http://mathbio.nimr.mrc.ac.uk/wiki/POPS and at the mirror site www.cs.vu.nl/~ibivu/programs/popswww. Major similarities/differences between the most commonly used programs listed above are compared in **Table 3.1.**

3.3   Representation of solvent accessibility

The ASA of residues in protein structures has been quantitatively obtained with the aid of several commonly available computer programs as described in the previous section. The pictorial representation of such ASA values provides an easy understanding of the location of each residue in the structure of a protein. It will also reveal the population of residues on the surface and interior core of a protein.

3.3.1 ASAView

ASAView is an algorithm and a database of schematic representations of the solvent accessibility of residues in a protein. In this program, a characteristic two-dimensional spiral plot of solvent accessibility has been implemented for providing a convenient graphical view of residues in terms of their exposed surface areas (Ahmad et al. 2004). Furthermore, the sequential plots are also displayed in the form of bar charts. Online plots of the proteins included in the entire PDB are provided for the whole protein as well as their chains separately.

Snapshots generated by ASAView for a DNA-binding protein (PDB code: 6CROA) are shown in **Figure 3.4**. The input options are illustrated in **Figure 3.4a**, and the output with a spiral view and a bar chart is shown in **Figures 3.4b and c**, respectively. In the spiral view (**Figure 3.4b**), the residues are placed in the order of their solvent accessibility. Most accessible residues come on the outermost ring of this spiral, and the buried residues are occurring in the innermost ring. The radius of the circles corresponds to the relative solvent accessibility. In the bar chart

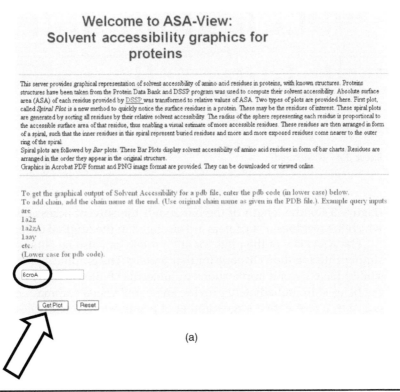

(a)

FIGURE 3.4 ASAView of a DNA-binding protein (PDB code: 6CROA): (a) input options, (b) spiral view—blue, red, green, gray, and yellow colors—indicates the positively charged, negatively charged, polar, nonpolar, and Cys residues, respectively. The size of the sphere shows the relative ASA, and (c) bar diagram: length of the bar represents the ASA.

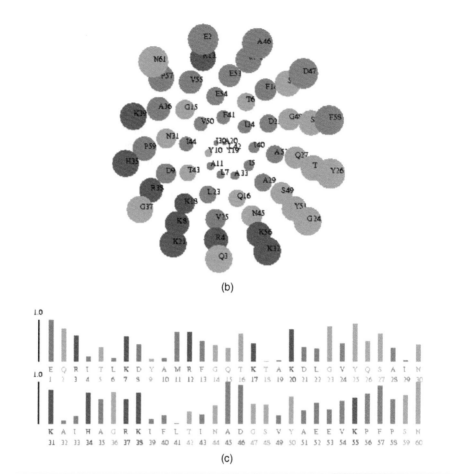

(b)

(c)

FIGURE **3.4**   *(Continued)*

(**Figure 3.4c**), the length of the bar shows the solvent accessibility of residues in which the residues of a protein are arranged in the original order as in the PDB.

The ASAView of the plots for any protein can also be obtained on the Web by simply entering the PDB code for that protein. The graphical plots of solvent accessibility have several applications in molecular biology. Especially, the spiral plot can be used to immediately provide an overall visual summary of the protein. For example, a plot with a large number of positively charged residues instantly tells that the given protein is charged as such. Similarly, the concentration of gray circles suggests the hydrophobic nature of proteins. Furthermore, the outward distribution of higher solvent accessible residues also provides the view of distribution of charged, hydrophobic, or polar residues in different ranges of solvent accessibility. The information about the residues with similar ASA may be helpful for further analyzing the relative number and nature of contacts in protein structure. ASAView is available at http://www.netasa.org/asaview/.

3.3.2 POLYVIEW

Porollo et al. (2004) developed a flexible visualization tool, POLYVIEW, for structural and functional annotations of proteins. A Web server has been developed for generating protein sequence annotations, including solvent accessibility. In this program, the relative solvent accessibility (RSA) has been represented with numerical values ranging from 0 to 9, with 0 corresponding to fully buried (0%–9% RSA) and 9 corresponding to fully exposed residue (90%–100% RSA), respectively. POLYVIEW is available at http://polyview.cchmc.org/.

3.4 Residue–residue contacts

During the process of protein folding, the amino acid residues along the polypeptide chain interact with each other in a cooperative manner to form the stable native structure. Protein structures are stabilized with hydrophobic, electrostatic, hydrogen bonding, disulfide bonds, and van der Waals interactions. These interactions involve the preference of residues to be in contact with each other and are generally termed as inter-residue interactions.

Tanaka and Scheraga (1975) categorized the inter-residue interactions into short-, medium-, and long-range interactions and proposed a hypothesis for protein folding by a three-step mechanism based on these interactions. The definition of short-, medium-, and long-range interactions in a protein is based on the amino acid residues, which are in contact with each other in the native structure, and their respective locations in the sequence.

3.4.1 Residue–residue contacts in protein structures

Each residue in a protein molecule is represented by its α-carbon atom. The center is fixed at the α-carbon atom of the first (N-terminal) residue, and the distances between this atom and the rest of the α carbon atoms (and moving the center to the C-terminal residue) in the protein molecule provide the information about the residue contacts in protein structures. An example is shown in **Figure 3.5**. In this figure, T is the central residue, and other residues, D, R, T, L, A, F, V and G are surrounded by Thr in the protein three-dimensional structure. The composition of the

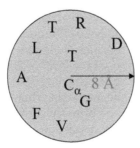

Figure 3.5 Definition of surrounding residues in protein structures. T is the central residue, and the residues that are occurring within the sphere of 8 Å radius are shown.

surrounding residues associated with this residue is calculated for a sphere of radius 8 Å. It has been shown that the influence of each residue over the surrounding medium extends effectively only up to 8 Å (Manavalan and Ponnuswamy, 1977), and this limit is sufficient to characterize the hydrophobic behavior of amino acid residues (Manavalan and Ponnuswamy, 1978; Ponnuswamy, 1993) and to accommodate both the local and nonlocal interactions (Gromiha and Selvaraj, 2004; Jiang et al. 2002). Furthermore, 8 Å limit has been used in several studies, so as to understand the folding rate of proteins (Debe and Goddard, 1999; Gromiha and Selvaraj, 2001a), protein stability upon mutations (Gromiha et al. 1999a), thermal stability of proteins (Gromiha, 2001; Gromiha and Thangakani, 2001), transition state structures of two-state protein mutants (Gromiha and Selvaraj, 2002), and to understand the relationship between hydrophobic clusters and long-range contacts in $(\alpha/\beta)_8$ barrel proteins (Selvaraj and Gromiha, 2003).

On the other hand, various contacting distances in space and different sets of atoms in protein structures have also been used in the literature for protein folding studies. Tudos et al. (1994) used the C_α atoms and the distance of 7 Å for defining long-range contacts. Kocher et al. (1994) used the centroid of each residue and the distance of 8 Å for deriving contact potentials. Jernigan and colleagues (Miyazawa and Jernigan, 1996; Bahar and Jernigan, 1997) represented the hydrophilic and hydrophobic contacts with the cutoff distance of 4 Å and 7 Å, respectively. Plaxco et al. (1998) proposed the concept of contact order, using all the atoms in a protein within the distance of 6 Å. Fariselli et al. (2001) used the cutoff distance of 8 Å between C_β atoms for representing inter-residue contacts. On the other hand, the information about the van der Waals radii of the atoms has also been used to define inter-residue contacts (Selbig, 1995; Dostyani et al. 1997).

3.4.2   Variation of sequence separation for defining residue–residue contacts

For a given residue, the composition of surrounding residues (say, within a sphere of 8 Å radius) is analyzed in terms of their location at the sequence level. The residues that are within a distance of two residues from the central residue are considered to contribute to short-range interactions, those within a distance of ±3 or ±4 residues to medium range, and those more than four residues away to long-range interactions (Ponnuswamy et al. 1973, 1980; Gromiha and Selvaraj, 1997; 2004). Miller et al. (2002) used the same cutoff of four residues to represent long-range contacts. However, the limit of residue separation has been varied for several studies. Gilis and Rooman (1997) considered the residues separated by more than 15 residues along the sequence as long-range contacts, while Tudos et al. (1994) proposed the limit of 20 residues. Dostyani et al. (1997) made the cutoff of 10 residues for defining stabilization centers in proteins. Gugolya et al. (1997) considered the interactions between residues that are separated by not more than 4 residues in the primary structure as short-range interactions, 5 to 20 residues as medium-range interactions, and more than 20 residues as long-range interactions. The limit of 12 residues shows the highest correlation between long-range order and protein folding rates in a set of 23 two-state proteins, and this limit varies for the proteins in different structural classes (Gromiha and Selvaraj, 2001a).

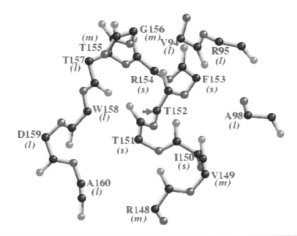

FIGURE 3.6   Representation of short-, medium-, and long-range contacts in protein structures. A typical example of surrounding residues around T152 of T4 lysozyme within 8 Å is shown): s: short-range contacts, m: medium-range contacts, and l: long-range contacts. Figure was adapted from Gromiha and Selvaraj (2004).

3.4.3   Definition of short-, medium-, and long-range interactions

The definition of the short-, medium-, and long-range interactions in protein structures is illustrated in **Figure 3.6**. In this figure, the surrounding residues of T152 (indicated by an arrow) in T4 Lysozyme (2LZM) within the sphere of the radius 8 Å are shown. The residues, I150, T151, F153, and R154 (marked as *s*), are close to the residue T152 (central residue) along N- and C-directions, and these residues contribute to short-range interactions; the residues, R148, V149, T155, and G156 (marked as *m*), are separated by three or four residues from the central residue (T152) in the amino acid sequence, and these residues contribute to medium-range interactions; other residues, T157, W158, D159, and A160, are separated by more than four residues, and V94, R95, and A98 are far in the sequence level from T152. Accordingly, these residues (marked as *l*) contribute to long-range interactions.

This classification and representation enable one to take into consideration of both local and nonlocal interactions involved in the three-dimensional structures of proteins.

3.5   Amino acid clusters in protein structures

The amino acid residues in protein structures interact with each other and form clusters. Heringa and Argos (1991) used a cutoff radius of 4.5 Å between side-chain atoms to delineate amino acid clusters and showed that most of the clusters are composed of three to four residues and are localized near the protein surface. On the other hand, Zehfus (1995) reported that an average of 65% of hydrophobic residues is involved in residue clusters, and each hydrophobic cluster contains at least seven residues. Karlin and colleagues (Karlin et al. 1994; Karlin and Zhu, 1996) proposed several methods for identifying residue clusters in protein structures based on average, minimum, and maximum distances of both all-atom (backbone and side-chain) and side-chain atom coordinates. The applications of this method have been demonstrated with the delineation of different types of clusters, namely, acidic clusters, cysteine clusters, iron–sulfur proteins and charge clusters, and clusters of multiple histidine residues. Chirgadze and Larionova (1999) suggested a different criterion for identifying charged clusters: Charged groups were included in a cluster if their charged N and O atoms were located at distances between 2.4 and 7 Å. This approach identified charged clusters in 86% of the considered 275 protein structures. Selvaraj and Gromiha (1998) identified the residue clusters in $(\alpha/\beta)_8$ barrel proteins based on the distance criteria of 8 Å limit between the C_α atoms of two residues in the protein structure. This procedure showed the presence of 14 identical amino acid clusters and a large number of physicochemically similar clusters based on the surrounding hydrophobicity, turn preference, bulkiness, refractive index, and antiparallel β-strand preference in a set of 36 $(\alpha/\beta)_8$ barrel proteins. Furthermore, a graph theoretical approach has been used to identify residue clusters within the distance of 6.5 Å in protein structures (Kannan and Vishveshwara, 1999; Kannan et al. 2001). These results show the influence of inter-residue interactions to the formation of residue clusters, which are important for the folding and stability of protein structures.

3.6   Contact potentials

The information about the preference of residue pairs to form medium- and long-range contacts has been used to understand the residue–residue cooperativity in protein folding.

3.6.1   Development of effective potentials

The most widely used approach for deriving effective potentials from an ensemble of experimentally determined protein structures consists of computing frequencies of sequence and structure features, and converting these frequencies into free energies (Sippl, 1990, 1995). Rooman and Wodak (1995) reviewed the two main approaches for developing database-derived potentials and their applications to protein structure prediction.

The first approach is in the context of standard statistical models for deriving scores based on the logarithm of frequencies (Bowie et al. 1991; Ouzounis et al. 1993; Wilmanns and Eisenberg, 1993). The second one is in the context of statistical mechanics for deriving potential terms expressed as $-kT \log$ (frequencies), where k is the Boltzmann's constant and T is the temperature (Sippl, 1990; Jones et al. 1992; Kocher et al. 1994). Zhang and Skolnick (1998) constructed an artificial protein structural database using contact and secondary structure propensity potentials (called as "true" potentials) and then derived new sets of potentials to see how they are related to the true potentials. They found that by using the Boltzmann distribution method, when the stability of the structures in the database lies within a certain range, both contact potentials and secondary structure propensities could be derived separately with remarkable accuracy.

Furthermore, several potentials have been derived based on the interactions between amino acid residues in a protein and are used to understand the protein-folding problem; to predict protein structure, stability, and fold recognition; and to design novel proteins (Flockner et al. 1995; Mirny and Shakhnovich, 1996; Reva et al. 1997; Gilis and Rooman, 1997; Furuichi and Koehl, 1998; Chiu and Goldstein, 1998; Miyazawa and Jernigan, 1999; Tobi et al. 2000; Russ and Ranganathan, 2002).

3.6.2 Backbone torsion potentials

The influence of amino acids on the backbone conformation of neighboring residues along the chain of a protein has been considered for developing backbone torsion potentials (Rooman et al. 1992; Kocher et al. 1994). They have reported two types of torsion potentials: residue-to-torsion potential and torsion-to-residue potential. The residue-to-torsion potential has been computed from the probabilities $P_{i-k,i-k}^{ai,ai}(t_k)$ and $P_{i-k,j-k}^{ai,aj}(t_k)$ that amino acid a at position i along the sequence and pairs of amino acids a at positions i and j, respectively, are associated with the torsion domain t at position k; i and k can be anywhere within a window of 17 sequence positions $[k-8, k+8]$ centered around k. The torsion-to-residue potential is computed from the probabilities $P_{i-k,i-k}^{ti,ti}(a_k)$ and $P_{i-k,j-k}^{ti,tj}(a_k)$ that torsion domains t at position i along the sequence and pairs of torsion domains t at positions i and j, respectively, are associated with the amino acid a at position k. Both the residue-to-torsion and torsion-to-residue potentials have been subdivided into two parts: a short-range part that includes only contributions from residues and torsion domains in the interval $[k-1, k+1]$ and a middle-range part that considers all the remaining contributions in the $[k-8, k+8]$ window. These torsion potentials represent best the interactions in protein surface (Gilis and Rooman, 1997), and it performs well for predicting the stability of surface mutations.

3.6.3 Residue–residue interaction potentials

Residue–residue potentials describe both local interactions (short and medium range) along the chain and interactions between residues that are far apart along the sequence but close in space (long range). They are computed from the propensities $P_{|i-j|}^{ai,aj}(d_{ij})$ of two residues a_i and a_j at positions i and j along the sequence, to be separated by a spatial distance d_{ij} (Kocher et al. 1994), and the pairs separated by 1 to 7 sequence positions have been considered as middle-range potentials. Pairs separated by more than 8 positions represent nonlocal (long-range) interactions along the chain. This distance potential is dominated by hydrophobic interactions and represents the main interactions that stabilize the protein core. Similar approach has also been proposed for devising a residue–residue potential function to calculate the conformational energy of proteins (Oobatake and Crippen, 1981).

3.6.4 Inter-residue contact potentials

Miyazawa and Jernigan (1985, 1996) estimated the effective inter-residue contact energies from the number of residue–residue contacts (within 6.5 Å) observed in crystal structures of globular proteins by means of a quasi-chemical approximation. This empirical energy function includes solvent effects and provides an estimate of the long-range component of conformational energies. The interaction energies e_{ij} (contact energy between the residues i and j) and e_{ij}' (the energy difference accompanying the formation of a contact pair ij from contact pairs ii and jj) are given in **Table 3.2** (Miyazawa and Jernigan, 1985; 1996). They have observed the following results: (i) The formation of Cys-X contacts from Cys-Cys and

TABLE 3.2 Contact energies derived from protein crystal structures

	Cys	Met	Phe	Ile	Leu	Val	Trp	Tyr	Ala	Gly	Thr	Ser	Asn	Gln	Asp	Glu	His	Arg	Lys	Pro
Cys	-5.44	-4.99	-5.80	-5.50	-5.83	-4.96	-4.95	-4.16	-3.57	-3.16	-3.11	-2.86	-2.59	-2.85	-2.41	-2.27	-3.60	-2.57	-1.95	-3.07
Met	0.46	-5.46	-6.56	-6.02	-6.41	-5.32	-5.55	-4.91	-3.94	-3.39	-3.51	-3.03	-2.95	-3.30	-2.57	-2.89	-3.98	-3.12	-2.48	-3.45
Phe	0.54	-0.20	-7.26	-6.84	-7.28	-6.29	-6.16	-5.66	-4.81	-4.13	-4.28	-4.02	-3.75	-4.10	-3.48	-3.56	-4.77	-3.98	-3.36	-4.25
Ile	0.49	-0.01	0.06	-6.54	-7.04	-6.05	-5.78	-5.25	-4.58	-3.78	-4.03	-3.52	-3.24	-3.67	-3.17	-3.27	-4.14	-3.63	-3.01	-3.76
Leu	0.57	0.01	0.03	-0.08	-7.37	-6.48	-6.14	-5.67	-4.91	-4.16	-4.34	-3.92	-3.74	-4.04	-3.40	-3.59	-4.54	-4.03	-3.37	-4.20
Val	0.52	0.18	0.10	-0.01	-0.04	-5.52	-5.18	-4.62	-4.04	-3.38	-3.46	-3.05	-2.83	-3.07	-2.48	-2.67	-3.58	-3.07	-2.49	-3.32
Trp	0.30	-0.29	0.00	0.02	0.08	0.11	-5.06	-4.66	-3.82	-3.42	-3.22	-2.99	-3.07	-3.11	-2.84	-2.99	-3.98	-3.41	-2.69	-3.73
Tyr	0.64	-0.10	0.05	0.11	0.10	0.23	-0.04	-4.17	-3.36	-3.01	-3.01	-2.78	-2.76	-2.97	-2.76	-2.79	-3.52	-3.16	-2.60	-3.19
Ala	0.51	0.15	0.17	0.05	0.13	0.08	0.07	0.09	-2.72	-2.31	-2.32	-2.01	-1.84	-1.89	-1.70	-1.51	-2.41	-1.83	-1.31	-2.03
Gly	0.68	0.46	0.62	0.62	0.65	0.51	0.24	0.20	0.18	-2.24	-2.08	-1.82	-1.74	-1.66	-1.59	-1.22	-2.15	-1.72	-1.15	-1.87
Thr	0.67	0.28	0.41	0.30	0.40	0.36	0.37	0.13	0.10	0.10	-2.12	-1.96	-1.88	-1.90	-1.80	-1.74	-2.42	-1.90	-1.31	-1.90
Ser	0.69	0.53	0.44	0.59	0.60	0.55	0.38	0.14	0.18	0.14	-0.06	-1.67	-1.58	-1.49	-1.63	-1.48	-2.11	-1.62	-1.05	-1.57
Asn	0.97	0.62	0.72	0.87	0.79	0.77	0.30	0.17	0.36	0.22	0.02	0.10	-1.68	-1.71	-1.68	-1.51	-2.08	-1.64	-1.21	-1.53
Gln	0.64	0.20	0.30	0.37	0.42	0.46	0.19	-0.12	0.24	0.24	-0.08	0.11	-0.10	-1.54	-1.46	-1.42	-1.98	-1.80	-1.29	-1.73
Asp	0.91	0.77	0.75	0.71	0.89	0.89	0.30	-0.07	0.26	0.13	-0.14	0.11	-0.24	-0.09	1.21	-1.02	-2.32	-2.29	-1.68	-1.33
Glu	0.91	0.30	0.52	0.46	0.55	0.55	0.00	-0.25	0.30	0.36	-0.22	-0.19	-0.21	-0.19	0.05	-0.91	-2.15	-2.27	-1.80	-1.26
His	0.65	0.28	0.39	0.66	0.67	0.70	0.08	0.09	0.47	0.50	0.16	0.26	0.29	0.31	-0.19	-0.16	-3.05	-2.16	-1.35	-2.25
Arg	0.93	0.38	0.42	0.41	0.43	0.47	-0.11	-0.30	0.30	0.18	-0.07	-0.01	-0.02	-0.26	-0.91	-1.04	0.14	-1.55	-0.59	-1.70
Lys	0.83	0.31	0.33	0.32	0.37	0.33	-0.10	-0.46	0.11	0.03	-0.19	-0.15	-0.30	-0.46	-1.01	-1.28	0.23	0.24	-0.12	-0.97
Pro	0.53	0.16	0.25	0.39	0.35	0.31	-0.33	-0.23	0.20	0.13	0.04	0.14	0.18	-0.08	0.14	0.07	0.15	-0.05	-0.04	-1.75

The upper half and diagonal elements represent e_{ij} and lower half indicates e'_{ij}. The energies are in RT units. Data from Miyazawa and Jernigan (1996).

X-X contacts represents a relatively large energy loss, because Cys-Cys often form disulfide bonds; (ii) the contact formation between negatively charged (Asp and Glu) and positively charged (Lys and Arg) residues is preferable due to electrostatic interactions; the magnitude of the interaction energies of Asp and Glu with His is smaller than that with Lys and Arg because of its less average charge; (iii) Tyr and Trp (to some extent) prefer contacts with polar residues because of the presence of polar atoms in their side chains, although they have hydrophobic characteristics as indicated by large negative values of e_{ij}; and (iv) the segregation of hydrophobic and hydrophilic residues can be directly seen from the values of e_{ij}'; e_{ij}' among hydrophobic residues (Met, Phe, Ile, Leu, and Val) takes small positive or negative values, indicating that these residues do not have strong specific preferences but are almost randomly mixed in protein structures. Hydrophilic residues (Thr, Ser, Asn, Gln, His, Arg, Lys, and Pro) for the most part prefer contacts with each other to those between the same types of residues; in the case of charged residues, the subtracted unfavorable electrostatic interactions would in part be responsible for this.

Zhang and Kim (2000) proposed that the residue contact energies strongly depend on the secondary structural environment and derived contact potentials in the context of secondary structural environment in proteins. These potentials have been used in threading and predicting the contacts in three-dimensional structures of proteins.

Parthiban et al. (2007) classified the atoms in amino acid residues into 40 types based on chemical behavior and connectivity, and developed atom pair potentials at different secondary structures and various ranges of solvent accessibility. The potentials have been successfully used to predict the stability of proteins upon single mutations.

3.6.5 Potentials based on distance criteria

The distance criterion has also been used to describe the residue pairs influenced by local (medium range) and nonlocal (long range) interactions. The inter-residue contacts in protein structures have been pictorially represented with the aid of contact maps. The contact map for a typical protein (Taka amylase) within the distance of 8 Å is shown in **Figure 3.7** (Gromiha and Selvaraj, 2004). In this figure, the diagonal residues show the short-range contacts, the residues close to the diagonal represent the medium-range contacts, and the residues far away from the diagonal indicate the long-range contacts.

For each medium- and long-range interaction, the average preference of surrounding residues has been computed for all the 20 amino acid residues. It is defined as

$$<N>_{ij} = \sum N_{ij} / \left(\sum N_i + \sum N_j \right), \tag{3.2}$$

where N_{ij} is the number of surrounding residues (contacts) of type j around residue i (400 combinations), and the summation is over all the residues in the considered proteins. ΣN_i and ΣN_j are respectively the total number of residues of type i and j (Gromiha and Selvaraj, 1999). Different sets of 20×20 matrices have been derived for medium- and long-range interactions for each structural class of globular proteins and membrane proteins (Gromiha and Selvaraj, 1999, 2001b). In principle, it

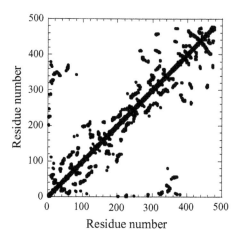

FIGURE **3.7**   Contact map in protein structures. The X and Y axes represent the amino acid residues, and the contacts between two residues are shown with filled circles. The diagonal elements indicate the short-range contacts, close to the diagonals represent medium-range contacts, and far from diagonal indicate the long-range contacts.

should be possible to convert the number of contacts into free energies using the standard procedures (Sippl, 1990, 1995; Miyazawa and Jernigan, 1996).

The analysis of residue pair preferences in globular protein structures showed that the charged and polar residues play a main role in forming medium-range contacts, although hydrophobic residues are also making significant contribution. In long-range contacts, hydrophobic residues dominate and the role of polar residues is minimal.

In transmembrane helical proteins, hydrophobic–hydrophobic residue pairs are predominant to form medium-range contacts, because the membrane-spanning helical segments are highly accommodated with a stretch of hydrophobic residues. In transmembrane strand proteins, the polar and the negatively charged residues have higher preference to form medium-range contacts.

The thermophilic proteins have a significant number of inter-residue contacts between residues of H-bond forming capability other than specific interactions, such as hydrophobic clusters that are observed in both mesophilic and thermophilic proteins (Gromiha, 2001). This reveals the influence of hydrogen bonds to the stability of thermophilic proteins (Vogt et al. 1997).

The structural analysis on slow- and fast-folding proteins showed that the short- and medium-range contacts between polar residues, such as NN, SQ, QE, QK, and QS, are dominant in fast-folding proteins. On the other hand, the long-range contacts in slow-folding proteins are mainly influenced with the hydrophobic residue pairs, such as AA, AG, GG, WL, MG, and CY. This might be due to the fact that the formation of hydrophobic core involves long-range interactions, which slows down the folding process (Huang and Gromiha, 2008).

3.7   Cation-π interactions in protein structures

In addition to noncovalent interactions, the cation-π interaction (Dougherty, 1996) is recognized as an important noncovalent binding interaction relevant to structural

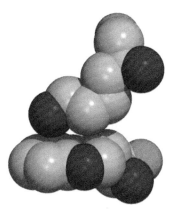

Figure 3.8   Representation of cation-π interactions in protein structures. An example between the residues Lys408 and Trp356 in sulfite reductase (1AOP) is shown.

biology. It is formed by the interaction of the positively charged residues, Lys and Arg, with the aromatic residues, Phe, Trp, and Tyr. Gallivan and Dougherty (1999) identified and evaluated cation-π interactions in protein structures. They reported that cation-π interactions are commonly occurring in protein structures in such a way that when a cationic side chain (Lys or Arg) is near to an aromatic side chain (Phe, Tyr, or Trp), the geometry is biased toward one that would experience a favorable cation-π interaction.

Cation-π interactions in protein structures have been identified with distance and energy criteria, (i) distance between the cationic group (the ammonium nitrogen (NZ) in Lys or the guanidinium carbon [CZ] in Arg) and the centers of all aromatic rings and (ii) the interaction energy between the positive charged and aromatic residues using OPLS (optimized potentials for liquid simulations) force field (see **Section 3.9.5**). An example is shown in **Figure 3.8**. This interaction is between Lys408 and Trp356 in a sulfite reductase (PDB code: 1AOP), and it is one of the strongest interactions with the energy of −9.4 kcal/mol. The total cation-π interaction energy ($E_{cat-\pi}$) has been divided into electrostatic (E_{es}) and van der Waals energies (E_{vdw}). The electrostatic energy (E_{es}) is calculated using the equation,

$$E_{es} = q_i q_j / r_{ij}, \tag{3.3}$$

where q_i and q_j are, respectively, the effective partial charges for the atoms i and j, and r_{ij} is the distance between them. The van der Waals energy between a pair of atoms i and j is given by

$$E_{vdw} = 4\varepsilon_{ij}\left(\sigma_{ij}^{12}/r_{ij}^{12} - \sigma_{ij}^{6}/r_{ij}^{6}\right), \tag{3.4}$$

where $\sigma_{ij} = (\sigma_{ii}\sigma_{jj})^{1/2}$ and $\varepsilon_{ij} = (\varepsilon_{ii}\varepsilon_{jj})^{1/2}$; σ and ε are, respectively, the van der Waals radius and well depth.

In identifying the cation-π interactions, all cation-π pairs (K or R with F, Y, or W) within 10 Å of each other are considered. If there is a gap large enough to insert a water molecule at closest contact, the structure is rejected, and the residues are considered "noninteracting." For the remaining "interacting pairs," the OPLS electrostatic energy, E_{es}, is evaluated. If $E_{es} \leq -2.0$ kcal/mol, the pair is counted as a cation-π interaction. If $E_{es} > -1.0$ kcal/mol, the structure is rejected.

If $-2.0 < E_{es} \leq -1.0$ kcal/mol, the structure is retained only if $E_{vdW} \leq -1.0$ kcal/mol (Gallivan and Dougherty, 1999).

The influence of cation-π interaction varies in different classes of proteins, such as globular, membrane, DNA binding, mesophilic, and thermophilic proteins (Wintjens et al. 2000; Gromiha et al. 2002a; Chakravarty and Varadarajan, 2002; Gromiha, 2003; Gromiha et al. 2004a,b; Gromiha and Suwa, 2005). Furthermore, cation-π interaction has been used to calculate the interaction energy between amino acid residues (Shacham et al. 2004). Recently, cation-π interactions are used to understand the recognition mechanism of protein–protein complexes (Gromiha et al. 2009).

The concept of cation-π interactions is simple to develop a computer program for delineating such interactions and their interaction energies. However, Gallivan and Dougherty (1999) developed the program, CAPTURE, which is available on the Web for academic users. It is available at http://capture.caltech.edu/. It has the options to run for any structure available in the PDB or upload the file in the PDB format. The output for the protein human lysozyme is shown in **Figure 3.9**. It lists the amino acid composition of the cation-π interaction forming residues and the number of cation-π interactions for all the possible six pairs. Furthermore, the residues involved in cation-π interactions along with their respective cation-π interaction energy are listed. This will be helpful to understand the strength of each cation-π interaction.

3.8   Noncanonical interactions

Babu (2003) developed a server, NCI, for the identification of noncanonical interactions in protein structures using geometric criteria. It takes the coordindates of protein structures in a PDB format and lists the noncanonical interactions, which include N–H . . . π, C(alpha)–H . . . π, C(alpha)–H . . . O=C and variants of them. In addition, the user can view the RasMol image highlighting the interactions in the protein structure and download the RasMol script. It is available at: http://www.mrc-lmb.cam.ac.uk/genomes/nci/.

3.9   Free energy calculations

The stability of protein structures can be estimated from the computation of different free energy contributions, such as hydrophobic free energy, electrostatic free energy, hydrogen bonding free energy, and van der Waals free energy. The calculation of these free energy terms reveals the importance of these interactions to the folding and stability of protein structures.

3.9.1   Hydrophobic free energy

Hydrophobicity is the major driving force in protein folding, and it can be estimated from solvent accessibility. It is calculated as (Eisenberg and McLachlan, 1986; Ponnuswamy and Gromiha, 1994)

$$G_{hy} = \sum \Delta \sigma_i [A_i(\text{folded}) - A_i(\text{unfolded})], \qquad (3.5)$$

```
**************** Report for the protein pdb1lz1.ent ****************

This protein has 130 amino acids of which:

        5       (3.8%) are Lysine
        14      (10.8%) are Arginine
        2       (1.5%) are Phenylalanine
        6       (4.6%) are Tyrosine
        5       (3.8%) are Tryptophan

This protein exists as a single chain

The protein has an apparent molecular weight of 14854 D



********************** Cation-Pi Summary **********************


Number of ARG/PHE interacting pairs:    0

Number of energetically significant ARG/PHE cation-pi interactions:      0

Number of ARG/TYR interacting pairs:    6

Number of energetically significant ARG/TYR cation-pi interactions:      0

Number of ARG/TRP interacting pairs:    4

Number of energetically significant ARG/TRP cation-pi interactions:      3

The following are the energetically significant ARG/TRP cation-pi interactions:

Cation  AA #    Chain   Pi      AA #    Chain   E(es)           E(vdw)
                                                (kcal/mol)      (kcal/mol)

ARG     98              TRP     64              -1.13           -1.50
ARG     107             TRP     112             -2.54           62.84
ARG     115             TRP     34              -5.43           -4.03

Number of LYS/PHE interacting pairs:    1

Number of energetically significant LYS/PHE cation-pi interactions:      1

The following are the energetically significant LYS/PHE cation-pi interactions:

Cation  AA #    Chain   Pi      AA #    Chain   E(es)           E(vdw)
                                                (kcal/mol)      (kcal/mol)

LYS     1               PHE     3               -4.81           -1.40
```

FIGURE **3.9** The output of the program CAPTURE showing cation-π interactions in human lysozyme. It has three energetically significant interactions between the pair of residues Arg and Trp.

where i stands for different atom types (carbon, neutral oxygen and nitrogen, charged nitrogen, charged oxygen, sulfur, etc.), and A_i(folded) and A_i(unfolded) represent, respectively, the ASA of each atom type in the folded and unfolded (extended) states of the protein.

The folded state ASA can be obtained with different programs such as Naccess, GETAREA, and ASC (see **Section 3.2**). The unfolding state ASA (A_i) is the value at extended state, which can be obtained with the average value of the residue in Gly-X-Gly or Ala-X-Ala conformation with the assumption that there is no steric hindrance due to the neighboring residues. The computed values with Ala-X-Ala conformation are given below: Ala-110.2; Asp-144.1; Cys-140.4; Glu-174.7; Phe-200.7; Gly-78.7; His-181.9; Ile-185.0; Lys-205.7; Leu-183.1; Met-200.1; Asn-146.4; Pro-141.9; Gln-178.6; Arg-229.0; Ser-117.2; Thr-138.7; Val-153.7; Trp-240.5; Tyr-213.7 (in Å^2).

$\Delta \sigma_i$ are the atomic solvation parameters obtained by fitting the free energy change for the transfer of amino acid residue to water (ΔG) and the accessible surface areas of residues in standard conformation (A_i). It is calculated as

$$\Delta G = \sum \Delta \sigma_i \, A_i. \qquad (3.6)$$

Generally any hydrophobicity scale can be used for the transfer free energy. However, experimentally observed scales, such as Tanford-Jones scale (Nozaki and Tanford, 1971; Jones, 1975) and Fauchere and Pliska scale (Fauchere and Pliska, 1983), are commonly used. From the theoretical scales, surrounding hydrophobicity scale (Manavalan and Ponnuswamy, 1978; Ponnuswamy, 1993), Kyte-Doolittle scale (Kyte and Doolittle, 1982), and Engelman scale (Engelman et al. 1986) are commonly used for understanding the hydrophobic character of amino acid residues and predicting the secondary structure and stability of globular and membrane proteins. Several hydrophobicity scales are available in the literature and are available in several databases, such as AAindex database (Kawashima et al. 1999) and SPLIT (Juretic et al. 1998). Comparison of hydrophobicity scales has also been reported (Cornette et al. 1987; Palliser and Parry, 2001).

The calculation of hydrophobic free energy for the amino acid residues in a protein and for the complete protein is implemented in several ASA prediction programs, or it can be computed using the result of ASA calculations. Furthermore, several atomic solvation parameters are available based on different groups of atoms (Eisenberg and McLachlan, 1986; Ooi et al. 1987; Vila et al. 1991; Wesson and Eisenberg, 1992; Ponnuswamy and Gromiha, 1994). **Table 3.3** lists the atomic solvation parameters proposed for different groups of atoms and using different hydrophobicity scales.

TABLE 3.3 Atomic solvation parameters for different groups of atoms

Atom type	Atomic solvation parameter (cal/mol/Å^2)				
	EM86	**Ooi87**	**Vila91**	**WE92**	**PG94**
C (any)	16			12	12
C (aliphatic)		8	216		
C (carbonyl or carboxyl)		427	−732		
C (aromatic)		−8	−678		
N (any)		−132	−312		
N (noncharged)	−6			−116	−6
N (charged)	−50			−186	−19
O (carbonyl or carboxyl)		−38	−262		
O (noncharged)	−6			−116	−6
O (charged)	−24			−175	−35
O (other)		−172	−910		
S (any)	21	−21	−281	−18	36

EM86: Eisenberg and McLachlan (1986).
Ooi87: Ooi et al. (1987).
Vila91: Vila et al. (1991).
WE92: Wesson and Eisenberg (1992).
PG94: Ponnuswamy and Gromiha (1994).

3.9.2   Electrostatic free energy

The major partners of electrostatic interaction in a protein chain are the charged side chains of the five residues, Lys, Arg, His, Glu, and Asp, and the neutral nitrogens and oxygens of the backbone and certain other side chains. These atomic groups interact mainly via hydrogen bonds, charge–charge, or ion-pairs and charge–helix dipole interactions. An ion-pair has been defined based upon the analysis of the distance distributions for like- and oppositely charged groups in proteins structures. Barlow and Thornton (1983) proposed the concept of ion pair if the distance between two charged groups is less than or equal to 4 Å and extensively surveyed the contribution of ion-pairs and salt bridges in protein structures. It has been reported that the electrostatic interactions between point charges are strongly context-dependent, with interactions between single ion-pairs on the protein surface generally contributing less than 1 kcal/mol and a buried ion-pair contributing around 3 kcal/mol to protein stability.

In an α-helix, the peptide dipoles align nearly parallel to the helix axis and produce approximately a field of continuous line dipole along the axis with a dipole density of 3.5D per 1.5 Å. Of particular interest is the fact that favorable dispositioned charges and helix dipoles impart stability to the folded state of globular proteins. The first quantitative measurement of the energy of a charge–dipole interaction in a helical segment of a native protein was performed by Fersht's group (Sali et al. 1988). From titration studies of histidine occurring at the C-terminus of an α helical segment in the protein barnase, they estimated an energy in the range of 1.4 to 2.1 kcal/mol for the interaction between the protonated histidine and the helix dipole. Nicholson et al. (1988) succeeded in increasing the thermostability of T4 lysozyme by introducing charged residues at sites designed to interact with α-helix dipoles.

3.9.3   Hydrogen bonding free energy

The entire secondary structural elements in proteins are held in their conformation mainly by hydrogen bonds involving the main chain NH/CO groups, and the polar groups of the side chains also form hydrogen bonds, whether these elements are fully or partially buried. Hydrogen bonds between the solvent/surface groups, counter-ions/protein groups/solvent, etc., are also a common feature. Hydrogen bonds are also observed to be an important factor to understand the stability of partially buried and exposed mutations (Gromiha et al. 2000; 2002b).

McDonald and Thornton (1994) developed a program, HBPLUS, for computing the hydrogen bonds in protein structures. It has the following features: (i) calculates the geometries of all hydrogen bonds, (ii) optionally lists neighbor interactions, (iii) calculates hydrogen positions, (iv) deals with hydrogens that can occupy more than one position, (v) optionally includes amino-aromatic H-bonds, (vi) supports full customization, e.g., H-bond criteria, and donor and acceptor atom types (vii) analyzes H-bonding near Asn, Gln, and His side-chains and suggests optimal conformations, (viii) supports .hbplusrc files, and (ix) outputs the PDB file, including extrapolated polar hydrogen positions. HBPLUS is freely available for academic purpose, and it can be obtained from the developers (www.biochem.ucl.ac.uk/bsm/hbplus/home.html).

3.9.4 Van der Waals free energy

Van der Waals free energy can be computed using Lennard-Jones 6–12 potential and is implemented in several force fields. In AMBER 6 force field (Cornell et al. 1995), it is computed as

$$G_{vw} = \left(A_{ij}/r_{ij}^{12} - B_{ij}/r_{ij}^{6} \right), \tag{3.7}$$

where $A_{ij} = \varepsilon_{ij}^{*}(R_{ij}^{*})^{12}$ and $B_{ij} = 2\varepsilon_{ij}^{*}(R_{ij}^{*})^{6}$; $R_{ij}^{*} = (R_{i}^{*} + R_{j}^{*})$ and $\varepsilon_{ij}^{*} = (\varepsilon_{i}^{*}\varepsilon_{j}^{*})^{1/2}$; R^{*} and ε^{*} are, respectively, the van der Waals radius and well depth, and these parameters are obtained from Cornell et al. (1995).

3.9.5 Force fields for computing free energy terms

Several force fields have been developed to calculate the interactions energy with different potentials. Essentially, all these programs include the terms for different energy terms, hydrophobic, hydrogen bonding, electrostatic, van der Waals, etc. In addition, the energy due to bond lengths, bond angles, and torsion angles may also be included. AMBER (Cornell et al. 1995), CHARMM (Brooks et al. 1983), OPLS (Jorgensen et al. 1996), GROMOS (Christen et al. 2005), and ECEPP (Arnautova et al. 2006) are the commonly used programs for the calculation of molecular mechanics and dynamics. These force fields are implemented in several commercial software, such as INSIGHT II, SYBYL, and MOE.

The AMBER force field has been developed by taking into account of bond lengths, bond angles, torsional angles, van der Waals, and electrostatic interactions. It is given by

$$E_{total} = \sum K_r(r - r_{eq})^2 + \sum K_\theta(\theta - \theta_{eq})^2 + \sum V_n/2[1 + \cos(n\phi - \gamma)]$$
$$+ \sum \left[(A_{ij}/r_{ij}^{12} - B_{ij}/r_{ij}^{6}) + q_i q_j / \varepsilon r_{ij} \right] \tag{3.8}$$

where r, θ, and ϕ are bond length, bond angle, and torsional angle, respectively. The values for K_r, r_{eq}, K_θ, and θ_{eq} have been derived by fitting structural and vibrational frequency data on small molecular fragments that make proteins; r_{ij} is the distance between the atoms i and j; q_i and q_j are the partial charges for the atoms i and j, respectively. A_{ij} and B_{ij} are van der Waals parameters as described in Equation 3.7. The details about the derivation of partial charges and the parameters for bond lengths, angles, torsional angles, van der Waals depth, and radius are available in Cornell et al. (1995).

Jorgensen et al. (1996) developed an OPLS all-atom force field for organic molecules and peptides. In OPLS force filed, the bond-stretching and angle-bending parameters have been adopted mostly from the AMBER all-atom force field. The torsional parameters were determined by fitting to rotational energy profiles obtained from ab initio molecular orbital calculations at the RHF/6-31G*//RHF/6-31G* level for more than 50 organic molecules and ions. The nonbonded parameters were developed in conjunction with Monte Carlo statistical mechanics simulations by computing thermodynamic and structural properties for 34 pure organic liquids, including alkanes, alkenes, alcohols, ethers, acetals, thiols, sulfides, disulfides, aldehydes, ketones, and amides.

CHARMM (Chemistry at HARvard Macromolecular Mechanics) is a highly flexible computer program, which uses empirical energy functions to model macromolecular systems (Brooks et al. 1993). The program can read or model build structures; energy minimize them by first- or second-derivative techniques; perform a normal mode or molecular dynamics simulation; and analyze the structural, equilibrium, and dynamic properties determined in these calculations.

GROMOS (Christen et al. 2005) is a package developed for the dynamical modeling of biomolecules using the methods of molecular dynamics, stochastic dynamics, and energy minimization. It has several features: (i) molecular dynamics (MD) simulation, stochastic dynamics (SD) simulation, and energy minimization; (ii) periodic boundary conditions, temperature and pressure control; (iii) long-range electrostatic interactions; (iv) charge-group based or atom-based cutoff for the nonbonded interactions; (v) nonphysical interactions; (vi) calculation of free energy changes based on the coupling parameter approach using thermodynamic integration, slow-growth, or one-step perturbation; and (vii) grid-based pairlist construction, enhanced sampling and path-integral simulation.

ECEPP-05 (Arnautova et al. 2006) is an all-atom force field developed for organic molecules and peptides with fixed bond lengths and bond angles. The van der Waals parameters were derived with global-optimization-based method: (i) An initial set of potential parameters is derived by fitting to ab initio interaction energies of dimers, and (ii) this initial set is refined to satisfy the criteria that the parameters should reproduce the observed crystal structures and enthalpies accurately (Arnautova et al. 2003). The values of the 1–4 nonbonded and electrostatic scale factors used in the force field were determined by computing the conformational energies of six model molecules, namely, ethanol, ethylamine, propanol, propylamine, 1,2-ethanediol, and 1,3-propanediol with different values of these factors. The partial atomic charges of these molecules were obtained by fitting to the electrostatic potentials calculated with the HF/6-31G quantum-mechanical method. To derive the torsional parameters for the peptide backbone, the partial atomic charges of the 20 neutral and charged amino acids were obtained by fitting to the electrostatic potentials of terminally blocked amino acids using the HF/6-31G quantum-mechanical method.

3.10 Amino acid properties derived from protein structural data

Protein structures have been used to derive several properties of amino acid residues. These properties reflect the behavior of amino acid residues in protein environment, which can be used for prediction algorithms.

3.10.1 Medium- and long-range contacts

The definition of medium- and long-range contacts has been described in **Section 3.4**. For each amino acid residue, the number of medium/long-range contacts has been computed within the radius of 8 Å, and the calculation has been repeated for all the residues in a protein and for a set of globular and membrane proteins. **Table 3.4** lists the average medium- and long-range contacts in different structural classes of globular and membrane proteins.

TABLE **3.4** Average medium- and long-range contacts in globular and membrane proteins

Residue	Globular		TMH		TMS	
	Medium	Long	Medium	Long	Medium	Long
Ala	2.1	3.9	3.2	2.8	0.8	6.2
Asp	1.8	2.9	2.5	2.1	0.8	4.6
Cys	1.9	5.6	3.4	3.6	1.3	7.7
Glu	2.1	2.7	2.3	2.3	0.7	5.4
Phe	2.0	4.5	3.2	2.5	0.6	5.5
Gly	1.5	4.3	2.9	3.2	0.7	5.9
His	2.0	3.8	3.0	2.5	0.9	4.5
Ile	1.8	5.6	3.4	2.6	0.8	6.0
Lys	2.0	2.8	2.3	2.2	0.7	5.7
Leu	2.2	4.6	3.4	2.3	0.6	6.4
Met	2.3	4.1	3.2	2.9	0.9	7.2
Asn	1.8	3.6	2.7	2.6	0.8	5.7
Pro	1.3	3.6	2.1	3.5	0.9	4.8
Gln	2.0	3.1	2.6	2.7	0.6	6.2
Arg	1.9	3.8	2.8	2.2	0.5	6.5
Ser	1.6	3.8	3.1	3.5	0.8	5.7
Thr	1.6	4.1	3.0	2.6	0.8	6.0
Val	1.6	5.4	3.3	2.6	0.6	6.6
Trp	1.9	4.8	3.2	2.6	0.4	6.0
Tyr	1.7	4.9	2.8	3.1	0.4	6.6

TMH: transmembrane helical; TMS: transmembrane strand.

3.10.2 Number of surrounding residues

For each residue, the distance between the specific residue and other residues in a protein has been computed and counted the number if the distance is below the cutoff distance. Manavalan and Ponnuswamy (1977) used the limit of 8 Å and reported the number of residues. The average number of surrounding residues for each of the 20 amino acid residues is listed in **Table 3.5**.

3.10.3 Surrounding hydrophobicity

The amino acid residues in a protein molecule are represented by their α-carbon atoms, and each residue is assigned with the hydrophobicity index obtained from thermodynamic transfer experiments (Nozaki and Tanford, 1971; Jones, 1975). The surrounding hydrophobicity (H_p) of a given residue is defined as the sum of hydrophobic indices of various residues, which appear within 8 Å radius limit from it (Manavalan and Ponnuswamy, 1978; Ponnuswamy, 1993).

$$H_p(i) = \sum_{j=1}^{20} n_{ij} h_j, \tag{3.9}$$

where n_{ij} is the total number of surrounding residues of type j around ith residue of the protein, and h_j is the experimental hydrophobic index of residue type j in kcal/mol (Nozaki and Tanford, 1971; Jones, 1975). The surrounding hydrophobicity indices reported in Manavalan and Ponnuswamy (1978) are presented in

TABLE 3.5   Characteristic features of amino acid residues in protein environment

Residue	N_s	H_p	ASA	G_t	B_r	$<R_A>$
Ala	6.05	12.97	113	0.20	0.38	3.70
Asp	4.95	10.85	151	−0.72	0.14	2.60
Cys	7.86	14.63	140	0.67	0.57	3.03
Glu	5.10	11.89	183	−1.09	0.09	3.30
Phe	6.62	14.00	218	0.67	0.51	6.60
Gly	6.16	12.43	85	0.06	0.38	3.13
His	5.80	12.16	194	0.04	0.31	3.57
Ile	7.51	15.67	182	0.74	0.56	7.69
Lys	4.88	11.36	211	−2.00	0.04	1.79
Leu	7.37	14.90	180	0.65	0.50	5.88
Met	6.39	14.39	204	0.71	0.42	5.21
Asn	5.04	11.42	158	−0.69	0.15	2.12
Pro	5.65	11.37	143	−0.44	0.18	2.12
Gln	5.45	11.76	189	−0.74	0.11	2.70
Arg	5.70	11.72	241	−1.34	0.07	2.53
Ser	5.53	11.23	122	−0.34	0.23	2.43
Thr	5.81	11.69	146	−0.26	0.23	2.60
Val	7.62	15.71	160	0.61	0.48	7.14
Trp	6.98	13.93	259	0.45	0.40	6.25
Tyr	6.73	13.42	229	−0.22	0.26	3.03

N_s: Average number of surrounding residues (Manavalan and Ponnuswamy, 1977).
H_p: Surrounding hydrophobicity in kcal/mol (Manavalan and Ponnuswamy, 1978).
ASA: Solvent accessibility in folded state of monomeric proteins in $Å^2$ (Miller et al. 1987).
G_t: Transfer free energy in kcal/mol (Miller et al. 1987).
B_r: Buriedness (Chothia, 1976).
$<R_A>$: Reduction in accessibility (Ponnuswamy et al. 1980).

Table 3.5. Furthermore, the scale has been computed for different secondary structures (Ponnuswamy et al. 1980) and different structural classes of globular (Cid et al. 1992) and membrane proteins (Ponnuswamy and Gromiha, 1993; Gromiha and Suwa, 2003).

3.10.4 Accessible surface area and transfer free energy

Miller et al. (1987) computed the ASA for different sets of monomeric and oligomeric protein structures and analyzed the tendency of each residue to be in the interior or on the surface of a protein (Miller et al. 1987). Furthermore, the information about the ASA of each residue has been converted into hydrophobic free energy for understanding the stability of a protein as follows: For each residue, the partition coefficient (f) has been computed from the ratio of the occurrence of that residue to be in the interior of the protein to that in the surface. The partition coefficient has been converted into transfer free energy using the expression (Janin et al. 1988),

$$\Delta G_t = -RT \ln f. \tag{3.10}$$

Table 3.5 includes the ASA of each amino acid residue in monomeric proteins and their hydrophobic free energies.

TABLE 3.6   Flexibility and thermodynamic parameters for the 20 amino acid residues

Residue	F	ΔG_h	ΔH_h	$-T\Delta S_h$	ΔCp_h
Ala	0.96	−0.54	−2.24	1.70	14.22
Asp	1.14	−2.97	−4.54	1.57	2.73
Cys	0.87	−1.64	−3.43	1.79	9.41
Glu	1.07	−3.71	−5.63	1.92	3.17
Phe	0.69	−1.06	−5.11	4.05	39.06
Gly	1.16	0.59	−1.46	0.87	4.88
His	0.80	−3.38	−6.83	3.45	20.05
Ile	0.76	0.32	−3.84	4.16	41.98
Lys	1.14	−2.19	−5.02	2.83	17.68
Leu	0.79	0.27	−3.52	3.79	38.26
Met	0.78	−0.60	−4.16	3.56	31.67
Asn	1.04	−3.55	−5.68	2.13	3.91
Pro	1.16	0.32	−1.95	2.27	23.69
Gln	1.07	−3.92	−6.23	2.31	3.74
Arg	1.05	−5.96	−10.43	4.47	16.66
Ser	1.13	−3.82	−5.94	2.12	6.14
Thr	0.96	−1.97	−4.39	2.42	16.11
Val	0.79	0.13	−3.15	3.28	32.58
Trp	0.77	−3.80	−8.99	5.19	37.69
Tyr	1.01	−5.64	−10.67	5.03	30.54

F: Flexibility in Å (Bhaskaran and Ponnuswamy, 1984).
ΔG_h: Hydration free energy change in kcal/mol (Oobatake and Ooi, 1993).
ΔH_h: Hydration enthalpy change in kcal/mol (Oobatake and Ooi, 1993).
$-T\Delta S_h$: Hydration entropy change in kcal/mol (Oobatake and Ooi, 1993).
ΔCp_h: Hydration heat capacity change cal/mol/K (Oobatake and Ooi, 1993).

3.10.5 Buriedness and reduction in accessibility

Buriedness is defined as the ratio between the number of residues (of type i) in the interior of the protein to the total number of residues (of type i) in a protein. This term reveals the tendency of each amino acid residue to be in the interior of the protein (Chothia, 1976). A related term is the reduction in accessibility, which is given by

$$<R_A> = A^0 - <A> /A^0,   \qquad (3.11)$$

where A^0 and $<A>$ represent, respectively, the accessible area in the unfolded (extended) and folded states of the protein (Ponnuswamy et al. 1980). **Table 3.5** shows the buriedness and reduction in the accessibility of the 20 amino acid residues in globular proteins.

3.10.6 Flexibility

Bhaskaran and Ponnuswamy (1984) developed a differential equation model for understanding the dynamic behavior of amino acid residues in globular proteins. They have calculated the root mean displacement for each of the 20 amino acid residues. This character has been related with the temperature factor of amino acid residues in globular proteins. In **Table 3.6**, the flexibility parameters are included for the 20 amino acid residues.

3.10.7 Thermodynamic parameters

Oobatake and Ooi (1993) computed the changes in hydration free energy, hydration enthalpy, hydration entropy, and hydration heat capacity change for the 20 amino acid residues using the three-dimensional structures of 113 proteins. The amino acid residues are portioned into seven groups, and the extended state ASA for these groups has been computed. These ASA values have been weighted with the hydration parameters, g_{ih}, h_{ih}, s_{ih}, and c_{ph}, to obtain the hydration free energy change, hydration enthalpy change, hydration entropy change, and hydration heat capacity change, respectively using the equations:

$$\Delta G_h = \sum g_{ih} A_i; \tag{3.12}$$

$$\Delta H_h = \sum h_{ih} A_i; \tag{3.13}$$

$$\Delta S_h = \sum s_{ih} A_i; \tag{3.14}$$

$$\Delta C_{ph} = \sum c_{ph} A_i. \tag{3.15}$$

The numerical values for these thermodynamic quantities have been presented in **Table 3.6**.

3.10.8 Other amino acid properties

Several other amino acid properties have been developed using three-dimensional structures of proteins. Furthermore, several properties of amino acid residues have been measured experimentally. Kidera et al. (1985) collected 188 amino acid properties and analyzed the relationship between them. Kawashima et al. (1999) developed a database, AAindex, which contains the numerical indices for various physicochemical and biochemical properties of amino acids and pairs of amino acids. Currently, it has 516 amino acid properties along with their references. It is available at http://www.genome.jp/dbget/aaindex.html. Furthermore, Juretic et al. (1998) compiled the amino acid indices of 88 hydrophobicity scales and used them for predicting membrane-spanning helices in membrane proteins. Cornette et al. (1987) collected 38 hydrophobicity scales for detecting amphipathic structures in proteins. Palliser and Parry (2001) listed the numerical values of 127 hydrophobicity scales and analyzed the performance of such scales for recognizing surface β-strands in proteins. Gromiha et al. (1999a,b; 2000) used a set of 49 physicochemical, energetic, and conformational properties of amino acid residues for understanding the stability of proteins (see **Section 2.5**).

3.11 Parameters for proteins

Several parameters have been also derived for a whole protein or protein chains, which can be related with protein stability, folding and unfolding rates, etc. Furthermore, the properties of all amino acid residues in a protein have also been studied.

3.11.1 Contact order

Plaxco et al. (1998) proposed the parameter, contact order (CO), which reflects the relative importance of local and nonlocal contacts to the native structure of a protein (Plaxco et al. 1998). It is defined as

$$CO = \sum \Delta S_{ij}/L.N, \tag{3.16}$$

where i and j are contacting residues in space within a distance of 6 Å and N is the total number of contacts, ΔS_{ij} is the sequence separation between the residues i and j, and L is the total number of residues in the protein. In a protein with low contact order, the residues interact with others are close in sequence. A high contact order implies that there are a large number of long-range interactions. Plaxco et al. (1998) found a striking correlation that has been observed between contact order and folding rate of small proteins.

3.11.2 Long-range order

Long-range order (LRO) for a protein has been defined from the knowledge of long-range contacts (contacts between two residues that are close in space and far in the sequence) in the protein structure (Gromiha and Selvaraj, 2001a). It is defined as

$$LRO = \sum n_{ij}/N; \; n_{ij} = 1 \quad if|i - j| > 12; \tag{3.17}$$
$$= 0 \, \text{otherwise,}$$

where i and j are two contacting residues in space within a distance of 8 Å and N is the total number of residues in a protein.

A good inverse relationship ($r = -0.78$) has been observed between LRO and experimental folding rate, $\ln(k)$ for a set of 23 considered proteins. The minimum distance of separation between two interacting residues has been varied from 1 to 50 residues and examined the correlation between LRO and folding rates. The results presented in **Figure 3.10** indicate that the minimum distance of 12 residues in defining LRO has the best correlation between LRO, and folding rates and significant correlation is obtained for the minimum residue separation of 10 to 15 residues (Gromiha and Selvaraj, 2001a).

3.11.3 Stabilization center

Stabilization centers (SC) in a protein are clusters of residues involved in long-range interactions (Dosztanyi et al. 1997). Two residues are considered to be in long-range interaction if they are separated by at least 10 residues in the sequence and at least one of their heavy-atom contact distances is less than the sum of their van der Waals radii of the two atoms, plus 1 Å. Two residues are part of stabilization centers if (i) they are involved in long-range interactions, and (ii) two supporting residues can be selected from both of their flanking tetrapeptides, which together with the central residues form at least seven out of nine contacts. When a residue satisfies these two criteria with one (or 2, 3, etc.) residue(s), the SC value of this residue is defined as 1 (or 2, 3, etc.). A public server is available for the identification of SC at http://www.enzim.hu/scide. As an example, the SCs in T4 lysozyme are shown in **Figure 3.11**. In this protein, the residues Ile17, Ile27, Ala74, Arg95, Ala98, Val103, Met120, Asn132, Thr152, and Phe153 are involved in stabilization centers.

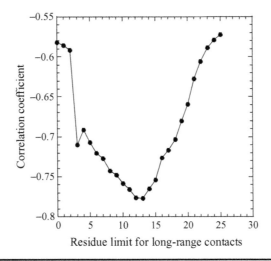

Figure 3.10   Plot connecting the correlation coefficient obtained between long-range contacts and folding rate of proteins, and the minimum limit to define long-range contacts. Figure was adapted from Gromiha and Selvaraj (2001).

3.11.4 Atom depth

Pintar et al. (2003) proposed the parameter, atom depth for describing the interior of the protein accurately. They described a simple and fast algorithm that measures the depth of each atom in a protein, defined as its distance from the closest solvent accessible atom. It can be easily computed from the 3D structure of a protein, thus complementing the information provided by the calculation of the solvent ASA and the buried surface area. The program reads a PDB file, which contains the atomic solvent accessibility in the B-factor field, and writes a file in the same

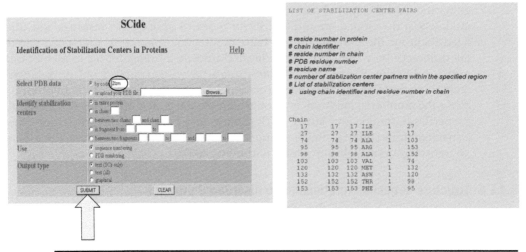

Figure 3.11   Identification of stabilization centers in protein structures. The results obtained for T4 lysozyme (2LZM) is shown. It has 10 stabilization centers as indicated in the figure.

format, where the B-factor field now contains the atom depth value. The algorithm is implemented in a standalone program written in C, and its source is freely available on the Web and on request from the developers (pintar@icgeb.org). Recently, a Web server, DPX, has been developed for calculating atom depth, and it is available at http://hydra.icgeb.trieste.it/dpx/.

3.11.5 Structural analysis of membrane proteins

The folding and assembly of membrane proteins are more complex than that of globular proteins due to the presence of lipid environment in the membrane. In helical membrane proteins, the *in vivo* assembly into membranes requires the assistance of a special cellular machine known as the translocon, and the process is associated with mechanical work, either from the translocation of the ribosome or from ATP1 hydrolysis (Wickner and Schekman, 2005). In contrast, transmembrane β-barrel membrane proteins must traverse through the translocon in the inner bacterial membrane and subsequently cross the periplasm to reach their final destination of folding and assembly. The process of folding and assembly of membrane proteins have been extensively reviewed (Popot and Engelman, 2000; White and von Heijne, 2008; Stanley and Fleming, 2008). The studies on energetics of membrane protein folding and stability reveal that several intrinsic properties resemble similar features in soluble and membrane proteins and the energetic differences mainly arise from specific intra- and intermolecular interactions within the membrane (Minetti and Remeta, 2006). The helix–helix interactions in membrane proteins are reported to play an important role to the folding and the stability of α-helical membrane proteins (MacKenzie and Fleming, 2008).

Using the available structures of membrane proteins, several investigations have been carried out to understand the behavior of amino acid residues in globular, α-helical, and β-barrel membrane proteins. Gromiha and Ponnuswamy (1996) derived numerical indices for different amino acid properties, including surrounding hydrophobicity, buriedness, the number of surrounding residues, hydrophobic gain ratio, and the spatial distribution of residues in membrane proteins. Mitaku et al. (2002) developed an amphiphilicity index of amino acid residues and utilized them for the transmembrane helix prediction. Yuan et al. (2003) compared the difference between signal peptides and transmembrane helices in membrane proteins. On the other hand, similar studies have been carried out on β-barrel membrane proteins to derive conformational parameters, surrounding hydrophobicity, the number of medium- and long-range contacts, and other physicochemical properties (Gromiha and Suwa, 2003; Jackups and Liang, 2005). Furthermore, the transmembrane strand proteins are more influenced with cation-π interactions than do transmembrane helical proteins (Gromiha, 2003).

3.11.6 Structure-sequence relationship in disordered proteins

It is generally believed that the function of a protein depends on its three-dimensional structure. Recently, it has been shown that many entire proteins and localized protein regions fail to fold into a 3D structure under physiological conditions and to perform functions. These proteins are named as intrinsically unstructured/disordered proteins/domains (Dunker et al. 2001). For example, the

N-terminal domain of P53 exists in a largely disordered structural state, yet they carry out basic cellular functions (Dawson et al. 2003). Hence, several investigations have been carried out to understand the relationship between protein disorder and function (Wright and Dyson, 1999; Dunker et al. 2002; Tompa, 2003).

Dosztanyi et al. (2005) reported that the amino acid sequence contains the information about protein disorder, and the amino acid composition/pairwise interaction energy between amino acid residues is an important parameter to identify the disordered domains. Consequently, the analysis of the amino acid composition on various aspects such as (i) folded and disordered proteins, (ii) folded and disordered domains, and (iii) short- and long-disordered domains compared with folded proteins revealed specific preferences of amino acid residues in disordered proteins/domains.

Sethi et al. (2008) reported that the content of aromatic residues is less in disordered proteins than the ordered ones. Shimizu et al. (2007) compared the frequency of amino acid residues in disordered and ordered proteins and showed that Gly, Pro, and Gln are dominant in disordered proteins, whereas an opposite trend was observed for Phe, Ile, Trp, and Tyr. This result indicates that polar residues mainly contribute for disordered proteins/domains. The variation on the propensity of amino acid properties based on physicochemical properties is presented in **Figure 3.12**. It clearly shows the dominance of polar, charged, and hydrophilic residues in disordered proteins and the dominance of aromatic, aliphatic, and hydrophobic residues in ordered proteins. The result reveals the importance of hydrophobic interactions to the folding and stability of ordered proteins (Ponnuswamy, 1993). Peng et al. (2006) compared the amino acid compositions of short (4–30 residues) and long

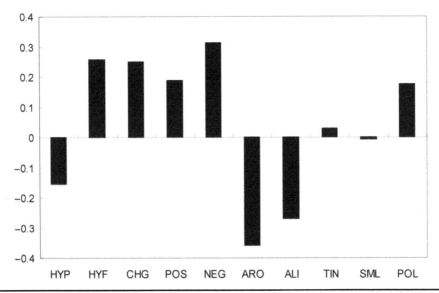

FIGURE 3.12   Propensity of amino acid residues based on physicochemical properties: HYP: hydrophobic, HYF: hydrophilic, CHG: charged, POS: positively charged, NEG: negatively charged, ARO: aromatic, ALI: aliphatic, TIN: tiny, SML: small, POL: polar. Figure was adapted from Shimizu et al. (2007).

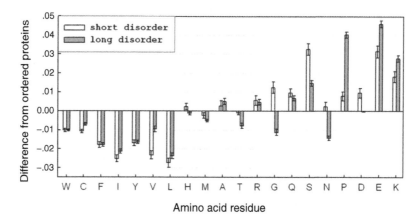

Figure 3.13 Comparison of amino acid compositions between short and long disordered regions. The y-axis represents the difference in amino acid compositions (fractions) from a reference dataset of ordered proteins. The error bars correspond to one standard deviation estimated using 5000 bootstrap samples. Figure was adapted from Peng et al. (2006).

(>30 residues) disordered regions and reported that both types of disordered regions exhibit similar overall compositional bias that characterizes intrinsic protein disorder. However, there were also some significant differences that short disordered regions are more depleted in C, I, V, and L, while long disordered regions are more enriched in K, E, and P but are less enriched in Q and S (**Figure 3.13**). In addition, long disordered regions are depleted in G and N, while short disordered regions are enriched in G and D.

Galzitskaya et al. (2006) calculated mean packing density for each amino acid residue in a database of 5829 protein structures and the average parameter for the 20 amino acid residues. It is the number of contacts between a pair of residues in which their heavy atoms distance is within the limit of 8 Å and excluding the neighboring residues. This information has been utilized to construct a packing density profile, which discriminates ordered and disordered regions.

3.12 Protein structure comparison

It has been suggested that proteins of similar structures have similar function. Hence, several methods have been developed to compare protein structures and domains. The structure-based multiple sequence alignment will also be helpful to understand the influence of similar/dissimilar structural regions.

3.12.1 DALI

Holm and Sander (1995) developed a program, DALI, for comparing protein structures. The database, FSSP, has been developed for the classification of 3D protein folds based on an all-against-all comparison of structures in the PDB. The results of the exhaustive pairwise structure comparisons are reported in the form of a fold tree generated by a hierarchical clustering and as a series of structurally representative sets of folds at varying levels of uniqueness. For each query structure from the representative set, there is a database entry containing structure–structure alignments with its structural neighbors in the representative set and its sequence homologs in the PDB. The DALI server accepts the coordinates of a query protein structure and compares them against those in the PDB. Furthermore, for any PDB structure, one can get the closely related structures using DALI. It provides the similarity score that is normalized with the domain size, percentage amino acid identity in aligned positions, root-mean-square deviation of C_α atoms in superimposition, number of structurally equivalent positions, and length of the structural neighbor protein. In favorable cases, comparing 3D structures may reveal biologically interesting similarities that are not detectable by comparing sequences. The DALI database is available at http://www.embl-ebi.ac.uk/dali/.

3.12.2 Combinatorial extension

Shindyalov and Bourne (1998) reported an algorithm, CE, based on the combinatorial extension, which builds an alignment between two protein structures. It involves a combinatorial extension of an alignment path defined by aligned fragment pairs rather than using dynamic programming and Monte Carlo optimization. Aligned fragment pairs are based on local geometry, rather than global features such as the orientation of secondary structures and the overall topology. Combinations of aligned fragment pairs that represent possible continuous alignment paths are selectively extended or discarded, thereby leading to a single optimal alignment. The method also provides the structural neighbors of each protein in the PDB. **Figure 3.14** shows the search example for 2lzm using CE and the output. It takes the PDB code as the input (**Figure 3.14a**) and lists its structural neighbors with

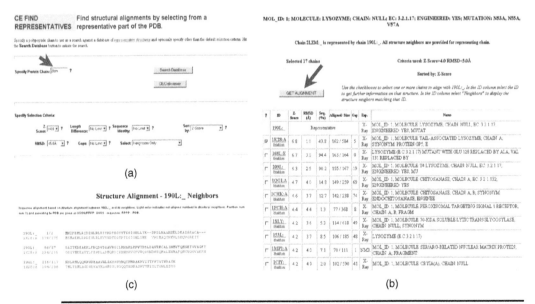

FIGURE 3.14 Structural alignments using CE. The identification of structural neighbors of 2LZM is illustrated in the figure: (a) input options, (b) identified structural neighbors, and (c) sequence alignment between two proteins.

several parameters, including the Z-score, root-mean-square deviation, percentage identity, length of aligned sequence, and the number of gaps (**Figure 3.14b**). It also shows the sequence alignment between the selected proteins (**Figure 3.14c**). CE is available at http://cl.sdsc.edu/ce.html.

3.12.3 PRIDE

Carugo and Pongor (2002) developed a method for comparing protein structures based on the distribution of $C_\alpha - C_\alpha$ distances between two residues, which are separated by 3 to 30 amino acid residues. The server, PRIDE, takes the structural data of the query protein as input and search with CATH and PDB databases. In PRIDE, structural alignment and secondary-structure assignment are not necessary for the calculation, which makes it fast enough to allow the scanning of large databases. It is available at http://hydra.icgeb.trieste.it/pride/.

3.12.4 MATRAS

Kawabata and Nishikawa (2000) developed the program, MATRAS, for comparing protein tertiary structures using the Markov transition model of evolution. In this method, the similarity score between structures i and j has been defined as log $P(j \to i)/P(i)$, where $P(j \to i)$ is the probability that structure j changes to structure i during the evolutionary process, and $P(i)$ is the probability that structure i appears by chance. MATRAS takes the PDB code or any file in the PDB format and sends the structure comparison results by e-mail. It has several features that one can search the similar positions within a protein, between two proteins, and a protein with several databases. It is freely available at http://biunit.aist-nara.ac.jp/matras/.

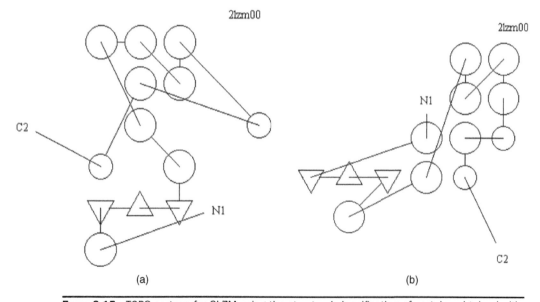

(a)   (b)

FIGURE 3.15   TOPS cartoon for 2LZM using the structural classification of proteins obtained with (a) SCOP and (c) CATH. The circles and triangles denote α-helices and β-strands, respectively.

3.12.5 TOPS

Michalopoulos et al. (2004) developed an enhanced database, TOPS, for describing protein structural topology. These topological descriptions reduce the protein fold to a sequence of Secondary Structure Elements (SSEs) and three sets of pairwise relationships between them, hydrogen bonds relating parallel and antiparallel β-strands, spatial adjacencies relating neighboring SSEs, and the chiralities of selected supersecondary structures, including connections in βαβ units and between parallel α-helices. It takes the file with the PDB format as input and displays the structure comparison results. Furthermore, TOPS shows the atlas of cartoons for any PDB structures with the representation of α-helices, β-strands, their connectivities, etc. TOPS can be found at http://www.tops.leeds.ac.uk/. An example is shown in **Figure 3.15**. It shows the TOPS cartoon for 2LZM using the structural classification based on SCOP (**Figure 3.15a**) and CATH (**Figure 3.15b**). In this figure, triangular and circular symbols represent β- strands and α-helices, respectively.

3.13 Exercises

1. Compare the secondary structural information given in DSSP and the PDB for the protein 2LZM.
   Hint: Download the DSSP file and get the secondary structure information. Get the same from the PDB with helix and strand. Compare them for each amino acid residue.
2. Compute the ASA for 2LZM per atom/atom type, residue and total ASA.
   Hint: Use GETAREA to obtain the parameters.
3. Compare the ASA for 2LZM using different programs.
   Hint: Get the result with DSSP, ASC, GETAREA, etc., and compare for each residue/atom.
4. Obtain the interior seeking residues of 2LZM.
   Hint: Use ASAview to view the ASA of 2LZM, and look into the spiral view to find the interior residues.
5. Analyze the patterns of the solvent accessibility for different types of proteins and at different secondary structures.
   Hint: Take datasets of all-α and all-β proteins, and compare the ASA for different residues and secondary structures.
6. How does the magnitude of ASA vary with different methods?
   Hint: Repeat the computations with different programs.
7. Identify the residues that are surrounded by Ile50 in 2LZM within the sphere of 8 Å.
   Hint: (i) Use C_α atoms of the protein, and compute the distance between Ile50 and other residues. If the distance is within 8 Å, take it as a contacting residue; (ii) use each heavy atom of Ile50, and compute the distance with the heavy atoms in other residues; if any heavy atom is within the distance of 8 Å, then take the residue as a contacting one.
8. Identify the contacting residues that are forming short-, medium-, and long-range contacts.

Hint: Obtain the contacting residues in the above question. Check the contacting residues of Ile50 in terms of sequence separation.

9. Obtain the short-, medium-, and long-range contacts in globular, transmembrane helical, and β-barrel membrane proteins, and compare them.
   Hint: Set up databases for three kinds of proteins, and compute the average short-, medium-, and long-range contacts in these classes of proteins.

10. Delineate the cation-π interactions in 2LZM.
    Hint: Use CAPTURE program.

11. Compare the hydrophobic free energy of proteins belonging to different structural classes.
    Hint: Prepare a dataset for different structural classes; compute ASA in folded and unfolded states for different types of atoms. Use Equation 3.5 to compute the hydrophobic free energy.

12. Discuss the variation of hydrophobic free energy with different hydrophobicity scales/atomic solvation parameters.
    Hint: Use different atomic solvation parameters given in **Table 3.3**.

13. Compare the cation-π/noncanonical interactions in DNA- and RNA-binding proteins.
    Hint: Prepare datasets of DNA- and RNA-binding proteins. Compute the cation-π/noncanonical interactions using CAPTURE/NCI.

14. Compute contact order and long-range order for 2LZM.
    Hint: Use equations 3.16 and 3.17.

15. Discuss the correlation between CO and LRO in different structural classes of proteins.
    Hint: Develop datasets based for structural classification of proteins and compute CO/LRO for all proteins belonging to different structural classes.

16. Obtain the stabilization centers in 4LYZ.
    Hint: Use Scide.

17. Analyze the stabilization center residues with respect to the secondary structure and the solvent accessibility.
    Hint: Prepare a nonredundant set of proteins; define SS and ASA (DSSP may be used). Compute stabilization center residues using SCide. Compare the results at different SSAA and/or residue types.

18. Compute the atom depth of 2LZM. Compare the results with ASA.
    Hint: Use DPX server. The last column is the atom depth.

19. Delineate the structurally similar proteins to human lysozyme with different methods.
    Hint: Use CE and other methods to obtain structurally similar proteins.

References

Ahmad S, Gromiha M, Fawareh H, Sarai A. ASAView: database and tool for solvent accessibility representation in proteins. BMC Bioinformatics. 2004;5:51.

Ahmad S, Gromiha MM, Sarai A. Real-value prediction of solvent accessibility from amino acid sequence. Proteins. 2003;50:629–635.

Arnautova YA, Jagielska A, Pillardy J, Scheraga HA. Derivation of a new force field for crystal-structure prediction using global optimization: nonbonded potential parameters for hydrocarbons and alcohols. J Phys Chem B. 2003;107(29):7143–7154.

Arnautova YA, Jagielska A, Scheraga HA. A new force field (ECEPP-05) for peptides, proteins, and organic molecules. J Phys Chem B. 2006;110(10):5025–5044.

Babu MM. NCI: a server to identify non-canonical interactions in protein structures. Nucleic Acids Res. 2003;31(13):3345–3348.

Bahar I, Jernigan RL. Inter-residue potentials in globular proteins and the dominance of highly specific hydrophilic interactions at close separation. J Mol Biol. 1997;266: 195–214.

Barlow DJ, Thornton JM. Ion-pairs in proteins. J Mol Biol. 1983;168:867–885.

Berman HM, Westbrook J, Feng Z, Gilliland G, Bhat TN, Weissig H, Shindyalov IN, Bourne PE. The Protein Data Bank. Nucleic Acids Res. 2000;28(1):235–242.

Bhaskaran R, Ponnuswamy PK. Dynamics of amino acid residues in globular proteins. Int J Pept Protein Res. 1984;24(2):180–191.

Bowie JU, Luthy R, Eisenberg D. A method to identify protein sequences that fold into a known three-dimensional structure. Science. 1991;253:164–170.

Brooks BR, Bruccoleri RE, Olafson BD, States DJ, Swaminathan S, Karplus M. CHARMM—A program for macromolecular energy, minimization and dynamics calculations. J Comp Chem. 1983;4(2):187–217.

Carugo O, Pongor S. Protein fold similarity estimated by a probabilistic approach based on C(alpha)-C(alpha) distance comparison. J Mol Biol. 2002;315(4):887–898.

Cavallo L, Kleinjung J, Fraternali F. POPS: A fast algorithm for solvent accessible surface areas at atomic and residue level. Nucleic Acids Res. 2003;31(13):3364–3366.

Chakravarty S, Varadarajan R. Elucidation of factors responsible for enhanced thermal stability of proteins: a structural genomics based study. Biochemistry. 2002;41(25): 8152–8161.

Chirgadze YN, Larionova EA. Spatial sign-alternating charge clusters in globular proteins. Protein Eng. 1999; 12:101–105.

Chiu TL, Goldstein RA. Optimizing energy potentials for success in protein tertiary structure prediction. Fold Des. 1998;3:223–228.

Chothia C. The nature of the accessible and buried surfaces in proteins. J Mol Biol. 1976;105:1–14.

Christen M, Hunenberger PH, Bakowies D, Baron R, Burgi R, Geerke DP, Heinz TN, Kastenholz MA, Krautler V, Oostenbrink C, Peter C, Trzesniak D, van Gunsteren WF. The GROMOS software for biomolecular simulation: GROMOS05. J Comput Chem. 2005;26(16):1719–1751.

Cid H, Bunster M, Canales M, Gazitua F. Hydrophobicity and structural classes in proteins. Protein Eng. 1992;5(5):373–375.

Cornell WD, Cieplak P, Bayly CI, Gould IR, Merz KM Jr, Ferguson DM, Spellmeyer D C, Fox T, Caldwell JW, Kollman PA. A second generation force field for the simulation of proteins, nucleic acids, and organic molecules. J Am Chem Soc. 1995;117:5179–5197.

Cornette JL, Cease KB, Margalit H, Spouge JL, Berzofsky JA, DeLisi C. Hydrophobicity scales and computational techniques for detecting amphipathic structures in proteins. J Mol Biol. 1987;195(3):659–685.

Dawson R, Müller L, Dehner A, Klein C, Kessler H, Buchner J. The N-terminal domain of p53 is natively unfolded. J Mol Biol. 2003;332(5):1131–1141.

Debe DA, Goddard WA. First principles prediction of protein folding rates. J Mol Biol. 1999;294:619–625.

Dosztányi Z, Csizmók V, Tompa P, Simon I. The pairwise energy content estimated from amino acid composition discriminates between folded and intrinsically unstructured proteins. J Mol Biol. 2005;347(4):827–839.

Dosztanyi Z, Fiser A, Simon I. Stabilization centers in proteins: identification, characterization and predictions. J Mol Biol. 1997;272:597–612.

Dougherty DA. Cation-pi interactions in chemistry and biology: a new view of benzene, Phe, Tyr, and Trp. Science. 1996;271(5246):163–168.

Dunker AK, Brown CJ, Lawson JD, Iakoucheva LM, Obradovic Z. Intrinsic disorder and protein function. Biochemistry. 2002;41(21):6573–6582.

Dunker AK, Lawson JD, Brown CJ, Williams RM, Romero P, Oh JS, Oldfield CJ, Campen AM, Ratliff CM, Hipps KW, Ausio J, Nissen MS, Reeves R, Kang C, Kissinger CR, Bailey RW, Griswold MD, Chiu W, Garner EC, Obradovic Z. Intrinsically disordered protein. J Mol Graph Model. 2001;19(1):26–59.

Eisenberg D, McLachlan AD. Solvation energy in protein folding and binding. Nature. 1986;319(6050):199–203.

Eisenhaber F, Argos P: Improved strategy in analytical surface calculation for molecular system-handling of singularities and computational efficiency. J Comp Chem. 1993;14:1272–1280.

Engelman DM, Steitz TA, Goldman A. Identifying nonpolar transbilayer helices in amino acid sequences of membrane proteins. Annu Rev Biophys Biophys Chem. 1986;15:321–353.

Fariselli P, Olmea O, Valencia A, Casadio R. Progress in predicting inter-residue contacts of proteins with neural networks and correlated mutations. Proteins. 2001; S5:157–162.

Fauchere JL, Pliska V. Hydrophobic parameters of amino acid side chains from the portioning of N-acetyl amino acid amides. Eur J Med Chem. 1983;18(4):369–375.

Flockner H, Braxenthaler M, Lackner P, Jaritz M, Ortner M, Sippl MJ. Progress in fold recognition. Proteins. 1995;23:376–386.

Fraczkiewicz R, Braun W. Exact and efficient analytical calculation of the accessible surface areas and their gradients for macromolecules. J Comp Chem. 1998;19: 319–333.

Furuichi E, Koehl P. Influence of protein structure databases on the predictive power of statistical pair potentials. Proteins. 1998;31:139–149.

Gallivan JP, Dougherty DA. Cation-p interactions in structural biology. Proc Natl Acad Sci USA. 1999;96:9459–9464.

Galzitskaya OV, Garbuzynskiy SO, Lobanov MY. FoldUnfold: web server for the prediction of disordered regions in protein chain. Bioinformatics. 2006;22(23):2948–2949.

Gianese G, Bossa F, Pascarella S. Improvement in prediction of solvent accessibility by probability profiles. Protein Eng. 2003;16:987–992.

Gilis D, Rooman M. Predicting protein stability changes upon mutation using database-derived potentials: solvent accessibility determines the importance of local versus non-local interactions along the sequence. J Mol Biol. 1997;272:276–290.

Gromiha MM. Important inter-residue contacts for enhancing the thermal stability of thermophilic proteins. Biophys Chem. 2001;91:71–77.

Gromiha MM. Influence of cation-pi interactions in different folding types of membrane proteins. Biophys Chem. 2003;103(3):251–258.

Gromiha MM, Ahmad S. Role of solvent accessibility in structure based drug design. Curr Comp Aided Drug Des. 2005;1:65–72.

Gromiha MM, Ponnuswamy PK. Hydrophobic distribution and spatial arrangement of amino acid residues in membrane proteins. Int J Pept Protein Res. 1996;48(5):452–60.

Gromiha MM, Selvaraj S. Influence of medium and long range interactions in different structural classes of globular proteins J Biol Phys. 1997;23:151–162.

Gromiha MM, Selvaraj S. Importance of long-range interactions in protein folding. Biophys Chem. 1999;77:49–68.

Gromiha MM, Selvaraj S. Comparison between long-range interactions and contact order in determining the folding rates of two-state proteins: application of long-range order to folding rate prediction. J Mol Biol. 2001a;310:27–32.

Gromiha MM, Selvaraj S. Role of medium and long-range interactions in discriminating globular and membrane proteins. Int J Biol Macromol. 2001b;29:25–34.

Gromiha MM, Selvaraj S. Important amino acid properties for determining the transition state structures of two-state protein mutants. FEBS Lett. 2002;526:129–134.

Gromiha MM, Selvaraj S. Inter-residue interactions in protein folding and stability. Prog Biophys Mol Biol. 2004;86(2):235–77.

Gromiha MM, Suwa M. Variation of amino acid properties in all-beta globular and outer membrane protein structures. Int J Biol Macromol. 2003;32(3–5):93–98.

Gromiha MM, Suwa M. Structural analysis of residues involving cation-pi interactions in different folding types of membrane proteins. Int J Biol Macromol. 2005;35:55–62.

Gromiha MM, Thangakani AM. Role of medium- and long-range interactions to the stability of the mutants of T4 lysozyme. Prep Biochem Biotech. 2001;31:217–227.

Gromiha MM, Oobatake M, Kono H, Uedaira H, Sarai A. Role of structural and sequence information in the prediction of protein stability changes: comparison between buried and partially buried mutations. Protein Eng. 1999a;12:549–555.

Gromiha MM, Oobatake M, Sarai A. Important amino acid properties for enhanced thermostability from mesophilic to thermophilic proteins. Biophys Chem. 1999b;82(1): 51–67.

Gromiha MM, Oobatake M, Kono H, Uedaira H, Sarai A. Importance of surrounding residues for protein stability of partially buried mutations. J Biomol Struct Dyn. 2000;18(2):281–295.

Gromiha MM, Thomas S, Santhosh C. Role of cation-π interaction to the stability of thermophilic proteins. Prep Biochem Biotech. 2002a;32:355–362.

Gromiha MM, Oobatake M, Kono H, Uedaira H, Sarai A. Importance of mutant position in Ramachandran plot for predicting protein stability of surface mutations. Biopolymers 2002b;64:210–220.

Gromiha MM, Santhosh C, Ahmad S. Structural analysis of cation-pi interactions in DNA binding proteins. Int J Biol Macromol. 2004a;34:203–211.

Gromiha MM, Santhosh C, Suwa M. Influence of cation-pi interactions in protein-DNA complexes. Polymer. 2004b;45:633–639.

Gromiha MM, Yokota K, Fukui K. Energy based approach for understanding the recognition mechanism in protein–protein complexes. Mol Biosystems. 2009 (in press). DOI:10.1039/B904161N.

Gugolya Z, Dosztanyi Z, Simon I. Interresidue interactions in protein classes. Proteins. 1997;27:360–366.

Heringa J, Argos P. Side-chain clusters in protein structures and their role in protein folding. J Mol Biol. 1991;220:151–171.

Holm L, Sander C. Dali: a network tool for protein structure comparison. Trends Biochem Sci. 1995;20(11):478–480.

Huang L-H, Gromiha MM. Analysis and prediction of protein folding rates using quadratic response surface models. J Comp Chem. 2008;29(10):1675–1683.

Hubbard SJ, Thornton JM. NACCESS, Computer Program. Department of Biochemistry and Molecular Biology. London: University College London; 1993.

Jackups R Jr, Liang J. Interstrand pairing patterns in beta-barrel membrane proteins: the positive-outside rule, aromatic rescue, and strand registration prediction. J Mol Biol. 2005;354(4):979–993.

Janin J, Miller S, Chothia C. Surface, subunit interfaces and interior of oligomeric proteins. J Mol Biol. 1988;204(1):155–164.

Jiang Z, Zhang, L, Chen, J, Xia, A, Zhao D. Effect of amino acid on forming residue-residue contacts in proteins. Polymer 2002;43:6037–6047.

Jones DD. Amino acid properties and side-chain orientation in proteins: a cross correlation approach. J Theor Biol. 1975;50:167–183.

Jones DT, Taylor WR, Thornton JM. A new approach to protein fold recognition. Nature. 1992:358,86–89.

Jorgensen WL, Maxwell DS, TiradoRives J. Development and testing of the OPLS all-atom force field on conformational energetics and properties of organic liquids. J Amer Chem Soc. 1996;118:11225–11236.

Juretic D, Lucic B, Zucic D, Trinajstic N. Protein transmembrane structure: recognition and prediction by using hydrophobicity scales through preference functions. In: Parkanyi C, ed. Theoretical and Computational Chemistry, Vol. 5. Theoretical Organic Chemistry. Amsterdam: Elsevier Science; 1998: 405–445.

Kabsch W, Sander C. Dictionary of protein secondary structure: pattern recognition of hydrogen-bond and geometrical features. Biopolymers. 1983;22:2577–2637.

Kannan N, Selvaraj S, Gromiha MM, Vishveshwara S. Clusters in alpha/beta barrel proteins: implications for protein structure, function, and folding: a graph theoretical approach. Proteins. 2001;43:103–112.

Kannan N, Vishveshwara S. Identification of side-chain clusters in protein structures by a graph spectral method. J Mol Biol. 1999;292:441–464.

Karlin S, Zhu ZY. Characterizations of diverse residue clusters in protein three-dimensional structures. Proc Natl Acad Sci USA. 1996;93:8344–8349.

Karlin S, Zuker M, Brocchieri L. Measuring residue associations in protein structures. Possible implications for protein folding. J Mol Biol. 1994;239:227–248.

Kawabata T, Nishikawa K. Protein structure comparison using the markov transition model of evolution. Proteins. 2000;41(1):108–122.

Kawashima S, Ogata H, Kanehisa M. AAindex: amino acid index database. Nucleic Acids Res. 1999;27(1):368–369.

Kidera A, Konishi Y, Oka M, Ooi T, Scheraga A. Statistical analysis of the physical properties of the 20 naturally occuring amino acids. J Prot Chem. 1985;4:23–55.

Kocher JP, Rooman MJ, Wodak SJ. Factors influencing the ability of knowledge-based potentials to identify native sequence-structure matches. J Mol Biol. 1994;235: 1598–1613.

Kyte J, Doolittle RF. A simple method for displaying the hydropathic character of a protein. J Mol Biol. 1982;157(1):105–132.

Lee B, Richards FM. The interpretation of protein structures: estimation of static accessibility. J Mol Biol. 1971;55:379–400.

MacKenzie KR, Fleming KG. Association energetics of membrane spanning alpha-helices. Curr Opin Struct Biol. 2008;18(4):412–419.

Manavalan P, Ponnuswamy PK. Hydrophobic character of amino acid residues in globular proteins. Nature. 1978;275(5681):673–674.

Manavalan P, Ponnuswamy PK. A study of the preferred environment of amino acid residues in globular proteins. Arch Biochem Biophys. 1977;184:476–487.

McDonald IK, Thornton JM. Satisfying hydrogen bonding potential in proteins. J Mol Biol. 1994;238:777–793.

Michalopoulos I, Torrance GM, Gilbert DR, Westhead DR. TOPS: an enhanced database of protein structural topology. Nucleic Acids Res. 2004;32(Database issue):D251–D254.

Miller EJ, Fischer KF, Marqusee S. Experimental evaluation of topological parameters determining protein-folding rates. Proc Natl Acad Sci USA. 2002;99(16):10359–10363.

Miller S, Janin J, Lesk AM, Chothia C. Interior and surface of monomeric proteins. J Mol Biol. 1987;196(3):641–656.

Minetti CA, Remeta DP. Energetics of membrane protein folding and stability. Arch Biochem Biophys. 2006;453(1):32–53.

Mirny LA, Shakhnovich EI. How to derive a protein folding potential? A new approach to an old problem. J Mol Biol. 1996;264:1164–1179.

Mitaku S, Hirokawa T, Tsuji T. Amphiphilicity index of polar amino acids as an aid in the characterization of amino acid preference at membrane-water interfaces. Bioinformatics. 2002;18(4):608–616.

Miyazawa S, Jernigan RL. Residue-residue potentials with a favorable contact pair term and an unfavorable high packing density term, for simulation and threading. J Mol Biol 1996;256:623–644.

Miyazawa S, Jernigan RL. An empirical energy potential with a reference state for protein fold and sequence recognition. Proteins. 1999;36:357–369.

Miyazawa S, Jernigan RL. Estimation of interresidue contact energies from protein crystal structures: Quasi-chemical approximation. Macromolecules. 1985;18:534–552.

Momen-Roknabadi A, Sadeghi M, Pezeshk H, Marashi SA. Impact of residue accessible surface area on the prediction of protein secondary structures. BMC Bioinformatics. 2008;9:357.

Mucchielli-Giorgi MH, Hazout S, Tuffery P. PredAcc: prediction of solvent accessibility. Bioinformatics. 1999;15(2):176–177.

Nicholson H, Becktel WJ, Matthews BW. Enhanced protein thermostability from designed mutations that interact with alpha-helix dipoles. Nature. 1988;336(6200):651–656.

Nozaki Y, Tanford C. The solubility of amino acids and two glycine peptides in aqueous ethanol and dioxane solutions. Establishment of a hydrophobicity scale. J Biol Chem. 1971;246(7):2211–2217.

Oobatake M, Crippen GM. Residue-residue potential function for conformational Analysis of proteins. J Phys Chem. 1981;85:1187–1197.

Oobatake M, Ooi T. Hydration and heat stability effects on protein unfolding. Prog Biophys Mol Biol. 1993;59(3):237–284.

Ooi T, Oobatake M, Nemethy G, Scheraga HA. Accessible surface areas as a measure of the thermodynamic parameters of hydration of peptides. Proc Natl Acad Sci U S A. 1987;84(10):3086–3090.

Ouzounis C, Sander C, Scharf M, Schneider R. Prediction of protein structure by evaluation of sequence-structure fitness. Aligning sequences to contact profiles derived from three-dimensional structures. J Mol Biol. 1993;232,805–825.

Palliser CC, Parry DA. Quantitative comparison of the ability of hydropathy scales to recognize surface beta-strands in proteins. Proteins. 2001;42(2):243–255.

Parthiban V, Gromiha MM, Hoppe C, Schomburg D. Structural analysis and prediction of protein mutant stability using distance and torsion potentials: role of secondary structure and solvent accessibility. Proteins. 2007;66(1):41–52.

Pascarella S, De Persio R, Bossa F, Argos P. Easy method to predict solvent accessibility from multiple protein sequence alignments. Proteins. 1998;32(2):190–199.

Peng K, Radivojac P, Vucetic S, Dunker AK, Obradovic Z. Length-dependent prediction of protein intrinsic disorder. BMC Bioinformatics. 2006;7:208.

Pintar A, Carugo O, Pongor S. Atom depth in protein structure and function. Trends Biochem Sci. 2003;28(11):593–597.

Plaxco KW, Simons KT, Baker D. Contact order, transition state placement and the refolding rates of single domain proteins. J Mol Biol. 1998;277:985–994.

Ponnuswamy PK, Gromiha MM. On the conformational stability of folded proteins. J Theor Biol. 1994;166:63–74.

Ponnuswamy PK, Gromiha MM. Prediction of transmembrane helices from hydrophobic characteristics of proteins. Int J Pept Protein Res. 1993;42(4):326–341.

Ponnuswamy PK. Hydrophobic characteristics of folded proteins. Prog Biophys Mol Biol. 1993;59(1):57–103.

Ponnuswamy PK, Prabakaran M, Manavalan P. Hydrophobic packing and spatial arrangement of amino acid residues in globular proteins. Biochim Biophys Acta. 1980;623:301–316.

Ponnuswamy PK, Warme PK, Scheraga HA. Role of medium-range interactions in proteins. Proc Natl Acad Sci USA. 1973;70:830–833.

Popot JL, Engelman DM. Helical membrane protein folding, stability, and evolution. Annu Rev Biochem. 2000;69:881–922

Porollo AA, Adamczak R, Meller J. POLYVIEW: a flexible visualization tool for structural and functional annotations of proteins. Bioinformatics. 2004;20:2460–2462.

Qin S, He Y, Pan XM. Predicting protein secondary structure and solvent accessibility with an improved multiple linear regression method. Proteins. 2005;61(3):473–480.

Reva B, Finkelstein AV, Sanner M, Olson AJ. Residue-residue mean-force potentials for protein structure recognition. Protein Eng. 1997;10:865–876.

Richmond TJ, Richards FM. Packing of alpha-helices: geometrical constraints and contact areas. J Mol Biol. 1978;119(4):537–555.

Rooman MJ, Wodak SJ. Are database-derived potentials valid for scoring both forward and inverted protein folding? Protein Eng. 1995;8:849–858.

Rooman MJ, Kocher JPA, Wodak SJ. Extracting information on folding from the amino acid sequence: accurate predictions for protein regions with preferred conformation in the absence of tertiary interactions. Biochemistry. 1992;31:10226–10238.

Rost B, Sander C. Conservation and prediction of solvent accessibility in protein families. Proteins. 1994;20:216–226.

Russ WP, Ranganathan R. Knowledge-based potential functions in protein design. Curr Opin Struct Biol. 2002;12:447–452.

Sali D, Bycroft M, Fersht AR. Stabilization of protein structure by interaction of alpha-helix dipole with a charged side chain. Nature. 1988;335(6192):740–743.

Selbig J. Contact pattern-induced pair potentials for protein fold recognition. Protein Eng. 1995;8:339–351.

Selvaraj S, Gromiha MM. Role of hydrophobic clusters and long-range contact networks in the folding of (a/b)8 barrel proteins. Biophys J. 2003;84:1919–1925.

Selvaraj S, Gromiha MM. An analysis of the amino acid clustering pattern in $(\alpha/\beta)_8$ barrel proteins. J Protein Chem. 1998;17:407–415.

Sethi D, Garg A, Raghava GP. DPROT: prediction of disordered proteins using evolutionary information. Amino Acids. 2008;35(3):599–605.

Shacham S, Marantz Y, Bar-Haim S, Kalid O, Warshaviak D, Avisar N, Inbal B, Heifetz A, Fichman M, Topf M, Naor Z, Noiman S, Becker OM. PREDICT modeling and in-silico screening for G-protein coupled receptors. Proteins. 2004;57(1): 51–86.

Shimizu K, Hirose S, Noguchi T. POODLE-S: web application for predicting protein disorder by using physicochemical features and reduced amino acid set of a position-specific scoring matrix. Bioinformatics. 2007;23(17):2337–2338.

Shimizu K, Muraoka Y, Hirose S, Tomii K, Noguchi T. Predicting mostly disordered proteins by using structure-unknown protein data. BMC Bioinformatics. 2007;8:78.

Shindyalov IN, Bourne PE. Protein structure alignment by incremental combinatorial extension (CE) of the optimal path. Protein Eng. 1998;11(9):739–747.

Sippl MJ. Calculation of conformational ensembles from potentials of mean force. An approach to the knowledge-based prediction of local structures in globular proteins. J Mol Biol. 1990;213:859–883.

Sippl MJ. Knowledge-based potentials for proteins. Curr Opin Str Biol. 1995;5:229–235.

Stanley AM, Fleming KG. The process of folding proteins into membranes: challenges and progress. Arch Biochem Biophys. 2008;469(1):46–66.

Tanaka S, Scheraga HA. Model of protein folding: inclusion of short-, medium-, and long-range interactions. Proc Natl Acad Sci. 1975;72:3802–3806.

Tobi D, Shafran G, Linial N, Elber R. On the design and analysis of protein folding potentials. Proteins. 2000;40:71–85.

Tompa P. The functional benefits of protein disorder. J Mol Str (Theochem). 2003; 666–667, 361–371.

Tudos E, Fiser A, Simon I. Different sequence environments of amino acid residues involved and not involved in long-range interactions in proteins. Int J Pept Protein Res. 1994;43:205–208.

Vila J, Williams RL, Vasquez M, Scheraga HA. Empirical solvation models can be used to differentiate native from near-native conformations of bovine pancreatic trypsin inhibitor. Proteins. 1991;10(3):199–218.

Vogt G, Woell S, Argos P. Protein thermal stability, hydrogen bonds, and ion pairs. J Mol Biol. 1997;269:631–643.

Wesson L, Eisenberg D. Atomic solvation parameters applied to molecular dynamics of proteins in solution. Protein Sci. 1992;1(2):227–235.

White SH, von Heijne G. How translocons select transmembrane helices. Annu Rev Biophys. 2008;37:23–42.

Wickner W, Schekman R. Protein translocation across biological membranes. Science. 2005;310(5753):1452–1456.

Wilmanns M, Eisenberg D. Three-dimensional profiles from residue-pair preferences: identification of sequences with beta/alpha-barrel fold. Proc Natl Acad Sci USA. 1993;90:1379–1383.

Wintjens R, Lievin J, Rooman M, Buisine E. Contribution of cation-pi interactions to the stability of protein-DNA complexes. J Mol Biol. 2000;302:395–410.

Wright PE, Dyson HJ. Intrinsically unstructured proteins: re-assessing the protein structure-function paradigm. J Mol Biol. 1999;293(2):321–331.

Yuan Z, Davis MJ, Zhang F, Teasdale RD. Computational differentiation of N-terminal signal peptides and transmembrane helices. Biochem Biophys Res Commun. 2003; 312(4):1278–1283.

Zehfus MH. Automatic recognition of hydrophobic clusters and their correlation with protein folding units. Protein Sci. 1995;4:1188–1202.

Zhang, C, Kim, S-H. Environment-dependent residue contact energies for proteins. Proc Natl Acad Sci USA. 2000;97:2550–2555.

Zhang L, Skolnick J. How do potentials derived from structural databases relate to "true" potentials? Protein Sci. 1998;7:112–122.

CHAPTER 4

Protein Folding Kinetics

Many small proteins are known to fold rapidly by simple two-state kinetics, involving only a single exponential function of time (Jackson, 1998). Folding is often described in terms of a two-state mass-action model, D↔N, where D and N are denatured (or unfolded, U) and native (or folded, F) states, respectively. The folding of a protein can also be illustrated with the Arrhenius diagram (**Figure 4.1**), which reveals the following ideas: (i) A single reaction coordinate exists from D to N, (ii) relatively distinct molecular structures appear in sequential order along the reaction coordinate, e.g., D → TS → N, and (iii) the single-exponential behavior in both forward and reverse directions has been traditionally interpreted in terms of rate-limiting steps, called transition states, TS (Schonbrun and Dill, 2003). In protein folding, there is no single microscopic reaction coordinate that every chain follows, and there may not be identifiable barriers of the traditional type because energy landscapes may be funnel shaped (Ozkan et al. 2001). Hence, the mechanism of protein folding has been addressed with mutational studies of folding rates and equilibrium constants (Fersht et al. 1986). This methodology, developed as Φ-value analysis, has been widely applied to many different proteins (Matouschek et al. 1989; Grantcharova et al. 1998). The experiments on protein engineering and Φ-value analysis also provide the information about the transition state structures at the level of individual residues (Matouschek et al. 1989; Itzhaki et al. 1995).

4.1   Φ-value analysis

The Φ-value analysis of a protein that folds via a two-state mechanism is illustrated in **Figure 4.2** (Nolting, 1999). A mutation causes a change of stability, $\Delta\Delta G_{F-U}$, between the folded (F) and unfolded states (U). In the transition state, #, the energy difference between mutant and wild type is $\Delta\Delta G_{\#-U}$. In this case, the fraction of energy difference or Φ-value is given by the following equation:

$$\Phi_{\#} = \Delta\Delta G_{\#-U}/\Delta\Delta G_{F-U}. \tag{4.1}$$

It depends on the amount of structure that has built up in the transition state, #, at the position of the mutation. $\Phi = 0$ corresponds to no structure formation in transition state as in the unfolded state, since the mutation does not shift the free energy difference between these two states. A Φ-value of 1 is interpreted that the residue has a native-like structure in the transition state, since the mutation shifts

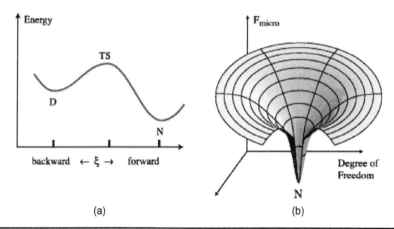

Figure 4.1 (a) Explanation of the transition state and (b) the funnel-shaped energy landscape in protein folding. ξ is a reaction coordinate. Figure was taken from Schonbrun and Dill (2003) and reproduced with permission from the National Academy of Sciences, USA (2003).

the free energy of the transition state by the same amount as the free energy of the folded state (Itzhaki et al. 1995).

Raleigh and Plaxco (2005) collected the experimental Φ-values of 296 well-characterized mutants in seven single-domain proteins and analyzed their occurrence at various ranges. The distribution of Φ-values obtained for a large dataset of 384 mutants in 15 proteins is shown in **Figure 4.3**. This figure shows that more than 82% of the 384 characterized residues in these proteins exhibit a Φ-value below 0.6, and only 7% produce a Φ-value in the range 0.8 to 1.2. Furthermore, the structural interpretation of Φ-values seems to be ambiguous due to the fact that about 86% of all Φ-values fall between zero and unity, which are termed as "partial Φ-values." There are two interpretations for partial Φ-values: (i) if a region of the polypeptide chain is partially native-like in all of the molecules in the transition state ensemble and (ii) if the transition state is heterogeneous; for example, a Φ-value of 0.5 will arise if a given position is fully native-like in half of the molecules and is unfolded in the remaining molecules. The debate on these two scenarios is yet to be solved, and it is quite possible that both mechanisms play a role (Raleigh and Plaxco, 2005).

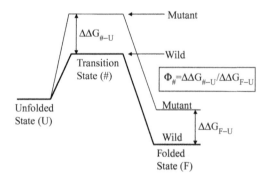

Figure 4.2 Φ value analysis of a protein that folds via a two-state mechanism. Figure was adapted from Nolting (1999).

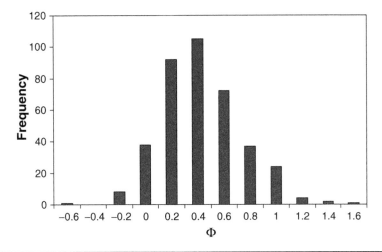

FIGURE 4.3 Distribution of 384 characterized mutants at various ranges of Φ values.

In addition, negative Φ-values are interpreted as a mutation stabilizes the folding transition state while it destabilizes the native state (Rios et al. 2005).

A set of single mutants, strategically distributed over the molecule, is used to map out the structure of the transition state at the resolution of single amino acid residues. This approach may analogously be applied on proteins, with more complicated energy landscapes that contain intermediates on the reaction pathway, and on proteins that have residual structure in their unfolded states.

4.1.1   Experimental studies

Several two-state proteins from different structural classes/folds, such as all-α, all-β, β-sandwich, and α/β have been characterized by the Φ-value analysis (Jackson, 1998). Acyl-coenzyme binding protein (ACBP) belongs to the all-α class of proteins. Eight conserved hydrophobic residues have been identified as critical for fast folding in ACBP. These residues are located in the interface between N- and C-terminal helices (Kragelund et al. 1999a). The SH3 domains of Src, α-spectrin, and Fyn belong to all-β class of proteins. The Sso7d protein also has a similar topology to that of the SH3 domains (Guerois and Serrano, 2000).

The transition state of the Src SH3 domain appears to be quite polarized, with one portion of the molecule highly ordered than the rest; a hydrogen bond network involving two β-turns and an adjacent hydrophobic cluster appear to be formed (Grantcharova et al. 1998). There is a good qualitative similarity among the transition state structures of Src SH3, α-spectrin, and Fyn SH3 (Martinez et al. 1998; Northey et al. 2002). In Fyn SH3, it has further been shown that residues at three positions are tightly packed in the transition state than the rest, and these residues are designated as the folding nucleus (Northey et al. 2002).

The transition-state structures of TNfn3 (third fibronectin type III domain of human tenascin), Fnfn10 (III domain of human fibronectin), and TI127 (human cardiac titin) that belong to the immunoglobulin fold have been characterized in detail (Hamil et al. 2000; Cota et al. 2001; Fowler and Clarke, 2001). The Φ-values indicate that the formation of folding nuclei in all these proteins is due to tertiary

TABLE **4.1**   Identified folding nuclei in 17 proteins

Protein name	PDB code	Folding nuclei	Reference
CI2	2ci2	A35, L68, I76	Itzhaki et al. 1995
Tenascin	1ten	I821, Y837, I860, V871	Hamill et al. 2000
CD2.d1	1hnf	L19, I21, I33, A45, V83, L96, W35	Lorch et al. 1999
CheY	3chy	D12, D13, D57, V10, V11, V33 A36, D38, A42, V54	Lopez-Hernandez and Serrano, 1996
ADA2h	1aye	I15, L26, F67, V54, I23	Villegas et al. 1998
Acp	1aps	Y11, P54, F84, Y25, A30, G45, V47	Chiti et al. 1999
U1A	1urn	I43, V45, L30, F34, I40, I14, L17, L26	Ternstrom et al. 1999
ACBP	2abd	F5, A9, V12, L15, Y73, I74, V77, L80	Kragelund et al. 1999a
FKBP12	1fkj	V2, V4, V24, V63, I76, I101, L50	Fulton et al. 1999
Fyn SH3	1fyn	I28, A39, I50	Northey et al. 2002
Src SH3	2src	E30, A45, I56	Grantcharova et al. 1998
α-spectrin	1shg	V23, V44, V53	Martinez et al. 1998
Sso7d	1bf4	V23, I30, A45, L59	Guerois and Serrano, 2000
WW	1pin	R14, S19, Y23, A31	Jäger et al. 2001
Titin	1tit	I2, I23, L36, I49, L58, F73	Fowler and Clarke, 2001
Fnfn	1ttf	L8, I20, I34, V50, A74, Y92	Cota et al. 2001
Villin	2vik	V7, I18, I23, M28, C44, I61, A77, T81, M84	Choe et al. 2000

The data for the first nine proteins were taken from Mirny and Shakhnovich (2001a).

interactions between residues in four strands, namely, B, C, E, and F. The residues with high Φ-values are clustered together in space in the folded structure.

The 64-residue chymotrypsin inhibitor 2 (CI2) belonging to the α/β class is one of the well-characterized proteins through the Φ-value analysis by more than 100 mutations under a variety of conditions (Otzen et al. 1994; Itzhaki et al. 1995). This analysis has revealed that CI2 folds around an extended nucleus that is composed of a contiguous region of structure (α-helix) and long-range interactions with groups distant in sequence (Fersht, 2000).

These experimental studies on the Φ-value analysis show the presence of several key residues, known as folding nucleus, important for initiating the folding and maintaining the stability. **Table 4.1** provides the list of folding nuclei identified in a set of 17 proteins by protein engineering experiments. Furthermore, Gromiha and Selvaraj (2002) listed the experimental Φ-values obtained for the mutants in CI2, FK506 binding protein (FKBP12), and SH3 domain of src (SH3). These data can be used to develop a theoretical model to predict the folding nuclei and to understand the transition-state structures in two-state proteins (Poupon and Mornon, 1999; Galzitskaya and Finkelstein, 1999; Gromiha and Selvaraj, 2002). Recently, Nolting et al. (2008) collected the Φ-values for 15 proteins and mutants to investigate the average inter-residue folding forces using the mutational data. The data for all the 15 proteins are listed in **Table 4.2**.

TABLE 4.2 Φ values and other related parameters for 15 proteins and mutants

Barstar					Barnase					Chymotrypsin inhibitor 2					Src SH3 domain					Spectrin R16 domain				
mt	$\Delta\Delta G_{F-U}$	$\Delta\Delta G_{\#-P}$	$F_{\#-P}$	$\Phi_{\#}$	mt	$\Delta\Delta G_{F-U}$	$\Delta\Delta G_{\#-P}$	$F_{\#-P}$	$\Phi_{\#}$	mt	$\Delta\Delta G_{F-U}$	$\Delta\Delta G_{\#-P}$	$F_{\#-P}$	$\Phi_{\#}$	mt	$\Delta\Delta G_{F-U}$	$\Delta\Delta G_{\#-P}$	$F_{\#-P}$	$\Phi_{\#}$	mt	$\Delta\Delta G_{F-U}$	$\Delta\Delta G_{\#-P}$	$F_{\#-P}$	$\Phi_{\#}$
IV5	1.04	0.24	1.31	0.63	IV4	0.81	0.00	0.00	-0.01	KA2	0.55	-0.17	-0.26	-0.31	TA9	0.64	-0.07	-0.28	-0.11	HA9	-0.5	-0.21	-0.42	
QG9	1.60	0.26	0.49	0.72	IA4	0.75	0.00	0.00	0.03	KM2	0.67	0.19	1.74	0.28	FA10	0.84	0.08	0.12	0.10	HG9	0.5	0.24	0.36	
IA13	1.35	0.09	0.17	0.45	NA5	1.59	0.16	0.56	0.09	TA3	0.85	0.22	0.87	0.26	FI10	1.65	-0.08	-0.77	-0.05	FL11	1.2	0.31	2.86	0.40
LV16	1.15	0.35	1.87	0.97	TA6	2.02	0.00	0.00	0.21	TG3	1.16	0.35	0.79	0.30	VA11	1.64	0.05	0.13	0.03	FA11	3.6	1.03	1.54	0.40
QG18	1.32	0.21	0.33	0.68	VA10	3.47	0.00	0.00	0.33	PA6	1.57	0.74	3.27	0.47	AG12	1.00	0.05	0.26	0.05	LA11	2.4	1.03	1.84	0.40
AG25	1.33	0.58	3.10	0.68	YA13	1.99	0.20	1.69	0.40	EA7	0.47	0.20	0.49		LA13	1.49	-0.04	-0.08	-0.03	RA13	0.3	0.20	0.25	
LV26	1.24	0.08	0.45	-0.04	VT10	3.00	0.30	0.41	0.49	LA8	2.68	-0.05	-0.10	-0.02	YA14	0.31	-0.02	-0.03		RG13	1.9	0.79	0.81	0.50
LV34	1.09	-0.08	-0.45	0.43	LA14	4.32	0.43	0.77	0.59	KA11	-0.42	-0.21	-0.31		DA15	0.43	-0.09	-0.40		MA15	2.1	0.92	1.66	0.60
AG36	1.13	0.54	2.89	0.70	TS16	1.36	0.14	0.69	0.80	SG12	0.80	0.44	1.80	0.55	YA16	2.27	0.07	0.09	0.03	DA16	0.2	0.04	0.18	
LA37	0.55	0.13	0.24	0.58	YA17	1.67	0.17	0.23	0.68	ED14	0.52	0.36	2.03	0.69	RA19	0.07	-0.05	-0.07		DG16	1.1	0.36	0.85	0.50
WF38	0.80	0.37	1.34	0.80	NA23	2.20	0.00	0.00	-0.04	EN14	0.70	0.68	5.34	0.97	TA20	-0.06	-0.08	-0.33		SG20	0.5	0.11	0.43	
LA41	2.00	0.73	1.32	0.45	IV25	0.83	0.00	0.00	-0.23	ED15	0.74	0.29	1.63	0.39	TA22	0.01	-0.03	-0.12		IV22	1.5	0.43	2.29	0.40
VA45	1.42	0.50	1.34	0.47	TA26	1.67	0.00	0.00	-0.02	EN15	1.07	0.42	3.28	0.39	DA23	0.56	0.07	0.31	0.13	IA22	3.5	0.83	1.49	0.40
LA49	0.90	0.46	0.83	0.47	EG29	2.05	0.00	0.00	-0.13	AG16	1.09	1.35	7.26	1.24	LA24	1.79	0.47	0.83	0.26	VA22	2.0	0.83	2.24	0.50
VG50	1.77	1.07	1.91	0.77	VA36	1.68	0.17	0.45	-0.06	KA17	0.49	-0.23	-0.35		FA26	1.97	0.79	1.18	0.40	EA24	-0.1	-0.13	-0.32	
LV51	0.60	0.00	-0.01	0.12	VT36	1.05	0.00	0.00	-0.03	KG17	2.32	0.60	0.71	0.26	KA27	0.44	-0.03	-0.04		EG24	1.1	0.04	0.07	0.20
FA56	2.08	0.32	0.47	0.35	VA45	1.17	0.12	0.31	-0.17	KA18	-0.21	-0.30	-0.45		KA28	0.09	-0.11	-0.16		KA25	0.0	0.20	0.29	
QG58	1.55	-0.03	-0.05	0.12	VT45	2.22	0.00	0.00	-0.06	KG18	0.99	0.57	0.67	0.58	EA30	1.94	1.20	2.92	0.62	VA29	0.1	0.85	2.28	
QG61	0.74	-0.19	-0.29	0.09	IV51	1.13	0.00	0.00	-0.18	VA19	0.49	-0.24	-0.65		RA31	0.32	0.07	0.09		TA39	-0.3	-0.17	-0.69	
LA62	-0.30	0.86	1.54	0.38	DA54	2.86	0.00	0.00	-0.16	IV20	1.30	0.70	3.77	0.54	LA32	2.26	1.24	2.23	0.55	TG39	0.1	0.11	0.24	
TA63	1.81	0.15	0.59	0.30	IA55	1.48	0.15	0.27	0.59	LA21	1.33	0.37	0.67	0.28	LV32	1.21	0.27	1.43	0.22	VA41	0.4	0.19	0.51	
AG67	1.75	0.33	1.77	0.52	IT55	0.90	0.09	0.30	0.42	LG21	1.38	0.29	0.39	0.21	QA33	0.21	0.23	0.50		NA43	-0.2	0.05	0.17	
EA68	1.10	0.57	1.40	0.41	NA58	2.00	0.00	0.00	0.94	QA22	0.02	-0.01	-0.02		IA34	0.32	1.29	2.31		NG43	0.4	0.24	0.51	
VA70	1.76	0.40	1.09	0.81	IV76	1.66	0.50	0.89	0.44	QG22	0.60	0.15	0.23	0.25	IV34	0.09	0.28	1.48		LA44	0.2	0.12	0.21	
QG72	1.19	0.60	0.92	0.72	NA77	0.98	0.00	0.00	0.01	DA23	0.96	0.27	1.14	0.28	VA35	0.77	0.59	1.59	0.77	KA46	-0.4	-0.11	-0.16	
VA73	1.01	0.00	0.01	0.77	NA84	1.45	0.00	0.00	-0.02	KA24	0.65	-0.03	-0.05	-0.05	NA36	0.20	-0.05	-0.19		KG46	0.5	0.47	0.55	
RE75	0.94	0.14	0.36	0.90	IV88	1.59	-0.16	-0.56	0.14	KG24	3.19	0.45	0.52	0.14	NA37	-0.07	-0.22	-0.76		HA48	1.4	0.92	1.85	1.00
AG77	2.03	1.62	8.70	0.63	IA88	1.64	0.33	1.76	0.91	PA25	1.76	0.21	0.94	0.12	WA42	1.29	0.32	0.34	0.25	RA50	-0.3	-0.01	-0.01	
AG79	1.25	0.37	2.00	0.63		4.02	1.61	2.88	0.92	EA26	0.32	0.07	0.17		WA43	1.20	0.18	0.19	0.15	RG50	0.6	0.67	0.69	1.00

(continued)

TABLE 4.2 Φ values and other related parameters for 15 proteins and mutants (continued)

Barstar

mt	$\Delta\Delta G_{F-U}$	$\Delta\Delta G_{\#-P}$	$F_{\#-P}$	$\Phi_{\#}$
TA85	1.71	2.82	0.72	0.51

Barnase

mt	$\Delta\Delta G_{F-U}$	$\Delta\Delta G_{\#-P}$	$F_{\#-P}$	$\Phi_{\#}$
LV89	0.47	-0.19	-1.01	
LT89	2.85	1.42	4.69	1.03
IV96	1.02	0.00	0.00	0.42
IA96	3.15	0.63	1.13	0.82
IA109	1.27	0.25	0.45	0.63
IV109	0.81	0.00	0.00	0.16

Chymotrypsin inhibitor 2

mt	$\Delta\Delta G_{F-U}$	$\Delta\Delta G_{\#-P}$	$F_{\#-P}$	$\Phi_{\#}$
IV29	1.11	0.49	2.62	0.44
IA29	3.90	1.09	1.96	0.28
IV30	-0.08	-0.11	-0.59	
IA30	2.12	1.21	2.16	0.57
IG30	3.52	1.30	1.75	0.37
IT30	1.34	0.84	2.78	0.63
LA32	2.37	0.78	1.40	0.33
LV32	0.50	0.02	0.11	
VT34	1.03	0.28	2.36	0.27
VA34	0.64	0.15	0.40	0.23
VG34	2.43	0.66	1.18	0.27
VA38	1.47	0.40	1.07	0.27
TA39	0.72	0.33	1.30	0.46
EA41	0.70	0.14	0.34	0.20
RA43	0.58	0.11	0.14	0.19
DA45	0.80	0.23	0.99	0.29
VA47	4.93	1.38	3.71	0.28
LA49	3.84	2.07	3.71	0.54
FL50	2.11	0.99	9.20	0.47
FV50	2.39	0.74	2.52	0.31
FA50	3.84	1.54	2.30	0.40
VA51	1.98	0.89	2.39	0.45
DA52	3.41	0.51	2.18	0.15
NA56	0.83	0.28	0.99	0.34
IV57	-0.19	-0.03	-0.16	
IA57	4.29	0.51	0.92	0.12
AG58	1.88	0.36	1.92	0.19
VT60	0.38	0.01	0.10	
VA60	1.51	0.26	0.69	0.17
VG60	3.24	0.23	0.41	0.07
PA61	3.34	0.37	1.63	0.11
VT63	1.15	0.15	1.27	0.13
VA63	1.45	0.36	0.97	0.25
VG63	3.50	0.49	0.88	0.14

Src SH3 domain

mt	$\Delta\Delta G_{F-U}$	$\Delta\Delta G_{\#-P}$	$F_{\#-P}$	$\Phi_{\#}$
WI43	0.77	-0.54	-1.41	-0.70
LA44	1.64	0.89	1.59	0.54
AG45	0.92	1.10	5.93	1.20
HA46	0.62	0.05	0.10	0.08
LA48	0.61	0.44	0.79	0.72
TA50	1.79	1.54	6.03	0.86
QA52	0.35	0.16	0.34	
TA53	1.11	0.75	2.96	0.68
YA55	1.52	0.85	1.17	0.56
IA56	1.84	1.31	2.34	0.71
PA57	1.36	0.33	1.45	0.24
NA59	0.14	0.00	0.00	
YA60	-0.23	0.04	0.06	
VA61	1.18	-0.07	-0.19	-0.06
AG62	0.53	-0.01	-0.05	-0.02
PA63	0.14	-0.05	-0.24	

Spectrin R16 domain

mt	$\Delta\Delta G_{F-U}$	$\Delta\Delta G_{\#-P}$	$F_{\#-P}$	$\Phi_{\#}$
LA51	2.7	0.79	1.42	0.40
AG53	1.1	0.27	1.43	0.20
LA55	3.7	0.99	1.77	0.40
AG57	1.1	0.08	0.41	0.10
HA58	2.3	0.62	1.25	0.40
IV62	0.5	0.41	2.22	
IA62	2.8	0.95	1.71	0.50
VA62	2.3	0.95	2.56	0.30
QA63	0.7	0.29	0.62	0.50
QG63	1.8	0.89	1.37	0.80
VA65	2.3	0.42	1.13	0.30
DA67	0.1	-0.04	-0.15	
DG67	1.3	0.39	0.91	0.50
KA71	0.1	0.06	0.08	
KG71	1.4	0.34	0.39	0.30
LA72	2.3	0.72	1.29	0.50
IV83	0.5	0.24	1.32	
IA83	2.0	0.65	1.17	0.50
VA83	1.4	0.65	1.75	0.40
QA85	0.0	-0.01	-0.02	
QG85	1.3	0.47	0.73	0.50
LA87	2.7	1.12	2.01	0.60
AG88	1.1	0.47	2.52	0.70
FL90	-0.6	-0.16	-1.49	
FA90	2.7	1.24	1.86	0.70
DA92	-0.7	-0.41	-1.76	
DG92	0.3	0.00	0.00	
KA95	0.4	0.30	0.45	
KG95	1.3	0.55	0.65	0.40
LA97	3.7	0.81	1.45	0.30
QA99	-0.4	-0.06	-0.14	
QG99	0.7	0.24	0.38	0.40
AG101	2.6	0.89	4.78	1.50
AG103	1.8	0.99	5.30	0.80
RA104	0.0	0.13	0.17	
DA106	-0.5	-0.21	-0.89	
DG106	1.3	0.83	1.97	0.90
LA108	1.4	0.95	1.71	1.00

Arc repressor					Apo-azurin					cspB					CTL9					FKBP12				
mt	$\Delta\Delta G_{F-U}$	$\Delta\Delta G_{\#-P}$	$F_{\#-P}$	$\Phi_{\#}$	mt	$\Delta\Delta G_{F-U}$	$\Delta\Delta G_{\#-P}$	$F_{\#-P}$	$\Phi_{\#}$	mt	$\Delta\Delta G_{F-U}$	$\Delta\Delta G_{\#-P}$	$F_{\#-P}$	$\Phi_{\#}$	mt	$\Delta\Delta G_{F-U}$	$\Delta\Delta G_{\#-P}$	$F_{\#-P}$	$\Phi_{\#}$	mt	$\Delta\Delta G_{F-U}$	$\Delta\Delta G_{\#-P}$	$F_{\#-P}$	$\Phi_{\#}$
MA1	0.52	0.05	0.09	0.10	IA7	3.12	0.31	0.56	0.10	LA2	0.12	0.07		0.13	LA72	3.77	1.09	1.95	0.29	VA2	2.43	1.34	3.60	0.55
KL2	-0.07	-0.02	-0.23		IA20	1.68	0.45	0.81	0.27	LE3	1.08	0.47	3.19	0.44	IA79	3.49	0.45	0.81	0.13	VA4	2.78	1.08	2.90	0.39
MA4	-0.01	-0.06	-0.10		VA22	1.44	0.26	0.70	0.18	KA5	1.39	1.09	1.63	0.78	IA93	3.19	0.26	0.47	0.08	IV7	0.92	0.15	0.81	0.16
KA6	-0.32	-0.15	-0.22		VA31	1.20	1.12	3.01	0.93	VT6	1.61	1.50	12.72	0.93	KA96	-0.61	-0.12	-0.18		TA21	1.60	0.70	2.74	0.44
MA7	0.60	0.03	0.05	0.05	LA33	0.96	0.87	1.56	0.91	KA7	1.10	0.99	1.48	0.90	IA98	4.28	0.39	0.70	0.09	TS21	1.44	0.79	4.03	0.55
PA8	-3.69	-0.53	-2.34		HG46	2.64	0.26	0.38	0.10	NA10	1.03	0.82	2.88	0.80	EG100	1.52	0.35	0.59	0.23	VA23	2.97	1.63	4.38	0.55
QA9	0.17	-0.01	-0.01		WA48	5.04	1.16	1.23	0.23	EA12	-0.05	-0.07	-0.17		EA100	0.98	-0.02	-0.05	-0.02	VA24	3.19	1.40	3.76	0.44
FA10	2.30	0.48	0.72	0.21	LA50	1.68	1.75	3.13	1.04	KA13	-0.31	-0.20	-0.30		LA102	2.37	0.45	0.81	0.19	TA27	1.97	0.75	2.94	0.38
NA11	0.07	-0.10	-0.34		VG60	3.12	0.56	1.00	0.18	FA15	1.82	0.97	1.45	0.53	LA108	2.33	0.05	0.09	0.02	TS27	1.49	0.94	4.79	0.63
LA12	2.51	0.87	1.56	0.35	IA81	2.40	1.25	2.24	0.52	FA17	1.18	0.10	0.15	0.09	LA110	3.21	0.16	0.29	0.05	FA36	3.54	-0.28	-0.42	-0.08
RA13	0.55	0.06	0.08	0.11	VA95	0.60	0.38	1.02	0.63	IV18	1.92	0.22	1.18	0.11	IA115	3.52	0.11	0.20	0.03	LA50	2.57	1.18	2.11	0.46
PA15	2.58	0.16	0.73	0.06	FA97	2.40	0.84	1.26	0.35	EA19	-0.10	-0.05	-0.12		LA117	0.82	0.11	0.20	0.13	VA55	2.13	0.26	0.70	0.12
RA16	-0.28	-0.24	-0.30		YA108	3.00	1.11	1.53	0.37	DA25	0.60	0.28	1.19	0.47	IA121	2.88	0.20	0.36	0.07	IA56	2.48	0.52	0.93	0.21
EA17	0.01	0.08	0.19		FA110	3.36	0.64	0.96	0.19	VT26	1.44	1.04	8.82	0.72	VA129	1.81	0.42	1.13	0.23	IT56	1.81	0.31	1.02	0.17
VA18	-0.62	0.06	0.15		HG117	2.40	0.24	0.37	0.10	AG32	0.79	0.11	0.60	0.14	VA131	1.93	1.14	3.06	0.59	ID56	3.16	0.25	0.77	0.08
LA19	0.81	0.35	0.62	0.43	LA125	0.72	0.58	1.04	0.81	IA33	2.28	0.03	0.05	0.01	LA133	2.94	1.85	3.31	0.63	RA57	0.81	0.34	0.43	0.42
DA20	0.36	0.04	0.17							LA41	2.33	0.83	1.49	0.36	VA137	2.48	1.14	3.06	0.46	RG57	2.29	0.21	0.21	0.09
LA21	3.48	1.17	2.10	0.34						QA45	0.53	0.13	0.28	0.26	LA141	3.56	1.60	2.87	0.45	EA60	2.13	0.28	0.68	0.13
RA23	0.75	0.27	0.35	0.36						FA49	3.57	1.02	1.53	0.28	VA143	3.25	0.72	1.93	0.22	EG60	2.84	0.17	0.28	0.06
KA24	0.63	0.14	0.21	0.22						FL49	0.70	0.11	1.02	0.16	VA145	1.92	0.36	0.97	0.19	EA61	0.84	0.08	0.19	0.10
VA25	-0.11	-0.09	-0.24							IA51	1.99	0.25	0.45	0.13						EG61	2.49	0.62	1.04	0.25
EA27	-0.05	0.05	0.13							AG60	2.28	0.31	1.66	0.14						VA63	2.97	1.46	3.92	0.49
EA28	1.35	0.36	0.88	0.27						VA63	3.05	0.43	1.16	0.14						TA75	2.60	0.88	3.45	0.34
NA29	2.50	0.45	1.59	0.18																IV76	0.76	0.43	2.31	0.57
RA31	3.54	0.93	1.17	0.26																IA76	3.81	1.94	3.47	0.51
VA33	2.31	0.73	1.97	0.32																IV91	0.38	0.30	1.61	
NA34	0.24	-0.33	-1.15																	IA91	1.54	0.00	0.00	0.00
YA38	3.98	0.96	1.33	0.24																LA97	3.56	0.57	1.02	0.16
QA39	0.11	-0.09	-0.19																	VA98	2.16	0.67	1.80	0.31
MA42	3.40	1.26	2.25	0.37																VA101	2.75	1.68	4.51	0.61
EA43	0.49	0.00	0.00																	LA106	2.32	0.79	1.42	0.34
KA46	0.87	0.01	0.02	0.01																				
KA47	1.64	0.00	0.00	0.00																				
EA48	2.79	0.34	0.83	0.12																				
RA50	2.16	0.32	0.40	0.15																				
IA51	1.80	0.36	0.64	0.20																				

(continued)

TABLE 4.2 Φ values and other related parameters for 15 proteins and mutants (continued)

α-Lactalbumin					IM7					IM9					Spectrin R17					Ubiquitin				
mt	$\Delta\Delta G_{F-U}$	$\Delta\Delta G_{\#-P}$	$F_{\#-P}$	$\Phi_{\#}$	mt	$\Delta\Delta G_{F-U}$	$\Delta\Delta G_{\#-P}$	$F_{\#-P}$	$\Phi_{\#}$	mt	$\Delta\Delta G_{F-U}$	$\Delta\Delta G_{\#-P}$	$F_{\#-P}$	$\Phi_{\#}$	mt	$\Delta\Delta G_{F-U}$	$\Delta\Delta G_{\#-P}$	$F_{\#-P}$	$\Phi_{\#}$	mt	$\Delta\Delta G_{F-U}$	$\Delta\Delta G_{\#-P}$	$F_{\#-P}$	$\Phi_{\#}$
VA8	0.88	−0.20	−0.54	−0.22	LA3	0.74	−0.32	−0.57	0.49	IV7	1.70	0.26	1.40	0.15	QG9	1.00	0.90	1.39	0.90	IV3	1.18	0.09	0.48	0.08
LA12	3.29	−0.65	−1.16	−0.20	IV7	1.46	0.02	0.11	0.53	AG13	0.79	0.78	4.20	0.97	FL11	2.10	1.05	9.74	0.50	IA3	3.18	1.07	1.92	0.34
VA27	1.90	−0.22	−0.59	−0.12	AG13	0.60	−0.05	−0.26	1.27	FA15	4.99	2.84	4.26	0.57	AG13	1.30	1.17	6.28	0.90	VA5	2.41	0.78	2.10	0.32
LA52	3.64	0.08	0.14	0.02	FA15	3.65	−0.49	−0.73	0.27	LA16	1.75	0.91	1.63	0.52	AG20	1.30	1.04	5.58	0.80	TA7	1.37	0.92	3.61	0.67
IV55	1.81	0.44	2.36	0.24	VA16	1.49	0.05	0.13	0.70	LA18	3.45	1.38	2.47	0.40	IV22	2.40	1.20	6.45	0.50	TA9	−0.06	−1.31	−5.13	
WA60	3.30	−0.44	−0.47	−0.13	LA18	3.02	−0.41	−0.73	0.50	VA19	3.24	1.04	2.79	0.32	IA22	1.30	0.65	1.16	0.50	IV13	1.16	−0.10	−0.54	−0.09
IV89	0.80	0.50	2.69	0.63	LA19	3.41	−0.31	−0.56	0.32	IV22	2.13	0.66	3.55	0.31	KG46	1.30	0.13	0.15	0.10	IA13	3.40	0.81	1.45	0.24
VA90	−0.61	−0.04	−0.11		IV22	2.13	0.45	2.42	0.24	TS27	0.84	0.10	0.51	0.12	AG50	1.00	0.10	0.54	0.10	LA15	3.85	1.34	2.40	0.35
KA93	2.35	−0.08	−0.12	−0.03	VA27	−0.50	−0.15	−0.40		LA33	2.42	0.65	1.16	0.27	FL51	1.40	0.28	2.60	0.20	VA17	1.65	0.54	1.45	0.33
IV95	1.57	0.09	0.48	0.06	AG28	0.19	0.22	1.19		LA36	2.37	0.59	1.06	0.25	VG57	1.00	0.00	0.00	0.00	TA22	1.74	0.58	2.27	0.33
LA96	1.51	−0.31	−0.56	−0.21	VA33	0.24	0.00	0.00		VA37	1.65	0.25	0.67	0.15	HA58	2.90	0.29	0.58	0.10	IV23	0.45	−0.07	−0.38	
LA105	0.99	−0.06	−0.11	−0.05	LA34	1.85	−1.21	−2.17	0.47	FL40	3.65	0.04	0.37	0.01	VA65	1.20	0.36	0.97	0.30	IA23	2.78	0.54	0.97	0.19
LA110	−0.32	−0.58	−1.04		VA36	0.10	0.13	0.35		LA52	3.67	0.11	0.20	0.03	AG67	1.60	0.32	1.72	0.20	IG23	3.58	1.22	1.64	0.34
WF118	0.63	0.19	0.69	0.30	LA37	2.85	−0.37	−0.66	0.48	IV53	1.49	0.10	0.54	0.07	AG85	0.90	0.72	3.87	0.80	VA26	3.37	1.02	2.74	0.30
					LA38	2.69	−1.12	−2.01	−0.01	IV67	1.25	0.51	2.74	0.41	MA87	2.20	1.10	1.97	0.50	KA27	2.56	−0.55	−0.82	−0.21
					FL41	1.89	−1.79	−16.61	0.24	VA68	2.37	0.55	1.48	0.23	LA90	4.10	2.05	3.67	0.50	AG28	0.52	0.61	3.27	1.17
					VA42	0.70	−0.35	−0.94	−0.24	VA71	2.85	1.03	2.77	0.36	SG95	1.00	0.60	2.45	0.60	IV30	1.21	0.20	1.07	0.17
					IV44	0.53	0.51	2.74	1.00	AG76	1.25	0.46	2.47	0.37	LA97	3.70	1.48	2.65	0.40	IA30	3.22	0.83	1.49	0.26
					TS51	0.96	−0.06	−0.31	−0.07	AG77	1.13	0.42	2.26	0.37	AG100	1.80	1.08	5.81	0.60	QA41	1.46	−0.49	−1.06	−0.34
					LA53	3.26	0.10	0.18	−0.01	FA83	5.04	1.56	2.34	0.31	AG106	1.10	0.33	1.77	0.30	LA43	4.36	−0.37	−0.66	−0.08
					IV54	2.64	0.16	0.86	0.13						LA108	3.20	0.00	0.00	0.00	LA50	2.73	−0.40	−0.72	−0.15
					IV68	0.60	−0.17	−0.91	0.85											LA56	4.05	0.36	0.64	0.09
					VA69	0.70	−0.03	−0.08	0.03											IV61	1.12	−0.09	−0.48	−0.08
					IV72	0.41	−0.03	−0.16												IA61	3.23	−0.20	−0.36	−0.06
					AG77	1.27	0.01	0.05	0.86											LA67	2.57	−0.19	−0.34	−0.07
					AG78	1.32	0.23	1.24	0.77											LA69	3.05	0.03	0.05	0.01

mt: mutation; $\Delta\Delta G_{F-U}$ (in kcal mol^{-1}) is the effect of mutation on the free energy change between unfolded state, U, and folded state, F. $\Delta\Delta G_{\#-P}$ (in kcal mol^{-1}) is the effect of mutation on the free energy change between major transition state, #, and preceding state, P. P is U for two-state folders and the intermediate state, I, for the three-state folders, respectively. $F_{\#-P}$ (in pN) is the folding force of # relative to P. $\Phi_{\#}$ is the Φ-value for #. Data were taken from Nolting et al. (2008) with permission from Elsevier.

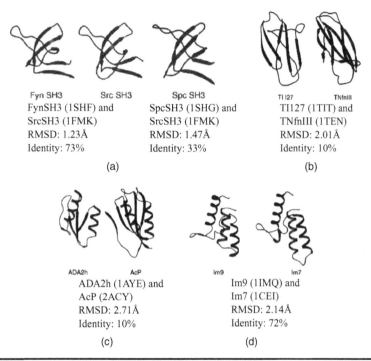

Fyn SH3   Src SH3   Spc SH3

FynSH3 (1SHF) and   SpcSH3 (1SHG) and
SrcSH3 (1FMK)   SrcSH3 (1FMK)
RMSD: 1.23Å   RMSD: 1.47Å
Identity: 73%   Identity: 33%

(a)

TI 127   TNfnIII

TI127 (1TIT) and
TNfnIII (1TEN)
RMSD: 2.01Å
Identity: 10%

(b)

ADA2h   AcP

ADA2h (1AYE) and
AcP (2ACY)
RMSD: 2.71Å
Identity: 10%

(c)

Im9   Im7

Im9 (1IMQ) and
Im7 (1CEI)
RMSD: 2.14Å
Identity: 72%

(d)

FIGURE 4.4 Structures of protein pairs on which Φ value analyses have been performed: (a) SrcSH3 with Fyn SH3 and Spc SH3, (b) TNfnIII and TI127, (c) ADA2h and AcP, and (d) Im7 and Im9. Figure was reproduced from Zarrine-Afsar et al. (2005) with permission from Elsevier.

4.1.2 Proteins with similar topologies share common transition state structures

Zarrine-Afsar et al. (2005) conducted a quantitative analysis on the influence of topology for determining the transition state structures of proteins. They have collected several sets of similar proteins and compared the Φ-values of aligned positions. An example for a set of five pairs of proteins (Src SH3 and Fyn SH3; Src SH3 and SpcSH3; TI127 and TNfnIII; ADA2h and AcP; and Im9 and Im7) is shown in **Figure 4.4**. These pairs of proteins have the sequence identity in the range of 10% to 73% and the root mean square deviation (rmsd) in the range of 1.23 Å to 2.71 Å. The correlation between the Φ-values of structurally aligned positions for all the pairs of proteins is shown in **Figure 4.5**. Remarkably, there is a good agreement between the Φ-values determined for these different pairs of proteins and correlation coefficients lying in the range of 0.70 to 0.92. This result indicates that proteins with similar topologies share common transition state structures.

4.2   Folding nuclei and Φ-values

The experimentally determined Φ-values upon amino acid substitutions have been used to predict critical residues in protein folding. The residues with high Φ-values form folding nuclei in protein structures, which have a critical set of interactions for folding and rapid assembly of their native states (Dobson, 2003; Dokholyan

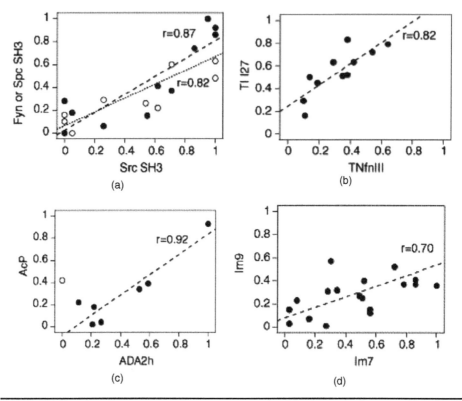

FIGURE 4.5 Correlation between experimental Φ values for different pairs of proteins: (a) SrcSH3 and Fyn (open circles)/Spc (closed circles) SH3, (b) TNfnIII and TI127, (c) ADA2h and AcP, and (d) Im7 and Im9. Figure was taken from Zarrine-Afsar et al. (2005) with permission from Elsevier.

et al. 2000). These are the residues, whose mutations affect the folding rate and stability in the transition state as strongly as that of the native protein (Matouschek et al. 1990). The availability of Φ-values for several proteins prompted researchers to identify the folding nuclei and map Φ-values on protein structures.

Galzitskaya and Finkelstein (1999) developed a method based on a search for free-energy saddle points on a network of protein-unfolding pathways to identify folding nuclei in three-dimensional protein structures. The identified folding nuclei in two typical proteins, chymotrypsin inhibitor (2CI2), and barnase are presented in **Figure 4.6**. The figures show a definite correlation between the chain fragments that form the globular native-like part with the lowest free-energy transition states and regions with high experimental Φ-values. Furthermore, the Φ-values obtained with theoretical model showed a good agreement with experimental observation for several residues.

Poupon and Mornon (1999) proposed a correspondence between conserved hydrophobic positions in amino acid sequences (topohydrophobic positions) and folding nuclei in protein structures. The conserved hydrophobic residues have distinct properties and are more buried than the nonconserved hydrophobic residues. These residues form a lattice of interacting residues in the inner core of the protein

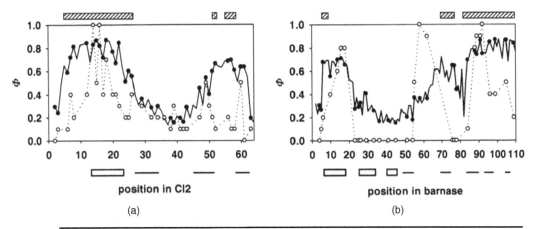

FIGURE 4.6 Correlation of experimental (open circles) and predicted Φ values (filled circles) in (a) chymotrypsin inhibitor, 2Cl2, and (b) barnase. The hatched rectangles on the top show the location of minimum free energy nucleus in the protein chain. The rectangles and lines show the native positions of the α-helices and β-strands in the chain. Figure (Galzitskaya and Finkelstein, 1999) was reproduced with permission from the National Academy of Sciences, USA.

and behave as folding nuclei. The authors tested the hypothesis on a set of protein structures and showed a relationship between topohydrophobic positions and folding nuclei.

Mirny and Shakhnovich (2001b) extended the approach on the conservation of amino acid residues based on physical-chemical properties and proposed a parameter, variability to identify the folding nuclei in protein structures. It is defined as

$$s(l) = -\sum_{i=1}^{6} p_i(l) \log p_i(l), \tag{4.2}$$

where $p_i(l)$ is the frequency of residues from class i in position l. They used six classes of residues to reflect physical-chemical properties of the amino acids and their natural pattern of substitutions: aliphatic (A, V, L, I, M, and C), aromatic (F, W, Y, and H), polar (S, T, N, and Q), basic (K and R), acidic (D and E), and special (reflecting their special conformational properties) (G and P). As a result of this classification, mutations within a class are ignored (e.g., V→L), while mutations that change the class are taken into account. **Figure 4.7** presents a variability profile for a set of nine proteins. The results show that the folding nuclei residues have low variability in amino acid sequences.

Shmygelska (2005) developed a simple, efficient, and robust algorithm to identify folding nuclei in protein structures. It is based on two aspects: (i) finding an ensemble of pathways with the lowest effective contact order and (ii) identifying contacts that are crucial for folding. This approach differs from the earlier methods, as it uses efficient graph algorithms and does not formulate restrictive assumptions about folding nuclei. The predictions provide additional details concerning the protein folding pathways. A computer program has been developed to screen folding nuclei in the native protein structures and the executable file for the proposed

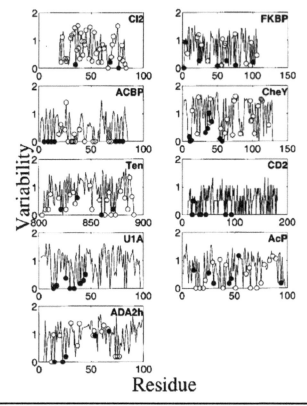

FIGURE 4.7 Variability profiles for nine different proteins. Circles indicate positions in which Φ values have been experimentally measured. Residues forming the folding nucleus are shown by filled circles. Figure was taken from Mirny and Shakhnovich (2001) with permission from Elsevier.

algorithm is available at http://www.cs.ubc.ca/~oshmygel/foldingnuclei.html. Recently, Faisca et al. (2008) showed that the critical residues (folding nuclei) have predominantly nonlocal contacts, and the mutation of the critical residues has a much stronger impact on the folding time than the geometry that is predominantly local.

4.3 Relationship between amino acid properties and Φ-values

The factors influencing the transition state structures of protein mutants have been analyzed by the relationship between amino acid properties (see **Section 3.10.8**) and Φ-values upon mutations using correlation coefficient approach (Gromiha and Selvaraj, 2002). The change in property values due to mutation is computed using the equation:

$$\Delta P(i) = P_{\mathrm{mut}}(i) - P_{\mathrm{wild}}(i), \tag{4.3}$$

where, $P_{\mathrm{mut}}(i)$ and $P_{\mathrm{wild}}(i)$ are, respectively, the property values of the ith mutant and wild type residue; i varies from 1 to N, N being the total number of mutants.

The computed differences of property values (ΔP) are related to the changes in experimental Φ-values using correlation coefficients. The correlation coefficients and regression equations were determined by standard procedures (Grewal, 1987). The results obtained with three typical proteins, FK506 binding protein (FKBP12), chymotrypsin inhibitor (CI2), and SH3 domain of src, are discussed below.

4.3.1   Buried mutations

In buried mutations, the three properties, volume, shape, and flexibility introduced by van Gunsteren and Mark (1992) for predicting the stability of protein mutants, show a very good correlation with Φ-values in FKBP12, which indicates that the transition state is more compact than the unfolded state (Main et al. 1999). In CI2, the conformational parameter, P_α (α-helical tendency), shows the highest correlation ($r = 0.93$) with experimental Φ-values, which suggests that the tendency for forming α-helix is very important in the transition state structures. Furthermore, the physical and thermodynamic properties show an appreciable correlation with Φ ($|r| > 0.6$), indicating that the mutations due to the reduction of methyl groups decrease the number of contacts in hydrophobic core, which is reflected in Φ-values (Itzaki et al. 1995).

4.3.2   Partially buried mutations

In partially buried mutations, the inclusion of sequence information improved the correlation between amino acid properties and Φ-values. It is computed using the equation,

$$P_{\text{seq}}(i) = \left[\sum_{j=i-k}^{j=i+k} P_j(i) \right] - P_{\text{mut}}(i), \tag{4.4}$$

where, $P_{\text{mut}}(i)$ is the property value of the ith mutant residue and $\Sigma P_j(i)$ is the total property value of the segment of $(2k + 1)$ residues ranging from $i - k$ to $i + k$ about the ith residue of wild type. The short- and medium-range energy (E_{sm}) is strongly correlated with the Φ-values, expressing the influence of medium-range interactions in the transition state structures of partially buried mutants. **Figure 4.8** shows the direct relationship between E_{sm} and Φ for all the considered proteins, FKBP12, CI2, and SH3. The correlation coefficients are, respectively, 0.72, 0.72, and 0.66.

4.3.3   Exposed mutations

In exposed mutations, the average long-range contact (N_l) has the strongest correlation with Φ($r = 0.79$) in all the three considered proteins. Other properties reflecting long-range interactions such as long-range non-bonded energy, average number of surrounding residues, and β-strand tendency show significant correlation with Φ-values. It is noteworthy that the hydrophobicity scales including the effect of surrounding residues have a good correlation, ranging from 0.65 to 0.75. These results are consistent with experimental observations that the long-range contacts play an important role in the formation of folding nucleus (Itzaki et al. 1995; Grantcharova et al. 1998).

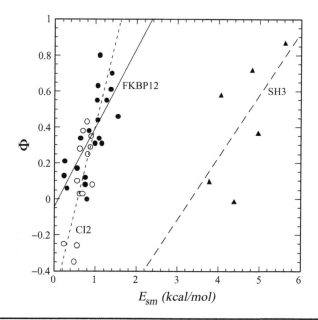

FIGURE 4.8 Relationship between experimental Φ values and E_{sm} in partially buried mutations of two-state proteins. The symbols, ●, ○, and ▲, represent FKBP12, CI2, and SH3 mutants, and their respective r-values are 0.72, 0.72, and 0.66 (Gromiha and Selvaraj, 2002).

4.4 Φ-value analysis with hydrophobic clusters and long-range contact networks

The Φ-values of 16 proteins have been analyzed using hydrophobic clusters and long-range contacts in protein structures (Selvaraj and Gromiha, 2004). The surrounding hydrophobicity of a given residue is defined as the sum of hydrophobic indices of various residues, which appear within the 8 Å radius limit from it (see **Section 3.10.3**). Segments that are composed of a group of residues with high surrounding hydrophobicity along the chain are considered as hydrophobic clusters; the residue with highest surrounding hydrophobicity within a cluster is taken as the key residue (nucleation site) in the hydrophobic cluster. Furthermore, for each residue, the residues that are close in space (within the distance of 8 Å) and far in sequence with the separation of more than four residues are termed as long-range contacts (Gromiha and Selvaraj, 2004; see **Section 3.4.3**).

4.4.1   Hydrophobic clusters

The importance of the hydrophobic cluster formation and the long-range contact network for understanding the transition state structures of proteins has been illustrated using one of the best-studied proteins, acyl coenzyme binding protein (PDB code: 2ABD). It belongs to the α-helical proteins fold, and its transition state structure has been characterized (Kragelund et al. 1999a,b). The surrounding hydrophobicity profile obtained for ACBP (2ABD) is shown in **Figure 4.9**. Four distinct

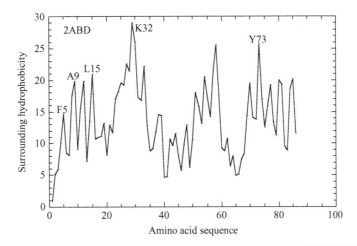

Figure 4.9 Surrounding hydrophobicity profiles for the acyl coenzyme binding protein (2ABD). Figure was adapted from Selvaraj and Gromiha (2004).

subdomains having high surrounding hydrophobicity values corresponding to the four helices A1 (A3-L15), A2 (D21-V36), A3 (K52-E62), and A4 (S65-Y 84) could be identified. Residues F5, A9, V12, L15, K32, Y73, V77, and L80 have been identified as folding nuclei using fluorescence measurements (Kragelund et al. 1999a, b). The approach based on hydrophobicity correctly identified F5, A9, V12, L15, Y73, V77, and L80 as key residues with the highest local maximum hydrophobicity. Further-more, the nearby residue of K32 (Q33) was also identified by the present approach. This result indicates the good agreement between the key residues identified us-ing surrounding hydrophobicity profile method and the experimentally reported folding nuclei (Selvaraj and Gromiha, 2004).

4.4.2 Long-range contact network

The long-range contact network between the residues that are identified as folding nuclei is shown in **Figure 4.10**. In this figure, residues in helix 1 (F5, A9, V12 and Leu15) form long-range contacts with residues in helix 4 (Y73 and L80). All these residues are thought to form productive interactions in the rate-limiting native like structure in this protein (Kragelund et al. 1999a).

4.4.3 Hydrophobic clusters and long-range contacts

The formation of hydrophobic clusters through long-range contacts in a typical protein, 1URN, is shown in **Figure 4.11**. The residues I12, L30, I43, I58, L69, and I84 are close to each other in space through long-range contacts. Interestingly, the residues L30 and I43 have high Φ-values. It is envisaged that the formation of such hydrophobic network of residues, through long-range interactions during the transition state, will facilitate the further downhill movement in the folding free energy landscape.

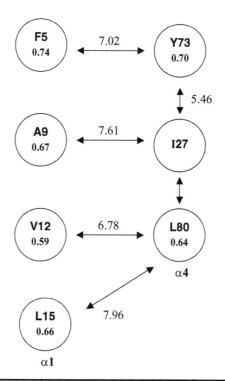

FIGURE 4.10 The long-range contact network pattern for the acyl coenzyme binding protein. Circles indicate α-helices. The residue information (name and number) and respective Φ values are given within the circles. The inter-residue distance (Cα-Cα) has been shown above the double head arrows. The secondary structural element to which a given residue belongs is indicated. Figure was adapted from Selvaraj and Gromiha (2004).

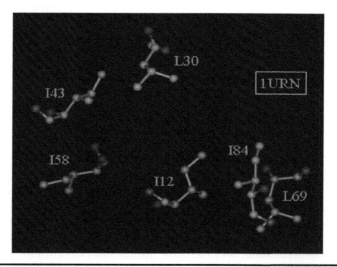

FIGURE 4.11 Hydrophobic clusters formed through long-range contacts in 1URN (Selvaraj and Gromiha, 2004).

4.5   Kinetic database for proteins

Fulton et al. (2007) developed a protein folding database, PFD, which contains the parameters that are important for understanding protein folding and stability. It includes annotated structural, methodological, kinetic, and thermodynamic data for more than 50 proteins, from 39 families. A user-friendly Web interface has been developed that allows searching, browsing, and information retrieval, while providing links to other protein databases. The database structure allows visualization of folding data. PFD can be accessed freely at http://pfd.med.monash.edu.au or http://www.foldeomics.org/pfd/. For each protein, information on the name of the species, family, molecular weight, contact order, free energy change, folding, and unfolding rates are given along with experimental conditions and links to other relevant sequences and structural databases. It has "advanced search" options, and one can search for the data with different conditions based on protein information, equilibrium data, kinetic data, mutant, measurement, and publication. **Figure 4.12a** shows the search with the protein name "chymotrypsin inhibitor 2," and the result is shown in **Figure 4.12b**.

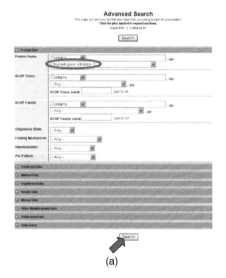

(a)

Select	ID	Protein	Species	Family	Class	Length	MW	CO	Eq. m_{D-N} (m)	Eq. ΔG_{D-N} (m)	k_u	k_f
☐	[ID: 10]	CI2	Barley	CI-2 family of serine protease inhibitors	α+β	66	7398.8	15	-8.3	32.5	0.000033 (0M)	314.19
☐	[ID: 310]	CI2 [view mutants (101)]	Barley	CI-2 family of serine protease inhibitors	α+β	66	7398.8	15	7.9	31.8	0.00012 (0M)	56.26
☐	[ID: 435]	CI2 [view mutants (101)]	Barley	CI-2 family of serine protease inhibitors	α+β	66	7398.8	15	7.9	31.8	0.02 (0M)	56.26

(b)

FIGURE 4.12   Usage of protein folding database: (a) search with protein name, chymotrypsin inhibitor 2 and (b) results obtained for the search.

Search in KineticDB

Field	Values	Action
Any:		search
Name: all ▾		Search
PDB:		Search
Authors: all ▾		Search
Buffer	any ▾	Filter
Kinetic type	any ▾	Filter
Denaturant	any ▾	Filter
SCOP	class: a ▾ fold: any ▾ sf: any ▾ family: any ▾	Filter

	N	Name	Mutation	PDB	ln k_f	ln k_u	ln k_{mf}	Kinetic type	Reference
☑	3	Villin headpeace subdomain	—	1VII	9.4	5.3	10.6	two	Wang M, Tang Y, Sato S, Vugmeyster L, McKnight CJ, Raleigh DP., Dynamic NMR line-shape analysis demonstrates that the villin headpiece subdomain folds on the microsecond time scale, JACS, 2003, **v. 125**, p. 6032-6033 ([12785814](#))
☑	4	Peripheral subunit-binding domain	WT	2PDD	9.8	5.4	9.8	two	Spector S, Raleigh DP, Submillisecond folding of the peripheral subunit-binding domain., JMB, 1999, **v. 293**, p. 763-768 ([10543965](#))
☑	5	GA module of albumin binding domain	K5I/K39V	1PRB	13.8	9.9	13.3	two	Wang T, Zhu Y, Gai F, Folding of a three-helix bundle at the folding speed limit., J Phys Chem B, 2004, **v. 108**, p. 3694-3697 ()
☑	6	B-domain of staphylococcal proteinA	WT	1BDD	11.7	4.2	5.8	two	Myers JK, Oas TG, Pre-organized secondary structure as an important determinant of fast protein folding., Nat Struct Biol, 2001, **v. 8**, p. 552-558 ([11373626](#))
☑	7	Engrailed homeodomain	WT	1ENH	10.5	7.6	8.1	multi	Gianni S, Guydosh NR, Khan F, Caldas TD, Mayor U, White GWN, DeMarco ML, Daggett V, Fersht AR, Unifying features in protein-folding mechanism., PNAS, 2003, **v. 100**, p. 13286-13291 ()
☑	8	TRF1 Myb domain	WT	1BA5	5.9	1.2	1.6	two	Gianni S, Guydosh NR, Khan F, Caldas TD, Mayor U, White GWN, DeMarco ML, Daggett V, Fersht AR, Unifying features in protein-folding mechanism., PNAS, 2003, **v. 100**, p. 13286-13291 ()

FIGURE 4.13 Search options in KineticDB. The result obtained with the search of data with "class a proteins" in SCOP database is shown.

Recently, Bogatyreva et al. (2009) proposed a database of folding kinetics, KineticDB, which contains the data for more than 90 unique proteins. It can be searched with the name of the protein, PDB code, authors, SCOP classification, etc. An example to search with "class a proteins" in SCOP database is shown in **Figure 4.13**. The KineticDB is available at http://kineticdb.protres.ru/db/index.pl.

4.6   Prediction of protein folding rates

Protein folding rate is a measure of slow/fast folding of a protein from its unfolded state to the native three-dimensional structure. Understanding the relationship between amino acid sequences and protein folding rates is an interesting and important task similar to the protein folding problem. It will help to understand the variations in protein folding kinetics, which may lead to several pathologies such as prion and Alzheimer diseases. Several methods have been proposed to understand/predict the folding rates of proteins from three-dimensional structures of proteins, secondary structural information, and amino acid sequence. These methods have different measures of accuracy and confidence levels.

4.6.1 Protein tertiary structure

Plaxco et al. (1998) proposed the concept of contact order (CO) using the information about the average sequence separation of all contacting residues in the native state of two-state proteins and found a significant correlation between CO and folding rates of two-state proteins (see **Section 3.11.1**). In an initial dataset of 12 proteins, CO shows a good correlation with protein folding rates, and the correlation coefficient is −0.81. Furthermore, CO has been effectively used for understanding protein folding kinetics. Gromiha and Selvaraj (2001) proposed the concept of LRO (see **Section 3.11.2**), and it showed a correlation of −0.78 with the folding rates of 23 two-state proteins. The classification of proteins based on three different structural classes improved the correlation to −0.72, −0.92, and −0.86 in all-α, all-β and mixed class proteins, respectively. Istomin et al. (2007) showed that LRO is the only parameter that shows good correlation in all structural classes of proteins.

Debe and Goddard (1999) have predicted the folding rates for 21 small, single-domain, topologically distinct proteins based on the first principles of protein folding and observed a good correlation with experimentally observed folding rates. Munoz and Eaton (1999) have proposed a simple statistical model to calculate the folding rates for 22 proteins from their three-dimensional structures and observed a correlation of 0.83 between predicted and experimental folding rates. Dinner and Karplus (2001) performed a statistical analysis to predict the protein folding rates and reported that both contact order and stability play important roles in determining the folding rate. Furthermore, neural networks-based models have been suggested to relate folding rates of proteins from the topological parameters, CO, LRO, and TCD (Zhang et al. 2003).

Miller et al. (2002) experimentally evaluated the role of structural topology to determine the protein folding rates and pathways. They have measured the folding rates for a set of circular permutants of the ribosomal protein S6 from *Thermus thermophilus* and estimated the correlation between folding rates and other topological parameters, CO (Plaxco et al. 1988), LRO (Gromiha and Selvaraj, 2001), and the fraction of short-range contacts (Mirny and Shakhnovich, 2001a). They observed that despite a wide range of relative contact order, the permuted proteins all fold with similar rates. On the other hand, LRO and the fraction of short-range contacts correlate very well with protein-refolding rates, including circular permutations of the ribosomal protein S6 from *Thermus thermophilus* (Miller et al. 2002).

Makarov et al. (2002) investigated the physical origin of protein folding rates and derived the formula $\ln(k) = \ln(N) + a + bN$, where N is the number of contacts in the folded state, and a and b are constants whose physical meaning is understood. They reported that this formula fit well the experimentally determined folding rate constants of the 24 proteins, with single values for a and b. Later, Makarov and Plaxco (2003) proposed the topomer search model, which quantitatively accounts for the broad scope of observed two-state folding rates. The model, which stipulates that the search for those unfolded conformations with a grossly correct topology is the rate-limiting step in folding, fits observed rates with a correlation coefficient of approximately 0.9 using two free parameters.

Dokholyan et al. (2002) showed that the topological properties of protein conformations determined their kinetic ability to fold. They used a macroscopic measure of the protein contact network topology, the average graph connectivity, by constructing graphs that are based on the geometry of protein conformations and reported that the average connectivity is higher for conformations with a high folding probability than for those with a high probability to unfold.

Capriotti and Casadio (2007) developed a server, K-fold, for discriminating the proteins that fold with two-state and multistate kinetics and for predicting protein folding rates using support vector machines. It takes the information about the three-dimensional structure of a protein as input and predicts the folding rate. It is freely available at http://gpcr.biocomp.unibo.it/cgi/predictors/K-Fold/K-Fold.cgi.

Total contact distance

Zhou and Zhou (2002) combined contact order and long-range order and devised a parameter, total contact distance (TCD). It is defined as

$$\text{TCD} = 1/n_r^2 \sum_{k=1}^{n_c} |i - j|, \tag{4.5}$$

where i and j are contacting residues; n_r and n_c are, respectively, number of residues and number of contacts in a protein. The summation is done for any cutoff residue separation (l_{cut}) and if $|i - j| > l_{\text{cut}}$. TCD is related to CO and LRO by a simple multiplication (TCD = CO × LRO) if LRO is calculated with the same l_{cut} value as CO. In a set of 28 two-state proteins, TCD showed a correlation of -0.88 with protein folding rates.

Cliquishness

Micheletti (2003) derived a parameter, cliquishness, using the concept of contact order (Plaxco et al. 1998) and an additional descriptor about the number of contacts. It is defined as

$$\text{Cliquishness} (i) = \sum \Delta_{ij} \Delta_{il} \Delta_{lj} / [N_c(N_c - 1)/2], \tag{4.6}$$

where N_c is the number of contacts for the residue, i. The cliquishness is properly defined only if the residue i is connected with at least two other residues.

Multiple contact index

Recently, Gromiha (2009) developed a novel parameter, the multiple contact index (MCI), to understand protein folding rates. The MCI is based on three parameters: (i) the distance between amino acid residues in space, (ii) the sequence separation between contacting residues and (iii) the number of residues that have multiple contacts. Using the three parameters, the MCI is defined as

$$n_{ci} = \sum n_{ij}; n_{ij} = 1 \text{ if } r_{ij} < 7.5 \,\text{Å}; |i - j| > 12 \text{ residues}; 0 \text{ otherwise};$$

$$\text{MCI} = \sum n_{\text{mi}}/N; n_{\text{mi}} = 1 \text{ if } n_{ci} \geq 4; 0 \text{ otherwise}, \tag{4.7}$$

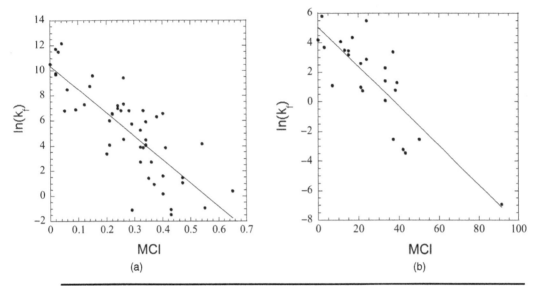

FIGURE 4.14 Relationship between the MCI and folding rates in (a) 50 two-state and (b) 25 three-state proteins. Figure was adapted from Gromiha (2009).

where n_c is the number of contacts for each residue and r_{ij} is the distance between the residues i and j. The MCI gives the total number of residues that have more than 4 contacts with the conditions that the contacting residues are within the limit of 7.5 Å and are separated with at least 12 residues. The MCI for three-state proteins is defined as below:

$$n_{ci} = \sum n_{ij}; n_{ij} = 1 \text{ if } r_{ij} < 6.5\,\text{Å}; |i - j| > 3 \text{ residues}; 0 \text{ otherwise.}$$

$$\text{MCI} = \sum n_{mi}; n_{mi} = 1 \text{ if } n_{ci} \geq 5; 0 \text{ otherwise.} \tag{4.8}$$

Note that no normalization has been done with chain length as it is a major determinant for the folding rates of three-state proteins (Galzitskaya et al. 2003). In a dataset of 50 two-state proteins, the MCI (Eqn. 4.7) showed a correlation of -0.80 with protein folding rates. On the other hand, the correlation was -0.83 between the MCI (Eqn. 4.8) and folding rates of 25 three-state proteins (Gromiha, 2009). **Figure 4.14** shows the relationship between the MCI and folding rates of 50 two-state and 25 three-state proteins.

The analysis on the propensity of residues to form multiple contacts showed that the aromatic and other hydrophobic residues prefer to form multiple contacts in two-state proteins, which reveals the formation of hydrophobic clusters and/or aromatic–aromatic interactions (**Table 4.3**). The propensity (P_{mc}) is computed as follows:

$$P_{mc}(i) = f_{mc}(i)/f_t(i), \tag{4.9}$$

where f_{mc} and f_t are the frequency of occurrence of amino acid residues that form multiple contacts and in the protein as a whole, respectively; i represents each of the 20 amino acid residues. The propensities have been normalized with the total

TABLE 4.3 Propensity of multiple contacts forming residues in two- and three-state proteins

Residue	Propensity		
	2-state proteins (50)	2-state proteins (27)	3-state proteins (25)
Ala	1.04	1.03	1.10
Asp	0.55	0.73	0.57
Cys	1.61	1.57	2.19
Glu	0.54	0.74	0.55
Phe	1.24	1.22	0.93
Gly	0.88	0.88	1.01
His	1.02	1.02	0.87
Ile	1.60	1.32	1.59
Lys	0.71	0.88	0.74
Leu	1.10	1.15	1.15
Met	0.79	1.08	0.88
Asn	0.89	0.81	0.29
Pro	0.94	0.98	0.70
Gln	0.89	0.75	0.66
Arg	0.82	1.05	0.67
Ser	0.76	0.94	1.10
Thr	1.04	1.01	1.30
Val	1.87	1.35	1.61
Trp	1.50	1.25	1.67
Tyr	1.34	1.10	1.39

Data were taken from Gromiha (2009); the number of proteins used in the computation is indicated in parenthesis.

number of multiple contacts forming residues and the total number of residues in all the considered proteins.

In three-state proteins, Cys has the highest preference, which indicates the formation of disulfide bridges and contacts with other residues. Interestingly, the polar residues Thr and Ser prefer to form multiple contacts in three-state proteins along with hydrophobic residues.

Furthermore, the analysis on the influence of residues with different number of contacts to the folding rates of two- and three-state proteins showed that the percentage of residues with up to seven contacts is higher in fast folding proteins than slow folding proteins (**Figure 4.15**). On the other hand, the number of residues with more than seven contacts is high in slow folding proteins.

4.6.2 Amino acid properties and secondary structure

Gromiha (2003) analyzed the influence of amino acid properties, secondary structure, and solvent accessibility for predicting the protein folding rates. The correlation coefficients obtained for selected amino acid properties, content and solvent accessibility at different secondary structures, and topological parameters are presented in **Table 4.4**. The properties, the number of medium-range contacts and α-helical tendency, showed moderate correlation with protein folding rates, which reveals the importance of local interactions to initiate protein folding.

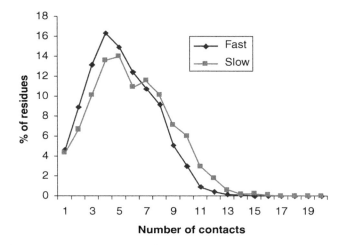

Figure 4.15   Percentage of residues with different number of contacts in 50 two-state proteins. Figure was adapted from Gromiha (2009).

The relationship between the secondary structure content and the protein folding rate shows that the α-helical content has moderate positive correlation of 0.52 (**Table 4.4**). This result also indicates the formation of α-helices prior to β-strands during the process of protein folding, as suggested by Sheinerman and Brooks (1998). The content of coil regions has negative correlation with protein folding rates.

Gong et al. (2003) calculated the secondary structure content for 24 two-state proteins and obtained coefficients that predict their folding rates. They showed that the predicted protein folding rates correlate strongly with experimentally determined ones and suggested that the folding rate of two-state proteins is a function of their local secondary structure content, consistent with the hierarchic model of protein folding.

4.6.3   Solvent accessibility

The correlation coefficients obtained between protein folding rates and various solvent accessibilities (obtained for the whole protein and at different secondary structures, helix, strand, turn, and coil) are given in **Table 4.4**. The results showed no significant correlation between them, and the highest correlation is 0.43 for the ASA at coil region (**Table 4.4**). Further examination on the effect of solvent accessibility by relating the number of residues at different ASA cutoff (0%–100% with an interval of 1%) and protein folding rates also showed no significant correlation at any specific cutoff ASA, which indicates that the number of residues in the interior/exterior of protein structures is not a major determinant for protein folding rates.

4.6.4   Chain length

Galzitskaya et al. (2003) demonstrated that chain length is the main determinant of the folding rate for proteins with the three-state folding kinetics. The logarithm of

TABLE 4.4 Correlation coefficient between selected amino acid properties, secondary structure content, solvent accessibility, and topological parameters with protein folding rates

No Variable	Correlation coefficient, r
Amino acid properties	
1. H_t, thermodynamic transfer hydrophobicity	0.05
2. H_p, surrounding hydrophobicity	0.08
3. pH_i, isoelectric point	0.07
4. M_w, molecular weight	0.00
5. P_α, α-helical tendency	0.35
6. P_β, β-strand tendency	−0.15
7. P_t, turn tendency	−0.27
8. P_c, coil tendency	−0.33
9. F, mean r.m.s. fluctuational displacement	−0.10
10. N_m, average medium-range contacts	**0.39**
11. N_l, average long-range contacts	−0.19
12. ΔASA, solvent accessible surface area for protein unfolding	0.08
13. ΔG, unfolding Gibbs free energy change	0.11
14. s, shape (position of branch point in a side-chain)	−0.04
Secondary structure content	
15. α-helix	**0.52**
16. β-strand	0.00
17. Turn	−0.03
18. Coil	−0.52
Accessible surface area (ASA)	
19. Average	0.28
20. at α-helix	0.32
21. at β-strand	0.00
22. at turn	−0.26
23. at coil	**0.43**
Topological parameters	
24. CO	−0.73
25. LRO	−0.81
26. TCD	**−0.88**

The highest correlation coefficient among amino acid properties, secondary structure content, ASA, and topological parameters is highlighted. Data were taken from Gromiha (2003).

the folding rates (k_f) strongly anticorrelates with their chain length L (the correlation coefficient being −0.80). At the same time, the chain length has no correlation with the folding rate for two-state folding proteins (the correlation coefficient is −0.07). Furthermore, they showed a significant difference of two- and three-state proteins that the two-state proteins have a strong anticorrelation between the folding rate and CO, whereas there is no such correlation for the three-state proteins. Furthermore, the folding rates of multistate proteins strongly correlate with their sizes (Ivankov et al. 2003).

4.6.5 Other parameters

Shao et al. (2003) proposed "helix parameter (HP)," which accounts the clustering of nonpolar residues in protein structures. They showed that the helix parameter is linearly correlated with the logarithms of folding rates of small two-state α-helical proteins. Zhang and Sun (2005) proposed the parameter, n-order contact distance, and showed a good correlation with protein folding rates. Dixit and Weikl (2006) considered the effect of chain connectivity due to crosslinks in protein chains by means of the graph theoretical concept of effective contact order. They reported that this method could predict the folding rates of two-state proteins with and without crosslinks. Prabhu and Bhuyan (2006) analyzed the folding rates of 45 two-state proteins in terms of length, secondary structure content, residue type, and free energy and developed empirical relations for predicting protein folding rates using these parameters.

4.6.6   Amino acid sequence

Recently, several methods have been proposed to predict the folding rates of proteins from amino acid sequence. They are mainly based on the predicted secondary structure, long-range contacts, and the combination of amino acid properties.

Ivankov and Finkelstein (2004) proposed a method for predicting folding rates of proteins from their amino acid sequences using the helical parameters predicted from their sequences. They have obtained the secondary structural information from DSSP and related protein folding rates with the combination of the number of helices and the number of residues in helical conformation. This approach showed the correlation of 0.81 between experimental and predicted protein folding rates. On the other hand, they have used PSIPRED (Jones, 1999; http://bioinf.cs.ucl.ac.uk/psipred/) for predicting the helical regions and related the helical parameters with protein folding rates. The predicted secondary structure information showed the correlation of 0.80 between the experimental and predicted folding rates. This result suggests that protein folding rates prediction obtained from the amino acid sequence (with the help of a secondary structure prediction done be PSIPRED) is not worse than the predictions done from 3D structures.

Punta and Rost (2005a) proposed an indirect method to predict protein folding rates from amino acid sequence: (i) developed a method, PROFcon, for predicting nonlocal contacts in protein structures from the amino acid sequence (see **Section 5.9**). This method combines information from alignments, predictions of secondary structure, and solvent accessibility (Punta and Rost, 2005b); (ii) the estimated nonlocal contacts have been utilized to compute long-range order (see **Section 3.11.2**), normalized with both chain length of the protein and square of the protein length, and (iii) the estimated long-range order is related with protein folding rates (Punta and Rost, 2005a). It has been shown that the estimated long-range order based on very noisy contact predictions was almost as accurate as the estimation based on known contacts for predicting the folding rates of two-state proteins.

Gromiha (2005) related the protein folding rates with physical-chemical, energetic, and conformational properties of amino acid residues and found that the classification of proteins into different structural classes shows an excellent correlation between amino acid properties and folding rates of two- and three-state proteins, indicating the importance of native-state topology in determining the protein folding rates. Furthermore, a simple linear regression model has been formulated for predicting the protein folding rates from amino acid sequences along with structural class information and obtained a good agreement between predicted and experimentally observed folding rates of proteins.

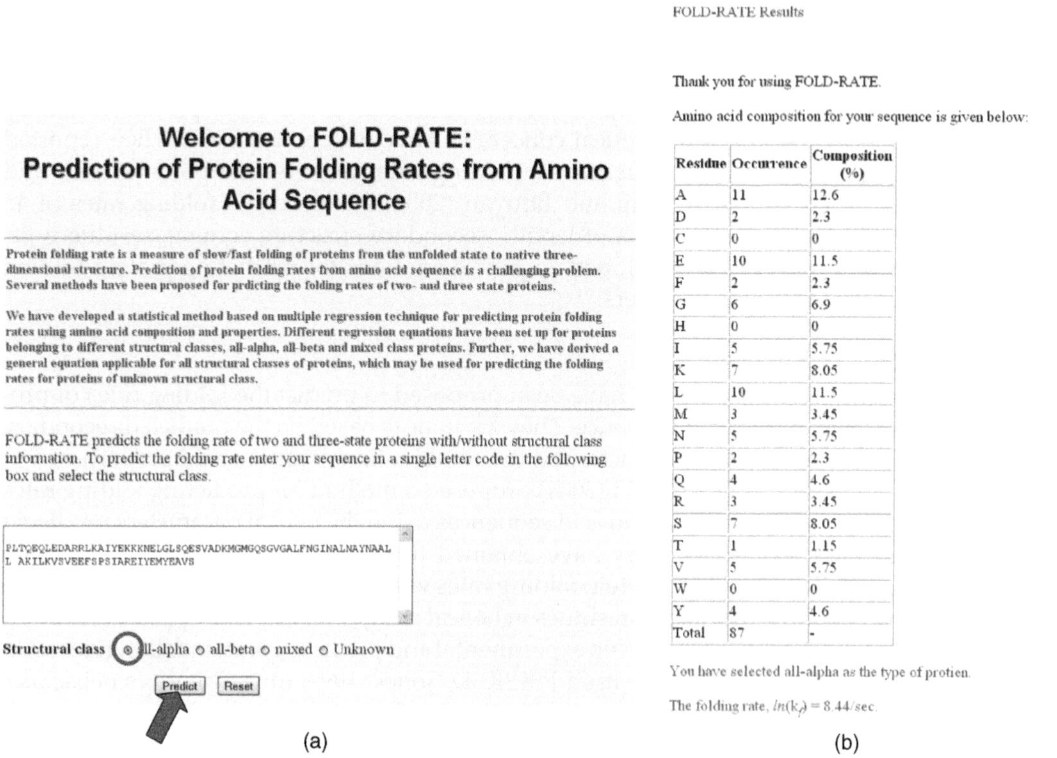

(a)

(b)

FIGURE 4.16   Web-based prediction of protein folding rates: (a) first page showing the input format (amino acid sequence in single letter code; an example is shown for λ repressor (1LMB) and structural class information (all-α) and (b) the query sequence, amino acid composition, type of the protein, and predicted folding rate are shown. The ln(k$_f$) value is predicted to be 8.44/sec.

A Web server, FOLD-RATE, has been developed for predicting protein folding rates from amino acid sequence, and it is available at http://psfs.cbrc.jp/fold-rate/ (Gromiha et al. 2006). An example for predicting protein folding rates is illustrated in **Figure 4.16**. **Figure 4.16a** shows the details to be provided in the input, i.e., the amino acid sequence of the protein and structural class information. The output is displayed in **Figure 4.16b**. It displays the amino acid composition of the query protein and the predicted folding rate. **Table 4.5** shows the experimental protein folding rates of 75 two- and three-state proteins.

The concept of the relationship between amino acid parameters and protein folding rates suggested by Gromiha and co-workers (Gromiha, 2005; Gromiha et al. 2006) has been used in several methods for predicting protein folding rates from amino acid sequence. Huang and Tian (2006) expressed the contribution of amino acid residues to protein folding rates based on amino acid properties (amino acid rigidity and dislike of amino acid for secondary structures) and showed a good correlation with protein folding rates. Ma et al. (2006) proposed a method based on amino acid composition for predicting protein folding rates. The analysis showed that the residues Ile, Ans, Gln, Thr, and Val are significantly

TABLE 4.5   Experimental folding rates in a set of 75 two- and three-state proteins

PDB code	Protein structural class	state	$\ln(k_f)$ experimental (/s)
1A6N	all-α	three	1.10
1BDD	all-α	two	11.75
1CEI	all-α	three	5.80
1EBD	all-α	two	9.68
1ENH	all-α	two	10.53
1HRC	all-α	two	8.76
1IMQ	all-α	two	7.31
1L8W	all-α	two	1.61
1LMB	all-α	two	8.50
1VII	all-α	two	11.52
1YCC	all-α	two	9.62
256B	all-α	two	12.20
2A5E	all-α	three	3.50
2ABD	all-α	two	6.55
2CRO	all-α	three	3.70
2PDD	all-α	two	9.80
1C8C	all-β	two	6.91
1C9O	all-β	two	7.20
1CBI	all-β	three	-3.20
1CSP	all-β	two	6.98
1EAL	all-β	three	1.30
1FNF-10	all-β	three	5.48
1FNF-9	all-β	two	-0.91
1G6P	all-β	two	6.30
1HNG	all-β	three	2.89
1HX5	all-β	three	0.74
1IFC	all-β	three	3.40
1LOP	all-β	two	6.60
1MJC	all-β	two	5.24
1NYF	all-β	two	4.54
1OPA	all-β	three	1.40
1PIN	all-β	two	9.44
1PKS	all-β	two	-1.05
1PNJ	all-β	two	1.10
1PSF	all-β	three	3.22
1SHF	all-β	two	4.50
1SHG	all-β	two	1.41
1SRL	all-β	two	4.04
1TEN	all-β	two	1.06
1TIT	all-β	three	3.47
1WIT	all-β	two	0.41
2AIT	all-β	two	4.20
1AON	mixed	three	0.80
1APS	mixed	two	-1.48
1AYE	mixed	two	6.80

(continued)

TABLE 4.5　Experimental folding rates in a set of 75 two- and three-state proteins (*continued*)

PDB code	Protein structural class	state	ln(k_f) experimental (/s)
1BNI	mixed	three	2.60
1BRS	mixed	two	3.40
1CIS	mixed	two	3.87
1COA	mixed	two	3.87
1DIV	mixed	two	6.58
1FKB	mixed	two	1.46
1GXT	mixed	three	4.38
1HDN	mixed	two	2.70
1HZ6	mixed	two	4.10
1PBA	mixed	two	6.80
1PCA	mixed	two	6.80
1PGB	mixed	two	6.00
1PHP C-TERMINAL	mixed	three	−3.45
1PHP N-TERMINAL	mixed	three	2.30
1POH	mixed	two	2.70
1qop α-subunit	mixed	three	−2.53
1qop β-subunit	mixed	three	−6.91
1RA9	mixed	three	−2.50
1RIS	mixed	two	5.90
1SCE	mixed	three	4.20
1UBQ	mixed	two	7.33
1URN	mixed	two	5.73
2ACY	mixed	two	0.92
2CI2	mixed	two	3.90
2HQI	mixed	two	0.18
2LZM	mixed	three	4.10
2PTL	mixed	two	4.10
2RN2	mixed	three	0.10
2VIK	mixed	two	6.80
3CHY	mixed	three	1.00

Data are taken from Gromiha et al. (2006).

correlated with protein folding rates of two-state proteins. Futhermore, they developed a parameter, composition index (CI), for predicting protein folding rates. It is given by

$$CI = C + \alpha W/LD, \tag{4.10}$$

where L is the sequence length of the protein and α is a constant parameter optimized with C and W/LD. The parameters, C, W, and D, are computed with the following equations:

$$C = (N_{Ala} + N_{Gln} - N_{Ile} - N_{Val} - N_{Cys})/L, \tag{4.11}$$

where, N_i represents the occurrence of residues of type i.

$$W = \sum w_i/L, \tag{4.12}$$

where w_i is the molecular weight of the residue i, and i varies for the 20 residues.

$$D = \sum d_i / L, \tag{4.13}$$

where d_i is the degeneracy of the residue i, and i varies for the 20 residues.

The composition index has been related with the folding rates of 37 two-state proteins and 25 three-state proteins, which showed a correlation of 0.73 and 0.71, respectively.

Huang and Gromiha (2008) developed a method based on quadratic response surface models for predicting the folding rates of 77 two- and three-state proteins, which showed a correlation of 0.90 between experimental and predicted protein folding rates using leave-one-out cross-validation method. It also discriminated the two- and three-state proteins with an accuracy of 90% using Baysean classification theory. A Web server has been developed for predicting protein folding rates, and it is available at http://bioinformatics.myweb.hinet.net/foldrate.htm. **Figure 4.17** shows the utility of the Web server for prediction. Recently, the parameters including the sequence, physiochemical properties of residues, and predicted secondary structure have been incorporated together to improve the accuracy of predicting protein folding rates (Jiang et al. 2009).

4.7   Relationship between Φ-values and folding rates

Releigh and Plaxco (2005) reviewed the experimental and theoretical analysis of Φ-values to understand the transition state structures of proteins and the relationship between Φ-values and folding rates. Φ-value analysis clearly shows that the side chains of many residues participate in energetic interactions in the transition state, and Φ-values provide a qualitative indication of a given side chain contributes to the overall barrier to folding. It is difficult to ascertain that the sum of these interactions defines absolute barrier heights because all of this structure is not simultaneously disrupted. The kinetic consequences of cosolvents, such as denaturants, that stabilize or destabilize many interactions simultaneously suggest, however, that the cumulative effect of the Φ-defined interactions is to modulate rates by 1 to 3 orders of magnitude (Plaxco et al. 2000). However, a growing body of evidence suggests that the specific interactions probed by Φ-value analysis play a perhaps surprisingly small role in defining the height of the folding barrier.

The most rapidly folding two-state protein folds a million times more rapidly than the slowest (Wittung-Stafshede et al. 1999; van Nuland et al. 1998), suggesting that the heights of even two-state folding barriers can vary by the maximum of 35kJ/mol. Several lines of experimental evidence suggest that the individual interactions probed by Φ-value analysis are, when compared to this 35 kJ/mol variation, a relatively minor determinant of folding rates. Results in favor of this argument include measurements of the decelerating effects of mutations that, unlike Φ-values, provide a quantitative measure of the extent to which a given interaction contributes to the folding barrier. It is thus striking that the vast majority of

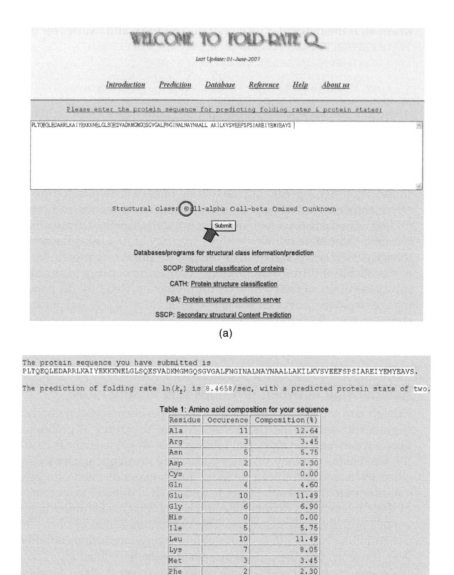

(a)

The protein sequence you have submitted is
PLTQEQLEDARRLKAIYEKKKNELGLSQESVADKMGMGQSGVGALFNGINALNAYNAALLAKILKVSVEEFSPSIAREIYEMYEAVS.

The prediction of folding rate $\ln(k_f)$ is 8.4658/sec, with a predicted protein state of two.

Table 1: Amino acid composition for your sequence

Residue	Occurence	Composition(%)
Ala	11	12.64
Arg	3	3.45
Asn	5	5.75
Asp	2	2.30
Cys	0	0.00
Gln	4	4.60
Glu	10	11.49
Gly	6	6.90
His	0	0.00
Ile	5	5.75
Leu	10	11.49
Lys	7	8.05
Met	3	3.45
Phe	2	2.30
Pro	2	2.30
Ser	7	8.05
Thr	1	1.15
Trp	0	0.00
Tyr	4	4.60
Val	5	5.75

(b)

FIGURE 4.17 (a) Snapshot showing the necessary items to be given as input for the prediction of protein folding rates using FOLD-RATE Q and (b) the predicted folding rate and the composition of amino acid residues for lambda-repressor (1LMB).

characterized mutations produce only small changes in folding rates (Releigh and Plaxco, 2005). The limited consequences of mutational studies suggest that the specific interactions probed by Φ-value analysis are the relatively minor determinants of observed folding rates, presumably because other, more global effectors (such as, perhaps, backbone topology) dominate barrier heights.

4.8   Exercises

1. Compare the distribution of Φ-values for different proteins given in **Table 4.2**.
   Hint: Make a histogram with Φ-values at different ranges.
2. Compare the folding nuclei in different proteins
   Hint: Check the availability of high Φ-values in **Table 4.2** for different proteins.
3. Map Φ-values for chymotrypsin inhibitor.
   Hint: Provide different color codes for different ranges of Φ-values and show in sequence and structure.
4. Compare the relationship between the number of contacts and Φ-values.
   Hint: Compute the number of medium- and long-range contacts (Chapter 3), and compare with Φ-values.
5. Are there any relationship between surrounding hydrophobicity and Φ-values?
   Hint: Compute surrounding hydrophobicity (Chapter 3) and relate with Φ-values.
6. Analyze the conservation score for the amino acid residues in 2CI2 with folding nuclei/Φ-values.
   Hint: Obtain conservation score using sequence/structure of CI2 (Chapter 2), and relate the score with ranges of Φ-values.
7. Obtain the data for folding rates in PFD.
   Hint: Display the data for all proteins in PFD.
8. Compare the protein folding rates data in different structural classes.
   Hint: Search KineticDB with different protein classes and analyze the data.
9. Compute the MCI for a two-state protein, 1NYF.
   Hint: Use Equation 4.7.
10. Compute the MCI for a three-state protein, 2LZM.
    Hint: Use Equation 4.8.
11. Compute cliquishness for the two proteins.
    Hint: Use Equation 4.6.
12. Compare the MCI and cliquishness for a set of proteins available in KineticDB
    Hint: Obtain the available proteins in KineticDB, and compute the parameters.
13. Predict the folding rate of 2LZM using FOLD-RATE, Fold Rate Q, and K-fold.
    Hint: Input the sequence for the first two methods and structural data for K-fold.
14. Estimate the relationship between $\Delta\Delta G$ and Φ-values in CI2.
    Hint: Compute the correlation using the data given in **Table 4.2**.
15. Compute the composition index for 2LZM.
    Hint: Use Equation 4.10.

References

Bogatyreva NS, Osypov AA, Ivankov DN. KineticDB: a database of protein folding kinetics. Nucleic Acids Res. 2009;37(Database issue):D342–D346.

Capriotti E, Casadio R. K-Fold: a tool for the prediction of the protein folding kinetic order and rate. Bioinformatics. 2007;23(3):385–386.

Chiti F, Taddei N, White PM, Bucciantini M, Magherini F, Stefani M, Dobson CM. Mutational analysis of acylphosphatase suggests the importance of topology and contact order in protein folding. Nat Struct Biol. 1999;6:1005–1009.

Choe SE, Li L, Matsudaira PT, Wagner G, Shakhnovich EI. Differential stabilization of two hydrophobic cores in the transition state of the villin 14T folding reaction. J Mol Biol. 2000;304:99–115.

Cota E, Steward A, Fowler SB, Clarke J. The folding nucleus of a fibronectin type III domain is composed of core residues of the immunoglobulin-like fold. J Mol Biol. 2001;305:1185–1194.

Debe DA, Goddard WA III. First principles prediction of protein folding rates. J Mol Biol. 1999;294:619–625.

Dinner AR, Karplus M. The roles of stability and contact order in determining protein folding rates. Nat Struct Biol. 2001;8:21–22.

Dixit PD, Weikl TR. A simple measure of native-state topology and chain connectivity predicts the folding rates of two-state proteins with and without crosslinks. Proteins. 2006;64:193–197.

Dobson CM. Protein folding and misfolding. Nature, 2003;426:884–890.

Dokholyan NV, Buldyrev SV, Shakhnovich EI. Identifying the protein folding nucleus using molecular dynamics. J Mol Biol. 2000;296:1183–1188.

Dokholyan NV, Li L, Ding F, Shakhnovich EI. Topological determinants of protein folding. Proc Natl Acad Sci USA. 2002;99:8637–8641.

Faísca PF, Travasso RD, Ball RC, Shakhnovich EI. Identifying critical residues in protein folding: insights from phi-value and P(fold) analysis. J Chem Phys. 2008;129(9):095108.

Fersht AR, Leatherbarrow RJ, Wells TNC. Quantitative analysis of structure–activity relationships in engineered proteins by linear free-energy relationships. Nature. 1986;322:284–286.

Fersht, AR. Transition-state structure as a unifying basis in protein-folding mechanisms: contact order, chain topology, stability, and the extended nucleus mechanism. Proc Natl Acad Sci USA. 2000;97:1525–1529.

Fowler SB, Clarke J. Mapping the folding pathway of an immunoglobulin domain: structural detail from Phi value analysis and movement of the transition state. Structure. 2001;9:355–366.

Fulton KF, Main ER, Daggett V, Jackson SE. Mapping the interactions present in the transition state for unfolding/folding of FKBP12. J Mol Biol. 1999;291:445–461.

Fulton KF, Bate MA, Faux NG, Mahmood K, Betts C, Buckle AM. Protein Folding Database (PFD 2.0): an online environment for the International Foldeomics Consortium. Nucleic Acids Res. 2007;35:D304–D307.

Galzitskaya OV, Finkelstein AV. A theoretical search for folding/unfolding nuclei in three-dimensional protein structures. Proc Natl Acad Sci USA. 1999;96:11299–11304.

Galzitskaya OV, Garbuzynskiy SO, Ivankov DN, Finkelstein AV. Chain length is the main determinant of the folding rate for proteins with three-state folding kinetics. Proteins. 2003;51:162–166.

Gong H, Isom DG, Srinivasan R, Rose GD. Local secondary structure content predicts folding rates for simple, two-state proteins. J Mol Biol. 2003;327:1149–1154.

Grantcharova VP, Riddle DS, Santiago JV, Baker D. Important role of hydrogen bonds in the structurally polarized transition state for folding of the src SH3 domain. Nat Struct Biol. 1998;5:714–720.

Grewal PS. Numerical Methods of Statistical Analysis. New Delhi: Sterling Publishers; 1987.

Gromiha MM. Importance of native state topology for determining the folding rate of two-state proteins. J Chem Inf Comp Sci. 2003;43:1481–1485.

Gromiha MM. A statistical model for predicting protein folding rates from amino acid sequence with structural class information. J Chem Inf Model. 2005,45:494–501.

Gromiha MM. Multiple contact network is a key determinant to protein folding rates. J Chem Inf Model. 2009;49(4):1130–1135.

Gromiha MM, Selvaraj S. Comparison between long-range interactions and contact order in determining the folding rate of two-state proteins: application of long-range order to folding rate prediction. J Mol Biol. 2001;310:27–32.

Gromiha MM, Selvaraj S. Important amino acid properties for determining the transition state structures of two-state protein mutants. FEBS Lett. 2002;526:129–134.

Gromiha MM, Selvaraj S. Inter-residue interactions in protein folding and stability. Prog Biophys Mol Biol. 2004;86:235–277.

Gromiha MM, Thangakani AM, Selvaraj S. FOLD-RATE: prediction of protein folding rates from amino acid sequence. Nucleic Acids Res. 2006;34:W70–W74.

Guerois R, Serrano L. The sh3-fold family: experimental evidence and prediction of variations in folding pathways. J Mol Biol. 2000;304:967–982.

Hamill SJ, Steward A, Clarke J. The folding of an immunoglobulin-like greek key protein is defined by a common-core nucleus and regions constrained by topology. J Mol Biol. 2000;297:165–178.

Huang JT, Tian J. Amino acid sequence predicts folding rate for middle-size two-state proteins. Proteins. 2006;63:551–554.

Huang LT, Gromiha MM. Analysis and prediction of protein folding rates using quadratic response surface models. J Comput Chem. 2008;29(10):1675–1683.

Istomin AY, Jacobs DJ, Livesay DR. On the role of structural class of a protein with two-state folding kinetics in determining correlations between its size, topology, and folding rate. Protein Sci. 2007;16(11):2564–2569.

Itzhaki LS, Otzen DE, Fersht AR. The structure of the transition state for folding of chymotrypsin inhibitor 2 analysed by protein engineering methods: evidence for a nucleation-condensation mechanism for protein folding. J Mol Biol. 1995;254: 260–288.

Ivankov DN, Finkelstein AV. Prediction of protein folding rates from the amino acid sequence-predicted secondary structure. Proc Natl Acad Sci USA. 2004;101: 8942–8944.

Ivankov DN, Garbuzynskiy SO, Alm E, Plaxco KW, Baker D, Finkelstein AV. Contact order revisited: influence of protein size on the folding rate. Protein Sci. 2003;12:2057–2062.

Jackson SE. How do small single proteins fold? Fold Des. 1998;3:R81–R91.

Jäger M, Nguyen H, Crane JC, Kelley JW, Gruebele MW. The folding mechanism of a β-sheet: the WW domain. J Mol Biol. 2001;311:373–393.

Jiang Y, Iglinski P, Kurgan L. Prediction of protein folding rates from primary sequences using hybrid sequence representation. J Comput Chem. 2009;30(5):772–783.

Jones DT. Protein secondary structure prediction based on position-specific scoring matrices. J Mol Biol. 1999;292(2):195–202.

Kragelund BB, Osmark P, Neergaard TB, Schiodt J, Kristiansen K, Kundsen J, Poulsen FM. The formation of a native-like structure containing eight conserved hydrophobic residues is rate-limiting in two-state protein folding of ACBP. Nature Struct Biol. 1999a;6:594–601.

Kragelund BB, Poulsen K, Andersen KV, Baldursson T, Kroll JB, Neergard TB, Jepsen J, Roepstorff P, Kristiansen K, Poulsen FM, Knudsen J. Conserved residues and their role in the structure, function, and stability of acyl-coenzyme A binding protein. Biochemistry. 1999b;38:2386–2394.

Lopez-Hernandez E, Serrano L. Structure of the transition state for folding of the 129 aa protein CheY resembles that of a smaller protein, CI-2. Fold Des. 1996;1:43–55.

Lorch M, Mason JM, Clarke AR, Parker MJ. Effects of core mutations on the folding of a beta-sheet protein: implications for backbone organization in the I-state. Biochemistry. 1999;38:1377–1385.

Ma BG, Guo JX, Zhang HY. Direct correlation between proteins' folding rates and their amino acid compositions: an ab initio folding rate prediction. Proteins. 2006;65:362–372.

Main ER, Fulton KF, Jackson SE. Folding pathway of FKBP12 and characterisation of the transition state. J Mol Biol. 1999;291:429–444.

Makarov DE, Keller CA, Plaxco KW, Metiu H. How the folding rate constant of simple, single-domain proteins depends on the number of native contacts. Proc Natl Acad Sci USA. 2002;99:3535–3539.

Makarov DE, Plaxco KW. The topomer search model: a simple, quantitative theory of two-state protein folding kinetics. Protein Sci. 2003;12:17–26.

Martinez JC, Pisabora MT, Serrano L. Obligatory steps in protein folding and the conformational diversity of the transition state. Nature Struct Biol. 1998;5:721–729.

Matouschek A, Kellis JT Jr, Serrano L, Fersht AR. Mapping the transition state and pathway of protein folding by protein engineering. Nature. 1989;340:122–126.

Matouschek A, Kellis JT Jr, Serrano L, Bycroft M, Fersht AR. Transient folding intermediates characterized by protein engineering. Nature. 1990;346(6283):440–445.

Micheletti C. Prediction of folding rates and transition-state placement from native-state geometry. Proteins. 2003;51:74–84.

Miller EJ, Fischer KF, Marqusee S. Experimental evaluation of topological parameters determining protein-folding rates. Proc Natl Acad Sci USA. 2002;99:10359–10363.

Mirny L, Shakhnovich E. Protein folding theory: from lattice to all-atom models. Annu Rev Biophys Biomol Struct. 2001a;30:361–396.

Mirny L, Shakhnovich E. Evolutionary conservation of the folding nucleus. J Mol Biol. 2001b;308(2):123–129.

Munoz V, Eaton WA. A simple model for calculating the kinetics of protein folding from three-dimensional structures. Proc Natl Acad Sci USA. 1999;96:11311–11316.

Nolting B. Protein folding kinetics. Heidelberg: Springer-Verlag; 1999.

Nölting B, Salimi N, Guth U. Protein folding forces. J Theor Biol. 2008;251(2):331–347.

Northey JG, Di Nardo AA, Davidson AR. Hydrophobic core packing in the SH3 domain folding transition state. Nature Struct Biol. 2002;9:126–130.

Otzen DE, Itzhaki LS, ElMasry NF, Jackson SE, Fersht AR. Structure of the transition state for the folding/unfolding of the barley chymotrypsin inhibitor 2 and its implications for mechanisms of protein folding. Proc Natl Acad Sci USA. 1994;91: 10422–10425.

Ozkan SB, Bahar I, Dill KA. Transition states and the meaning of Phi-values in protein folding kinetics. Nat Struct Biol. 2001;8(9):765–769.

Plaxco KW, Simons KT, Ruczinski I, Baker D. Topology, stability, sequence, and length: defining the determinants of two-state protein folding kinetics. Biochemistry. 2000;39(37):11177–11183.

Plaxco KW, Simons KT, Baker D. Contact order, transition state placement and the refolding rates of single domain proteins. J Mol Biol. 1998;277:985–994.

Poupon A, Mornon JP. Predicting the protein folding nucleus from sequences. FEBS Lett. 1999;452:283–289.

Prabhu NP, Bhuyan AK. Prediction of folding rates of small proteins: empiricalrelations based on length, secondary structure content, residue type, and stability. Biochemistry. 2006;45:3805–3812.

Punta M, Rost B. Protein folding rates estimated from contact predictions. J Mol Biol. 2005a;348:507–512.

Punta M, Rost B. PROFcon: novel prediction of long-range contacts. Bioinformatics. 2005b;21:2960–2968.

Raleigh DP, Plaxco KW. The protein folding transition state: what are Phi-values really telling us? Protein Pept Lett. 2005;12(2):117–122.

Rios MA, Daneshi M, Plaxco KW. Experimental investigation of the frequency and substitution dependence of negative phi-values in two-state proteins. Biochemistry. 2005;44(36):12160–12167.

Schonbrun J, Dill KA. Fast protein folding kinetics. Proc Natl Acad Sci USA. 2003;100(22):12678–12682.

Selvaraj S, Gromiha MM. Importance of hydrophobic cluster formation through long-range contacts in the folding transition state of two-state proteins. Proteins. 2004; 55:1023–1035.

Shao H, Peng Y, Zeng ZH. A simple parameter relating sequences with folding rates of small alpha helical proteins. Protein Pept Lett. 2003;10:277–280.

Sheinerman FB, Brooks CL III. Molecular picture of folding of a small alpha/beta protein. Proc Natl Acad Sci USA. 1998;95:1562–1567.

Shmygelska A. Search for folding nuclei in native protein structures. Bioinformatics. 2005;21:i394–i402.

Ternstrom T, Mayor U, Akke M, Oliveberg M. From snapshot to movie: phi analysis of protein folding transition states taken one step further. Proc Natl Acad Sci USA. 1999;96:14854–14859.

van Gunsteren WF, Mark AE. Prediction of the activity and stability effects of site-directed mutagenesis on a protein core. J Mol Biol. 1992;227:389–395.

van Nuland NA, Chiti F, Taddei N, Raugei G, Ramponi G, Dobson CM. Slow folding of muscle acylphosphatase in the absence of intermediates. J Mol Biol. 1998;283(4):883–891.

Villegas V, Martinez JC, Aviles FX, Serrano L. Structure of the transition state in the folding process of human procarboxypeptidase A2 activation domain. J Mol Biol. 1998;283:1027–1036.

Wittung-Stafshede P, Lee JC, Winkler JR, Gray HB. Cytochrome b562 folding triggered by electron transfer: approaching the speed limit for formation of a four-helix-bundle protein. Proc Natl Acad Sci USA. 1999;96(12):6587–6590.

Zarrine-Afsar A, Larson SM, Davidson AR. The family feud: do proteins with similar structures fold via the same pathway? Curr Opin Struct Biol. 2005;15(1):42–49.

Zhang L, Sun T. Folding rate prediction using n-order contact distance for proteins with two- and three-state folding kinetics. Biophys Chem. 2005;113:9–16.

Zhang L, Li J, Jiang Z, Zia A. Folding rate prediction on neural network model. Polymer 2003;44:1751–1756.

Zhou H, Zhou Y. Folding rate prediction using total contact distance. Biophys J. 2002;82:L458–L463.

CHAPTER 5

Protein Structure Prediction

Deciphering the native conformation of a protein from its amino acid sequence, termed as "protein folding problem," is one of the long-standing challenges in molecular and computational biology. Several methods have been proposed to predict the structural class, secondary structure content, location of secondary structures, and modeling tertiary structures. Progress has been shown for the past several decades, and still the accuracy is modest.

The prediction of protein secondary structures is an intermediate goal for determining its tertiary structure. The analysis on the performance of several secondary structure prediction methods in different structural classes showed that all the methods predict the secondary structure of all-α proteins more accurately than other classes (Gromiha and Selvaraj, 1998). Hence, the successful prediction of the structural class and schemes developed for each class may help to improve the accuracy levels of secondary structure predictive schemes in globular proteins. Consequently, Rost and Sander (1993a) developed a method for predicting helices in all-α proteins with high accuracy.

5.1   Protein structural class

The concept of "amino acid composition" plays a major role in predicting the structural class of globular proteins, and it is defined as the statistical preference of each of the amino acid residues occurring in protein molecules. The amino acid composition ($Comp_i$) for an amino acid "i" in a particular structural class is computed using the formula $Comp_i = n_i / N$, where n is the number of amino acid residues of type i—i varies from 1 to 20—and N is the total number of residues (see **Section 2.3.2**). The composition of amino residues in four different structural classes is presented in **Table 5.1**. Using the amino acid composition, several methods have been proposed to discriminate protein structural classes, and these methods are claming to achieve the accuracy close to 100%.

Klein (1986) proposed a method based on discriminant analysis to predict the structural class of globular proteins. This method uses the distribution of vectors of attributes characterizing the sequences of different structural classes. Chou's group performed a detailed analysis on the preference of amino acid residues in different structural classes of globular proteins and proposed various methods to

TABLE 5.1   Amino acid composition for the four structural classes of globular proteins

Amino acid	All-α	All-β	$\alpha + \beta$	α/β
Ala	11.72	6.28	8.31	9.65
Asp	5.70	5.10	6.27	6.32
Cys	0.71	1.99	2.11	1.03
Glu	6.90	5.29	5.56	6.18
Phe	4.23	4.40	3.78	3.68
Gly	6.63	8.49	6.80	8.76
His	3.01	1.33	2.56	2.27
Ile	4.06	4.22	4.58	6.13
Lys	8.05	5.71	6.19	6.21
Leu	10.94	7.24	7.77	7.88
Met	2.52	1.57	1.93	2.29
Asn	3.52	5.10	4.67	4.19
Pro	2.86	4.72	5.06	4.08
Gln	4.18	4.48	4.02	3.92
Arg	4.16	3.94	4.75	4.27
Ser	5.97	8.71	6.73	5.60
Thr	4.72	8.11	6.21	5.25
Val	6.19	7.02	6.58	7.49
Trp	1.10	1.75	1.46	1.32
Tyr	2.81	4.55	4.65	3.47

Data were taken from Kumarevel et al. (2000).

discriminate them. Chou and Zhang (1992) introduced a simple correlation coefficient method to predict the structural class of proteins and examined the validity of the method with both training and test sets of proteins. In this method, the amino acid compositions of four structural classes have been computed. For a new protein, compute the composition and the correlation with the precalculated composition of the four structural classes. The new protein is of all-α type if the correlation is the highest with the composition of all-α class proteins.

The predictive accuracy has been improved by component coefficient method (Zhang and Chou, 1992), expressing the protein as a vector of 20 dimensional space, in which its 20 components are defined by the composition of 20 amino acid residues. Similar approach on the weighting to the 20 constituent amino acid residues (Zhou et al. 1992) also produced better results. The component coupled method based on amino acid composition improved the accuracy further (Chou and Maggiora, 1998). The progress in the prediction of protein structural classes has been reviewed in Chou (2000, 2005).

Gromiha and Ponnuswamy (1995) analyzed the hydrophobic characteristics of proteins and developed a method based on the amphipathic character of α-helices and β-strands (see **Section 2.4**) and their conformational preferences for predicting protein structural classes.

Eisenhaber et al. (1996) analyzed the success rates of the secondary structural class prediction with different methods and addressed the necessity of another class of irregular type, which have been proposed by Nakashima et al. (1986). They have also developed a Web server (http://www.bork.embl-heidelberg.de/SSCP/) for predicting the structural class based on amino acid composition and compositional couplings between two amino acid residues. Furthermore, methods have been developed based on amino acid index (Bu et al. 1999) and Bayes decision rule (Wang and Yuan, 2000).

Kumarevel et al. (2000) proposed a method based on amino acid distribution along the sequence for predicting the structural classes of proteins. In this method, the compositions of residue pairs have been computed, and 20×20 matrices have been developed for each of the four structural classes. For a new protein, (i) compute the preference of residue pairs and (ii) get the deviation from the precalculated residue pair preference of each of the four structural classes. The new protein would be predicted to belong to the structural class for which the deviation is the lowest.

Zhou and Assa-Munt (2001) analyzed the concepts of few advanced algorithms, such as the least Mahalanobis distance algorithm, the component-coupled algorithm, and the Bayes decision rule for predicting the structural classes of proteins. They showed that the Bayes decision rule introduced for the protein structural class prediction is completely the same as the earlier component-coupled algorithm. Furthermore, the least Mahalanobis distance algorithm is an approximation of the component-coupled algorithm, also named as the covariant-discriminant algorithm (Chou and Elrod, 1999). This clarification may help to use these powerful algorithms effectively and correctly interpret the results in the protein structure prediction.

Luo et al. (2002) used the compositions of several dipeptides, tripeptides, tetrapeptides, pentapeptides, and hexapeptides for classifying protein structures. Jin et al. (2003) introduced a measure of information discrepancy to the prediction of protein structural classes. This approach is based on the comparisons of subsequence distributions, and they have considered the effect of residue distribution in protein structures. In a dataset of 1401 sequences with no more than 30% redundancy, the overall correctness rates of the resubstitution test and the jackknife test are reported to be 99.4% and 75.02%, respectively.

Chou and Cai (2004) calculated the functional domain composition to predict the structural class of a protein or a domain according to the following classification: all-α, all-β, α/β, $\alpha + \beta$, μ (multidomain), σ (small protein), and ρ (peptide). This method has the advantage that both the sequence-order-related features and the function-related features are naturally incorporated in the predictor. They have performed the jackknife cross-validation test on a dataset that consists of proteins and domains with only less than 20% sequence identity to each other and reported the overall success rate of 98%. Later, Cai et al. (2006) proposed a method based on LogitBoost for classifying proteins. It performs classification using a regression scheme as the base learner, which can handle multiclass problems and is particularly in coping with noisy data. Shen et al. (2005) introduced "supervised fuzzy clustering approach" that is featured by utilizing the class label information during the training process and used it for the class prediction.

Wang and Yuan (2000) developed a dataset of 674 globular proteins belonging to different structural classes (155 all-α proteins, 156 all-β proteins, 184 $\alpha + \beta$ proteins, and 179 α/β proteins) and reported that methods claiming 100%

accuracy for the structural class prediction, predicted only with the accuracy of 60% with this dataset. This result suggests that the success rates reported by different investigators are contradictory, and it mainly depends on the dataset.

5.2   Secondary structure content

Several methods have been proposed for predicting the secondary structure content of a protein based on amino acid composition and residue-pair composition (Eisenhaber et al. 1996; Chou, 1999).

Muskal and Kim (1992) presented a method based on two computer-simulated neural networks placed in "tandem" to predict the secondary structure content of globular proteins. The first of the two networks, NET1, predicts a protein's helix and strand content using the protein's amino acid composition, molecular weight, and heme presence. This network has more adjustable parameters (network weights) than learning examples, which caused problems with memorization. To overcome this problem, they designed a second network, NET2, which learned to determine when NET1 was in a state of generalization. These two networks predicted the contents of α-helical and β-strand segments within the error of 5.0% and 5.6%, respectively.

Eisenhaber et al. (1996) suggested a couple of methods: (i) The amino acid composition of an unknown protein is represented by the best (in a least square sense) linear combination of the characteristic amino acid compositions of the three secondary structural types computed from a learning set of tertiary structures and (ii) generalization of the first one and takes into account also possible compositional couplings between any two sorts of amino acids. Its mathematical formulation results in an eigenvalue/eigenvector problem of the second moment matrix describing the amino acid compositional fluctuations of secondary structural types in various proteins of a learning set. In a set of more than 400 proteins, this method predicted that the fractions of helix, sheet, and coil of the query protein are within the absolute error of 13.7%, 12.6%, and 11.4%, respectively. A Web server, SSCP, has been developed to predict the secondary structure content of a protein and hence the structural class. It is available at www.bork.embl-heidelberg.de/SSCP/. **Figure 5.1** shows an example for providing the input sequence and the output given by the SSCP server. It takes the amino acid sequence as the input (**Figure 5.1a**) and shows the percentage of residues in helical, strand, and coil regions along with the predicted protein structural class (all-α, all-β, $\alpha + \beta$, α/β, and irregular) in the output (**Figure 5.1b**).

Zhang et al. (2001) developed a multiple linear regression method to predict the content of α-helix and β-strand in a globular protein from its amino acid sequence and structural class information. In this method, the main attributes are the amino acid composition and the autocorrelation functions derived from the hydrophobicity profile of the primary sequence. They have examined the influence of amino acid residues and auto correlation functions for predicting the secondary structure content and selected a part of them to reduce the prediction error. Lin and Pan (2001) suggested a similar method using the information about the amino acid composition, the autocorrelation function, and the interaction function of side-chain mass.

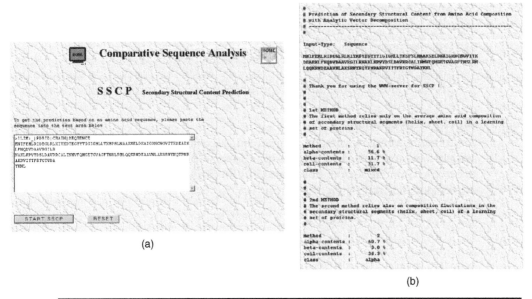

FIGURE 5.1 Prediction of the secondary structure content and the structural class using SSCP server: (a) input form containing the query amino acid sequence and (b) results obtained with the program with two methods.

Cai et al. (2003) classified the secondary structures into α-helix, β-strand, β-bridge, 3-10-helix, π-helix, H-bonded turn, bend, and random coil, and computed the composition of residue pairs in each of these secondary structures. Furthermore, the information about residue pair composition has been fed into a neural network for predicting the content of each of the secondary structures in a protein.

Ruan et al. (2005) proposed a novel approach for predicting the content of secondary structures in a protein from its primary sequence based on a composition moment vector, which is a measure that includes information about both the composition of a given primary sequence and the position of amino acids in the sequence. In contrast to the composition vector, this approach provides functional mapping between the primary sequence and the helix/strand content. This concept has been tested with over 11,000 protein sequences from Protein Data Bank (PDB), and they showed that a neural network method using a composition moment vector could predict with an average accuracy of 91.5% for the helix and 94.5% for the strand contents. It has been shown that the new measure, composition moment vector, reduced the error about 40% compared with the composition vector results.

Furthermore, several machine learning techniques have been used for discriminating protein structural classes and folding types (Niu et al. 2006; Chen et al. 2006; Dong et al. 2006; Taguchi and Gromiha, 2007; Kurgan et al. 2008).

5.3   Secondary structural regions

For the past four decades, there have been rapid advancements in the protein secondary structure prediction. Several methods have been proposed to predict the secondary structures, and the success rate is limited with their own advantages and disadvantages. In earlier days, methods were developed using statistical analysis, hydrophobicity analysis, pattern recognition, physicochemical properties, and sequence alignment. Recently, several machine learning techniques have been proposed, and the methods based on neural networks, support vector machines (SVMs), etc., along with alignment profiles significantly improved the accuracy.

5.3.1 Measures of accuracy

In protein secondary structure prediction algorithms, two measures have been widely used to assess the quality of prediction. They are the three-state prediction accuracy (Q3) and segment overlap (SOV or Sov). Q3 is a measure of the overall percentage of correctly predicted residues, to observed ones. It is given by

$$Q3 = \sum_{i=H,E,C} \frac{\text{predicted}(i)}{\text{observed}(i)} \times 100, \tag{5.1}$$

where H, E, and C represent respectively the residues in helix, strand, and coil.

The definition of SOV, proposed by Zemla et al. (1999) is widely used to assess the secondary structure prediction accuracy. Consider the terms s1 and s2 denote segments of secondary structure in conformational state i (i.e., helix, H; strand, E; or coil, C). Segments s1 and s2 correspond to the two secondary structure assignments being compared. The first assignment is considered as "observed," obtained from experiments and the second one as "predicted," obtained from different prediction methods. Let (s1, s2) denote a pair of overlapping segments, S(i), the set of all the overlapping pairs of segments (s1, s2) in state i, i.e,

$$S(i) = \{(s1, s2): s1 \cap s2 \neq \varnothing, s1 \text{ and } s2 \text{ are both in conformational state } i\}$$

S'(i) is the set of all segments s1 for which there is no overlapping segment s2 in state i, i.e,

$$S'(i) = \{s1: \forall s2, s1 \cap s2 = \varnothing, s1 \text{ and } s2 \text{ are both in conformational state } i\}$$

For state i, the segment overlap measure is then defined as

$$\text{Sov}(i) = 100 \times \frac{1}{N(i)} \sum_{S(i)} \frac{\min \text{ov}(s1, s2) + \delta(s1, s2)}{\max \text{ov}(s1, s2)} \times \text{len}(s1) \tag{5.2}$$

with the normalization value $N(i)$ defined as

$$N(i) = \sum_{S(i)} \text{len}(s1) + \sum_{S'(i)} \text{len}(s1). \tag{5.3}$$

The sum in Equation (5.2) and the first sum in Equation (5.3) are taken over all the segment pairs in state i which overlap by at least one residue, the second sum in Equation (5.3) is taken over the remaining segments in state i found in the reference assignment, len(s1) is the number of residues in segment s1, min ov(s1, s2) is the length of the actual overlap of s1 and s2, i.e., for which both segments have residues in state i, max ov(s1, s2) is the total extent for which either of the segments s1 and s2 has a residue in state i, and δ(s1, s2) is defined as

$$\delta(s1, s2) = \min\{(\max \text{ ov}(s1, s2) - \min \text{ ov}(s1, s2));$$
$$\min \text{ ov}(s1, s2); \text{int}(\text{len}(s1)/2); \text{int}(\text{len}(s2)/2)\}, \quad (5.4)$$

where $\min\{x1; x2; x3; \ldots; xn\}$ is the minimum of n integers.

The measure defined in equations (5.2–5.4) is easily extended to evaluate multi-state secondary structure assignments. In particular, for the three-state case of helix (H), strand (E), and coil (C),

$$\text{Sov}(i) = 100 \times \frac{1}{N} \sum_{i \in \{H,E,C\}} \sum_{S(i)} \frac{\min \text{ ov}(s1, s2) + \delta(s1, s2)}{\max \text{ ov}(s1, s2)} \times \text{len}(s1), \quad (5.5)$$

where the normalization value N is a sum of $N(i)$ over all three conformational states:

$$N = \sum_{i \in \{H,E,C\}} N(i). \quad (5.6)$$

The quality of match of each segment pair is taken as a ratio of the overlap of the two segments (min ov[s1, s2]) and the total extent of that pair (max ov[s1, s2]). The definition allows to improve this ratio by extending the overlap by the value of δ(s1, s2). The normalization procedure assures that Sov values are always within the range of 0 to 100 and thus can be used in the percentage scale to allow a direct comparison with other prediction evaluation measures, for example, Q3.

To illustrate the calculation of Sov, consider a prediction given in **Figure 5.2a** and evaluate the strand assignment, i.e., calculate Sov(E). In the observed structure, the first strand β_{o1} belongs to the set S'(E) because it does not produce any overlapping pair, while the second strand produces two of them: (β_{o2}, β_{p1}) and (β_{o2}, β_{p2}). The value of Sov(E) is calculated as follows:

$$\text{Sov}(E) = 100 \times \frac{1}{6+6+3} \times \left(\frac{1+1}{10} + \frac{2+1}{6} \right) \times 6 = 28.0 \quad (5.7)$$

As in Sov (Eqn. 5.5), all three conformational states are assigned equal weight, and coil regions are treated in the same way as strands or helices. Thus the value of Sov calculated for all three conformational states for this prediction is equal to 39.4.

Furthermore, SOV can be computed on the Web at http://proteinmodel.org/AS2TS/SOV/sov.html (**Figure 5.2b**). It takes the observed and predicted secondary structure along with the amino acid sequence and lists the Q3 (Eqn. 5.1) and SOV for all the secondary structures and overall score.

5.3.2 Statistical analysis

Statistical methods are based on studies of the database of proteins of known primary and secondary structures. In these methods the prediction accuracy lies in the range of 55% to 65%. Chou and Fasman (1974) proposed the first method for

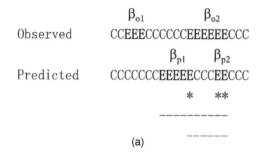

SOV: Segment Overlap Measure

Paste or type your input data into the box below:

```
>OSEQ (observed assignments)
CCEEECCCCCCEEEEEECCC
>PSEQ (predicted assignments)
CCCCCCCEEEEECCCEECCC
>AA   (amino-acid sequence; OPTIONAL)
MQTRSIGVQWERTYASDFGH
```

Start processing:

[Process] [Clear Form]

```
SECONDARY STRUCTURE PREDICTION ACCURACY EVALUATION.   N_AA =    20
```

		ALL	HELIX	STRAND	COIL
Q3	:	50.0	100.0	33.3	63.6
SOV	:	39.4	100.0	28.0	54.9

(b)

FIGURE 5.2   Sov(E) calculation: (a) illustration; * and - correspond respectively, to min ov and max ov in the overlapping segment pairs from observed and predicted structures. Figure was adapted from Zemla et al. (1999); (b) computation of Sov on the Web with input features (observed and predicted secondary structures as well as the amino acid sequence) and output results.

predicting the secondary structures of globular proteins based on (i) the calculation of the statistical propensities of each amino acid residue that forms either α-helix or β-strand and (ii) the classification of each residue into 12 classes depending on the likelihood of forming α-helix or β-strand. These statistical preferences of amino acid residues have been used to predict the probable helical and strand segments in a protein. The propensity of an amino acid residue i in α-helical conformation has been defined as the percentage of residue i in α-helical conformation to the percentage of all residues in the same conformation. It is written as

$$\text{propensity}_\alpha(i) = \% \text{ of residue } i \text{ in } \alpha\text{-helix}/\% \text{ of all residues in } \alpha\text{-helix.} \quad (5.8)$$

$$\% \text{ of residue } i \text{ in } \alpha\text{-helix} = n_\alpha(i)/N(i) \quad (5.9)$$

$n_\alpha(i)$ = number of residues of type i in α-helix

$N(i)$ = number of residues of type i in the whole dataset

$$\% \text{ of all residues in } \alpha\text{-helix} = n_\alpha/N \quad (5.10)$$

n_α = total number of residues in α-helix

N = total number of residues in the whole dataset

The classification of residues is based on strong helix formers (Hα), helix formers (hα), weak helix formers (Iα), weak helix breakers (iα), helix breakers (bα), strong helix breakers (Bα), and similar six classes for β-strands (Hβ, hβ, Iβ, iβ, bβ, and Bβ). The propensity values and the residues belonging to these 12 classes are shown in **Table 5.2**. The following rules have been proposed to predict the α-helical and β-strand regions (Chou and Fasman, 1974).

TABLE 5.2 Chou–Fasman parameters

Residue	P_α	Residue	P_β	Residue	P_t
Glu	Hα 1.53	Hβ Met	1.67	Asn	1.68
Ala	1.45	Val	1.65	Gly	1.68
Leu	1.34	Ile	1.60	Ser	1.56
His	hα 1.24	hβ Cys	1.30	Pro	1.54
Met	1.20	Tyr	1.29	Asp	1.26
Gln	1.17	Phe	1.28	Tyr	1.25
Trp	1.14	Gln	1.23	Cys	1.17
Val	1.14	Leu	1.22	Trp	1.11
Phe	1.12	Thr	1.20	Lys	1.01
Lys	Iα 1.07	Trp	1.19	Arg	1.00
Ile	1.00	Iβ Ala	0.97	Thr	1.00
Asp	iα 0.98	iβ Arg	0.90	Phe	0.71
Thr	0.82	Gly	0.81	His	0.69
Ser	0.79	Asp	0.80	Met	0.67
Arg	0.79	bβ Lys	0.74	Ile	0.58
Cys	0.77	Ser	0.72	Ala	0.57
Asn	bα 0.73	His	0.71	Gln	0.56
Tyr	0.61	Asn	0.65	Leu	0.53
Pro	Bα 0.59	Pro	0.62	Glu	0.44
Gly	0.53	Bβ Glu	0.26	Val	0.30

Data were taken from Chou and Fasman (1974).

Hα(β): Strong helix (strand) former
hα(β): Helix (strand) former
Iα(β): Weak helix (strand) former
iα(β): Weak helix (strand) breaker
bα(β): Helix (strand) breaker

Bα(β): Strong helix (strand) breaker
P_α: α-helical propensity
P_β: β-strand propensity
P_t: turn propensity

Helix

H1. Locate clusters of four helical residues (h_α or H_α) out of six residues along the polypeptide chain. Weak helical residues (I_α) count as 0.5 h_α (i.e., three h_α and two I_α residues out of six could also nucleate a helix). Helix formation is unfavorable if the segment contains 1/3 or more helix breakers (b_α or B_α), or less than 1/2 helix formers.

H2. Extend the helical segment in both directions until terminated by tetrapeptides with the average helical parameter, $\langle P_\alpha \rangle < 1.00$.

H3. Pro cannot occur in the inner helix or at the C-terminal helical end.

H4. Pro, Asp and Glu prefer the N terminal helical end. His, Lys and Arg prefer the C-terminal helical end.

General rule for helix: any segment of six residues or longer in a native protein with $\langle P_\alpha \rangle \geq 1.03$ as well as $\langle P_\alpha \rangle > \langle P_\beta \rangle$, and satisfying conditions H1 to H4 is predicted as a helix.

Strand

S1. Locate clusters of three strand residues (h_β or H_β) out of five residues along the polypeptide chain. β-strand formation is unfavorable if the segment contains 1/3 or more β-strand breakers (b_β or B_β), or less than 1/2 β-strand formers.

S2. Extend the strand segment in both directions until terminated by tetrapeptides with the average strand parameter, $\langle P_\beta \rangle < 1.00$.

S3. Glu occurs rarely in the β region. Pro occurs rarely in the inner β region.

S4. Charged residues occur rarely at the N-terminal β-strand, and infrequently at the inner β-strand region and C-terminal β-strand. Trp occurs mostly at the N-terminal β-strand and rarely at the C-terminal β-strand.

General rule for strand: any segment of five residues or longer in a native protein with $\langle P_\beta \rangle \geq 1.05$ as well as $\langle P_\beta \rangle > \langle P_\alpha \rangle$, and satisfying conditions S1 to S4, is predicted as a β-strand.

Conflict situation: a region containing overlapping helical and strand assignments is considered as a helix (or strand) if average propensity of α-helix (β-strand) is greater than that of β-strand (α-helix). A Web server has been developed to predict the secondary structures of a protein based on the Chou–Fasman method, and the prediction results are available at http://fasta.bioch.virginia.edu/fasta/chofas.htm. **Figure 5.3** shows the input sequence and predicted secondary structure using the Chou–Fasman method.

Garnier et al. (1978) introduced a method, GOR by treating the primary sequences and the sequence of secondary structures as two messages related by a translation process, using information theory. The secondary structure assignment using the improved GOR method is available at http://npsa-pbil.ibcp.fr/cgi-bin/npsa_automat.pl?page=npsa_gor4.html. This method was improved by Gibrat et al. (1987) by recalculating all the information terms using a larger database and using side chain-side chain interactions. In the nearest neighbor methods, the secondary structure of a new primary sequence is assigned to be the same as that of the closest primary sequence to it of the known three-dimensional structures.

FIGURE 5.3 Prediction of secondary structure using the Chou–Fasman method.

Several methods have been put forward with different numbers of neighboring residues, ranging from 1 to 50 residues (Levin et al. 1986; Nishikawa and Ooi, 1986; Salzberg and Cost, 1992; Yi and Lander, 1993; Mugilan and Veluraja, 2000). These methods are relatively simple, but they are not directly related to chemical or physical theory.

5.3.3 Sequence alignment

One of the powerful methods for predicting the secondary structures of globular proteins is the alignment of the sequence to a homologue of a known three-dimensional structure. In this method, for the two given sequences A and B of lengths *m* and *n*, all possible overlapping segments having a particular length from A are compared to all segments of B (Barton, 1996). For each pair of segments, the amino acid pair scores are accumulated over the length of the sequence. Different types of scores, such as identity scoring, genetic code scoring, chemical similarity scoring, and scoring scheme based on the observed substitution, have been used in the computation.

Russell and Barton (1993) proposed a method based on a multiple sequence alignment, and the predictive accuracy rises up if the sequence similarity is high enough to ensure reliable alignment (Rost and Sander, 1994a). Mehta et al. (1995) improved the accuracy of the secondary structure prediction by using a multiple sequence alignment of a large dataset with 2500 proteins from 70 families. Furthermore, the local alignment has also been used to predict protein secondary structures (Salamov and Solovyev, 1997). Cuff and Barton (2000) improved the predictive accuracy up to 76% with the application of a multiple sequence alignment profile. A Web server, Jpred, has been developed for predicting the secondary structures of proteins using a multiple sequence alignment and is available at http://www.compbio.dundee.ac.uk/~www-jpred/ (Cole et al. 2008).

5.3.4   Hydrophobicity profiles

Hydrophobicity analysis has remained at the central focus for understanding protein folding and stability and, especially, secondary structures of proteins, interior and exterior regions, antigenic sites, periodicities in residue distributions, and membrane associated regions. Rose (1978) computed the first hydrophobicity profile for a set of proteins to predict the turn regions in them. Hydrophobicity profile is a graph showing the local hydrophobicity of the amino acid sequence as a function of position. To display a hydrophobicity plot, one has to choose a hydrophobicity scale and an averaging procedure. It is also possible to get the profile without any average for single residue hydrophobicity plot. As an example, hydrophobicity profile of T4 lysozyme using surrounding hydrophobicity scale (Ponnuswamy and Gromiha, 1993) is shown in **Figure 2.14b**.

Cid et al. (1992) generated hydrophobicity profiles for several proteins using the surrounding hydrophobicity scale (Ponnuswamy et al. 1980) and derived general rules for identifying secondary structures. The hydrophobicity profiles for α-helices, buried and exposed β-strands and turns are shown in **Figure 5.4**. The following facts have been taken into account for drawing the profiles: (i) Turns occur at those sites in the polypeptide chain where the hydrophobicity is at a local minimum, (ii) helices are generally located nearer the surface of the protein and tend to have hydrophilic and hydrophobic surfaces at opposite sides. This fact originates alternating regions with low and high hydrophobicity with a periodicity of every 3 to 4 residues, corresponding to the α-helix periodicity of 3.6 amino acid residues, (iii) β-strands have a tendency to be buried in the interior of the protein, which shows up by a clustering of hydrophobic amino acid residues in the region

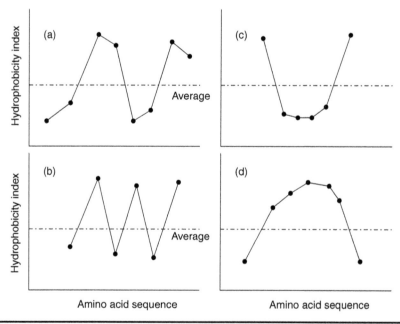

FIGURE 5.4   The basic four hydrophobicity profiles: (a) exposed α-helical structure, (b) exposed β-strand, (c) β-turn and (d) buried β-strand. Figure was adapted from Cid et al. (1992).

of the sequence where a β-strand occurs, and (iv) few β-strands lying on the protein surface will show the presence of alternating hydrophobic and hydrophilic amino acid residues. The advantages of this method are as follows: (i) it is simple, (ii) it predicts β-turns and loops with great accuracy, (iii) when predicting β-strands, it can distinguish between exposed and buried β-strands, and (iv) it provides an independent criterion to differentiate between helical and β-structures in regions where they present similar probabilities according to the Chou and Fasman (1974) method. The main disadvantages are that (i) it cannot differentiate between buried helical structures and buried β-strands, and (ii) the database does not include all types of proteins, like membrane proteins or glycoproteins, and hence its application is restricted.

Gromiha and Ponnuswamy (1995) followed the method of Cid et al. (1982) and constructed different hydrophobicity profiles of different window sizes for all-α, all-β and mixed class proteins. These profiles have been combined with the preference of amino acid residues at N − 1, N, C, and C + 1 positions of α-helices and β-strands to predict the secondary structures of globular proteins. These hydrophobicity-based methods can provide the physical basis of secondary structures, and the accuracy is modest.

5.3.5 Neural network and alignment profiles

Neural network is one of the most powerful tools to predict the secondary structures of globular proteins. It is a network of nonlinear processing units that have adjustable connection strengths, and the secondary structure prediction is mainly based on feed-forward networks using the back propagation learning rule. The goal of the method is to find a good input–output mapping that can then be used to predict the test set.

The artificial neural network (ANN) consists of a number of nodes (also called neurons) and zero or more hidden layers. An example is shown in **Figure 5.5**, where

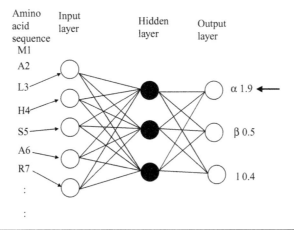

Figure 5.5 A simple artificial neural network for predicting secondary structure. Residues, i − 2, i − 1, i, i + 1 and i + 2, are used for the prediction of residue i (S5). The output is, for each type of the secondary structure, a number describing the strength of prediction that the residue belongs to this class. Residue five is predicted to be a helix.

there is one hidden layer, consisting of three nodes. Each node in the input layer gets a value for a feature; as shown in **Figure 5.5**, it is a value describing an amino acid type. Each node in a layer is connected to all nodes in the preceding layer. In this case, the layers are fully connected. A weight is associated with each connecting line. For each node, an activation value is calculated based on the node's input value. Denote a node by j, and let the nodes on its predecessor layer be i, $i = 1$, $2, \ldots k$. Furthermore, let the weight between nodes i and j be w_{ji} and the activation value for node i be x_i. Then a value $S_j = \Sigma w_{ji} a_i$ is calculated ($i = 0$ corresponds to an extra node with $a_0 = 1$, used as a threshold). The activation level at node j is then calculated as a "soft threshold function" $f(S_j)$, and this value is sent on all connection lines out from j. A common function is $f(S_j) = 1/(1 + e^{-S_j})$.

Training the network means finding values for the weights associated with each line in the ANN so that the network gives correct predictions as often as possible. For this purpose, a supervised learning approach is used. In a supervised approach, a training set is constructed as a set of labeled instances, i.e., a set of residues along with their correct labels (in **Figure 5.5,** each label being α, β, 1). During the training phase of the ANN, the weights associated with each of the lines in the network are adjusted. Each instance, y in the training set is presented to the ANN, the ANN produces an output using its current weights and the output is compared with the correct label of y, and an error is calculated for each of the nodes in the output layer. Now the weights of the lines into the output layer can be adjusted to reduce the error for this one instance. Then the so-called back-propagation algorithm is used to propagate the errors to the nodes in the layers before and to adjust the weights of the lines into this layer. The back propagation algorithm can then be used recursively to propagate the errors to all layers and adjust the weights until all weights in the ANN have been adjusted. All instances in the training set are presented to the ANN many times, and the order in which the instances are presented is randomized. Different criteria can be used to decide when to end the training phase.

A danger of training an ANN is that the weights become very well adjusted to the training set, producing a very low error on this set, but fail to generalize to new dataset. To test the performance of a prediction system, a test set must be defined that is independent of the training set (does not contain any of the sequences in the training set).

Several investigators used neural networks to secondary structure prediction, and their approaches differ mainly in the size of the window of amino acid residues and the dataset of proteins. Qian and Seijnowski (1988) used a network with one hidden layer, a window size of 13 and 21 inputs for each position and predicted the three-state secondary structure at an accuracy of 64%. Subsequently, neural network based methods have been proposed to predict the β and γ-turns in globular proteins (McGregor et al. 1989; Shepherd et al. 1999; Kaur and Raghava, 2003, 2004). Chandonia and Karplus (1996) used primary and secondary neural networks to secondary structure prediction for 681 proteins and obtained an accuracy of 75%. Petersen et al. (2000) reported an accuracy of 80% in predicting protein secondary structures.

Rost's group is working on the prediction of secondary structures using neural networks for about two decades, and they have proposed several improvements. Rost and Sander (1993b) trained a two-layered, feed-forward neural network on a nonredundant dataset of 130 proteins and claimed a three-state accuracy of 71%.

A Web server, PHD, has been developed for predicting protein secondary structures, which is widely used by several researchers (Rost, 1996). Predication results are available at http://cubic.bioc.columbia.edu/predictprotein/. Later, it has been suggested that the accuracy of protein secondary structure prediction improves with the growth of alignment profiles (Przybylski and Rost, 2002). The accuracy of PHD using pair-wise alignments increased to 72%, and using larger databases and PSI-BLAST it raised accuracy to 75%. Further analysis indicated that more than 60% of the improvement originated from the growth of current sequence databases; about 20% resulted from detailed changes in the alignment procedure (substitution matrix, thresholds, and gap penalties), and 20% of the improvement resulted from carefully using iterated PSI-BLAST searches. The prediction results using alignments are also available at the same site. Pollastri et al. (2002a) used ensembles of bidirectional recurrent neural network architectures and PSI-BLAST-derived profiles for predicting the secondary structures in three and eight states. The three-state prediction showed an accuracy of 78% correct prediction. The program, Sspro, is implemented on a Web server, which is available at http://promoter.ics.uci.edu/BRNN-PRED/.

Cuff and Barton (2000) developed a neural network prediction algorithm that works by applying multiple sequence alignments, along with PSIBLAST and Hidden Markov Model (HMM) profiles. Consensus techniques are applied that predict the final secondary structure more accurately. This method shows an accuracy of 76%, and the program is available at http://www.compbio.dundee.ac.uk/www-jpred/.

Jones (1999b) proposed a two-stage neural network to predict protein secondary structure based on the position-specific scoring matrices (PSSMs) generated by PSI-BLAST. The results are found to be better than other methods as evidenced from the Critical Assessment of Techniques for Protein Structure Prediction experiment (CASP3), where the method was evaluated by stringent blind testing. Using a new testing set based on a set of 187 unique folds and three-way cross-validation based on structural similarity criteria rather than sequence similarity criteria, the PSIPRED showed an average Q3 score of between 76.5% and 78.3% depending on the precise definition of an observed secondary structure. A Web server has been developed for predicting the secondary structures of proteins and online services are available at http://bioinf.cs.ucl.ac.uk/psipred/ (McGuffin et al. 2000). **Figure 5.6** shows the utility of PSIPRED for predicting a typical amino acid sequence. The server takes the input sequence and sends the result by e-mail. Furthermore, the source code can be downloaded from http://bioinf.cs.ucl.ac.uk/memsat/memsat-svm/.

Lin et al. (2005) developed a Web server, YASPIN, for predicting protein secondary structures in seven states using hidden neural networks and optimized the output using HMMs (a computation structure for describing the subtle patterns that define families of homologous sequences; HMMs are powerful tools for detecting distant relatives. They include the possibility of introducing gaps into the generated sequence, with position-dependent gap penalties and carry out the alignment and assignment of probabilities together). YASPIN does not use an alignment algorithm directly but uses the information encoded in the PSSM that can be generated using PSI-BLAST (Altschul et al. 1997; Altschul and Koonin, 1998) or any other

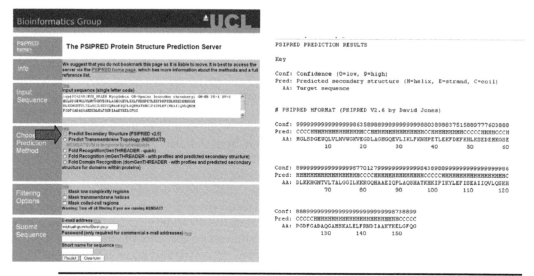

FIGURE 5.6 Utility of PSIPRED for predicting the protein secondary structure. The output shows the predicted secondary structure along with the confidence level.

alignment program. The prediction results are shown for three-state predictions, and the services are available at http://ibivu.cs.vu.nl/programs/yaspinwww/.

5.3.6   Support vector machines

Machine learning techniques are popular in several biological applications, including protein secondary structure prediction. The applications of neural networks have been discussed elaborately in the previous section. Recently, several investigations have been carried out using SVMs for predicting the secondary structures of proteins. It is a learning algorithm (Cristianini and Shawe-Taylor, 2000), which from a set of positively and negatively labeled training vectors learns a classifier that can be used to classify new unlabeled test samples. SVM learns the classifier by mapping the input training samples $\{x_1, \ldots, x_n\}$ into a possibly high dimensional feature space and by seeking a hyperplane in this space that separates the two types of examples with the largest possible margin, i.e., distance to the nearest points. If the training set is not linearly separable, SVM finds a hyperplane, which optimizes a trade-off between good classification and large margin. The implementation of SVM can be done with the freely downloadable SVM-light package by Joachims (1999). It is available at http://svmlight.joachims.org/.

Hua and Sun (2001) developed the first method based on SVMs for predicting protein secondary structures. They have started to construct several binary classifiers and then assembled a tertiary classifier for three secondary structure states (helix, sheet, and coil) based on these binary classifiers. The SVM method achieved a good performance of segment overlap accuracy SOV = 76.2% and per-residue accuracy (Q3) of 73.5% through sevenfold cross validation on a database of 513 non-homologous protein chains with multiple sequence alignments. The applications of SVM include effective avoidance of over-fitting, the ability to handle large feature spaces, and information condensing of the given data set. Ward et al. (2003) applied

SVMs to predict the protein secondary structure and reported that the accuracy of the prediction is similar to that of neural networks. They reported an accuracy of 77% for the three-state prediction. Kim and Park (2003) used both SVMs and PSI-BLAST profiles for predicting the secondary structures of proteins and obtained an average accuracy of 78%.

5.4   Discrimination of transmembrane helical proteins and predicting their membrane-spanning segments

Membrane proteins have become attractive targets for pharmacological agents, and about 20% to 30% of protein sequences in genomes are identified as transmembrane helical proteins. On the other hand, the known three-dimensional structures of membrane proteins are limited, and the location of transmembrane helical segments in membrane proteins would help to model their tertiary structures. Hence, several methods have been proposed for predicting the membrane-spanning helical segments in membrane proteins. These methods include positive inside rule, hydrophobicity analysis, HMM, neural network, etc.

These methods are mainly based on the following concepts (Chen and Rost, 2002): (i) TM helices are predominantly apolar and between 12 and 35 residues long, (ii) globular regions between membrane helices are typically shorter than 60 residues (Wallin and von Heijne, 1998; Liu and Rost 2001), (iii) most TMH proteins have a specific distribution of the positively charged amino acids arginine and lysine, known as the "positive-inside-rule" (von Heijne, 1986; von Heijne, 1989). Connecting "loop" regions on the inside of the membrane has more positive charges than "loop" regions on the outside, (iv) long globular regions (>60 residues) differ in their composition from those globular regions subject to the "inside-out-rule." These simple facts have been used in several prediction methods developed over the past two decades.

5.4.1   Hydrophobicity profile method

Kyte and Doolittle (1982) developed a hydrophobicity profile method for locating the membrane-spanning helical segments from amino acid sequences. To identify membrane regions, they implemented a moving-window approach in which the hydrophobicity indices of the amino acid residues have been summed over a stretch of residues and took the average. It has been reported that the window length of 19 residues (a central residue and 9 residue on each sides of the central one) performs the best to discriminate globular and membrane proteins and to detect membrane-spanning helical segments. They have fixed a threshold of 1.6 and if the average hydrophobicity for a segment exceeds this limit, the segment has been suggested as a transmembrane helix. The hydrophobicity plot for the Kyte–Doolittle method is available at http://fasta.bioch.virginia.edu/fasta/grease.htm. Eisenberg et al. (1984) developed the helical hydrophobic moment as a measure of the amphiphilicity of α-helix. This hydrophobic moment differed between transmembrane and globular helices, and could thus be explored to predict transmembrane regions (Eisenberg et al. 1982). Klein et al. (1985) combined a quadratic discriminant function with the hydrophobicity analysis of Kyte and Doolittle for discriminating membrane proteins. A window of 17 adjacent residues has been used in this approach, and proteins with values less than 0 were classified as integral membrane proteins. Nakai and Kanehisa (1992) applied the same concept of filtering the simple scales through a quadratic discriminant function in their method. The rationale of discriminant function is that it first tentatively evaluates the number of putative membrane helices using a low threshold of 0.5. Then it refines the predicted number by using a more stringent threshold of –2.0. After the transmembrane regions are predicted, they used the concept of modified positive-inside rule (von Heijne, 1986; Hartmann et al. 1989) to predict the protein's topology, which in the realm of membrane proteins refers to the orientation of its N-terminus with respect to the lipid bilayer. A Web server (ALOM) has been developed to predict the membrane-spanning helical segments using discriminant analysis and the topology using the concept of Hartmann et al. (1989), and it is available at http://psort.nibb.ac.jp/form.html.

Ponnuswamy and Gromiha (1993) developed a surrounding hydrophobicity scale applicable to both globular and membrane proteins and proposed a "surrounding hydrophobicity profile" method for predicting transmembrane helices from the amino acid sequence of a protein. This profile is simply the plot of the surrounding hydrophobicity indices of the residues against their sequence numbers. In this plot, the hydrophobic and hydrophilic parts are distinguished by a horizontal line representing the average hydrophobicity value, which is obtained from the surrounding hydrophobicity values for all the amino acid residues in the sample set of proteins. The surrounding hydrophobicity profile thus constructed projects the transmembrane helices as a sequence of peaks and valleys above the average middle line (or with a few valleys crossing down the average line), and the other parts as peaks and valleys frequently crossing the middle line, of falling below the middle line. The criterion of a continuous sequence of 20 to 24 points above the average line with a maximum of two nonadjacent exceptions was used to determine the length of a predicted transmembrane helix.

Followed by these methods, several researchers have developed their own hydrophobicity scales and proposed different methods for predicting transmembrane helical segments (Engelmen et al. 1986). Jayasinghe et al. (2001) attempted to improve hydropathy analysis by improving the hydropathy scales. The commonly used hydrophobicity scales neglect the thermodynamic constraints imposed by α-helices on transmembrane stability, and hence they have derived a whole-residue hydropathy scale from the Wimley-White experiments that took into account the backbone constraints. They have developed a Web server (Membrane Protein Explorer, MPEx) by means of hydropathy plots using thermodynamic principles for identifying the membrane-spanning helical segments, and it is available at http://blanco.biomol.uci.edu/mpex/. Deber et al. (2001) developed a method, TMFinder, using a hydrophobicity scale based on the HPLC retention time of peptides with nonpolar phase helicity. It measured the propensity of an amino acid to be in an α-helical state based on circular dichroism.

5.4.2   Positive-inside rule

Gunnar von Heijne (1986) observed that the positively charged amino acids are important elements in targeting peptides that direct proteins into mitochondria, nuclei, and the secretory pathways of both prokaryotic and eukaryotic cells. On the basis of this, he introduced the concept of "positive-inside rule," which observes that regions of polytopic (multispanning) membrane proteins facing the cytoplasm are generally enriched in Arg and Lys residues, whereas translocated regions are largely devoid of these residues, implying that the distribution of positively charged amino acids may also be a major determinant of the transmembrane topology of integral membrane proteins. Using this concept, several methods have been proposed for predicting the membrane-spanning segments. Von Heijne (1989) described a derivative of Escherichia coli leader peptidase, a polytopic inner-membrane protein that switches from sec-gene-dependent membrane insertion with a N_{out}–C_{out} transmembrane topology to sec-gene-independent insertion with a N_{in}–C_{in} topology in response to the addition of four positively charged lysines to its N-terminus. Hartmann et al. (1989) altered this rule slightly by omitting the region flanking the first helix from the compilation for predicting the orientation of eukaryotic membrane proteins.

5.4.3 Combining hydrophobicity and positive-inside rule

A considerably more complex scheme for post-processing hydrophobicity scales was implemented in TopPred (von Heijne, 1992). TopPred predicts the complete topology of membrane proteins by using hydrophobicity analysis, automatic generation of possible topologies, and ranking these topologies by the positive-inside rule. First, the method introduces a particular sliding trapezoid window to detect segments of outstanding hydrophobicity using the hydrophobicity scale proposed by Engelman et al. (1986). The two bases of the trapezoid were chosen to be 11 and 21 residues long, and the shape of a trapezoid was used to combine the favorable noise-reduction of a triangular window (Claverie and Daulmiere, 1991) with a more physically relevant rectangular window, which represents the central nonpolar region of the lipid bilayer. Next, TopPred explores the positive-inside rule. This rule simply states the observation that positively charged residues (Arg and Lys) are more abundant on the inside of membranes (von Heijne, 1986). Generally, this fact allows for the membrane protein topology prediction. However, TopPred has an option to choose the thresholds for considering a segment as membrane helix that yielded the optimal difference between the number of positively charged residues at the inside and at the outside. All these refinements implemented in TopPred led to a major improvement in prediction accuracy (von Heijne, 1992). Prediction results are available at http://www.sbc.su.se/~erikw/toppred2/.

5.4.4 Statistical method

Amino acid preferences for membrane and nonmembrane proteins and/or that for membrane-spanning α-helices and other regions can also be used for prediction. Rather than using the observation that hydrophobic residues are abundant in transmembrane helices, a more general strategy has been conceived to infer from the known membrane helices, which amino acids have the highest preference for that state. Such a simple statistical evaluation was already the base for the first methods predicting secondary structure for globular proteins (Schulz, 1988; Fasman, 1989). Hofmann and Stoffel (1993) analyzed the statistical preferences of amino acid residues using the sequence data available in

SWISS-PROT database. On the basis of these preferences, a method, TMpred, has been proposed for predicting transmembrane helices using a combination of several weight matrices for scoring. The prediction results are available at http://www.ch.embnet.org/software/TMPRED_form.html.

Gromiha (1999) developed a set of conformational parameters for the 20 amino acid residues in transmembrane helical proteins using the frequency of occurrence of amino acid residues in the transmembrane helical part of membrane proteins, and that in the whole complex in a set of 70 membrane proteins. On the basis of these conformational parameters, a simple algorithm has been formulated using a set of primary and secondary rules to predict the transmembrane α-helices in membrane proteins. Furthermore, a computer program has been developed that takes the amino acid sequence as input and gives the predicted transmembrane α-helices as output.

5.4.5 Combining statistical preferences and hydrophobicity scales

Hydrophobicity-based methods still appear to be effective in predicting transmembrane segments. One of the drawbacks was that such methods fail to discriminate accurately between membrane regions and highly hydrophobic globular segments. Juretic and colleagues integrated multiple scales for amino acids for the prediction of transmembrane regions and developed the method SPLIT (Juretic et al. 1993; Juretic et al. 1998). They derived amino acid preferences for the "state" membrane helix from a dataset of integral membrane proteins with a partially known secondary structure. They also extracted preferences for β-strand, turn, and nonregular secondary structure based on sets of soluble proteins of the known structure. The comparison with hydrophobicity plots suggested that the preference profiles were more accurate, exhibited higher resolution, and had less noise. Shorter, unstable or movable membrane helices were often missed by the hydrophobicity analyses in proteins with transport functions. In contrast, they were predicted by the combination of preferences. For instance, the N-terminal TM helices of voltage-gated ion channels and glutamate receptors were correctly identified by SPLIT. A Web server has been developed for predicting transmembrane helices and is available at http://split.pmfst.hr/split/.

Pasquier et al. (1999) proposed an algorithm, PRED-TMR, which uses a standard hydrophobicity analysis with an emphasis on the detection of potential helix ends. Using the propensities of amino acid residues at the termini of transmembrane helices, PRED-TMR compiles scores for the termini of each putative segment. The best prediction has been achieved by a scoring function obtained from the two termini scores, a hydropathy score and a length constraint. The prediction results are available at http://o2.biol.uoa.gr/PRED-TMR/.

5.4.6 Physicochemical parameters

Hirokawa et al. (1998) developed a method, SOSUI, for discriminating transmembrane helical proteins and predicting their membrane-spanning segments using the combination of a variety of physicochemical parameters. Specifically, the parameters, Kyte–Doolittle hydropathy index, amphiphilicity, relative and net charges, and protein length have been used to detect transmembrane α-helices. It has been reported that this method could discriminate the transmembrane helical

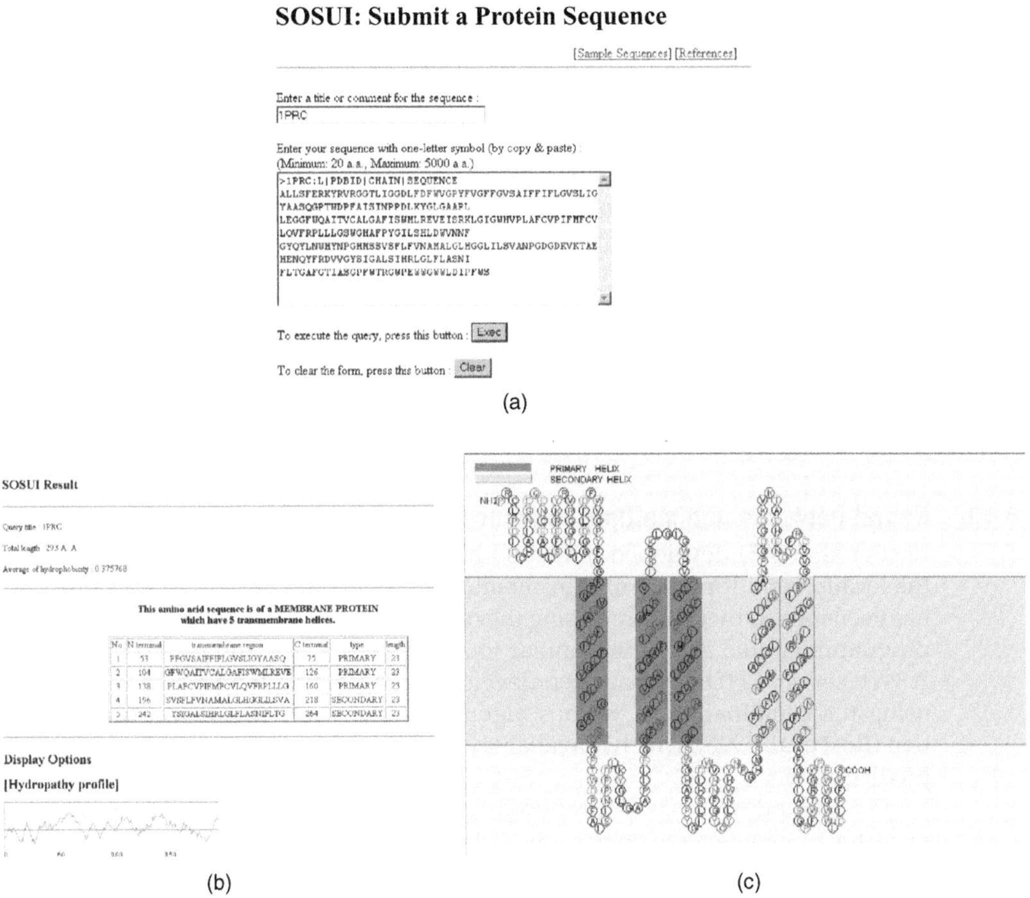

SOSUI: Submit a Protein Sequence

[Sample Sequences] [References]

Enter a title or comment for the sequence :

1PRC

Enter your sequence with one-letter symbol (by copy & paste)
(Minimum: 20 a.a., Maximum: 5000 a.a.)

>1PRC:L|PDBID|CHAIN|SEQUENCE
ALLSFERKYRVRGGTLIGGDLFDFWVGPYFVGFFGVSAIFFIFLGVSLIG
YAASQGPTHDPFAISINPPDLKYGLGAAPL
LEGGFWQAITVCALGAFISWMLREVEISRKLGIGWHVPLAFCVPIFMFCV
LQVFRPLLLGSWGHAFPYGILSHLDWVNNF
GYQYLNWHYNPGHMSSVSFLFVNAMALGLHGGLILSVANPGDGDEVKTAE
HENQYTRDVVGYSIGALSIHRLGLFLASNI
FLTGAFGTIASGPFWTRGWPEWWGWWLDIPFWS

To execute the query, press this button : [Exec]

To clear the form, press this button : [Clear]

(a)

SOSUI Result

Query title : 1PRC

Total length : 273 A.A

Average of hydrophobicity : 0.375768

This amino acid sequence is of a MEMBRANE PROTEIN
which have 5 transmembrane helices.

No.	N terminal	transmembrane region	C terminal	type	length
1	53	FFGVSAIFFIFLGVSLIGYAASQ	75	PRIMARY	23
2	104	GFWQAITVCALGAFISWMLREVE	126	PRIMARY	23
3	138	PLAFCVPIFMFCVLQVFRPLLLG	160	PRIMARY	23
4	196	SVSFLFVNAMALGLHGGLILSVA	218	SECONDARY	23
5	242	TSIGALSIHRLGLFLASNIFLTG	264	SECONDARY	23

Display Options

[Hydropathy profile]

FIGURE 5.7   Prediction of transmembrane helices and topology using SOSUI: (a) input showing the query amino acid sequence (L chain of 1PRC), (b) identified transmembrane regions and the hydropathy profile, and (c) topology of the membrane protein.

proteins at high accuracy. On the other hand, it predicts the signal peptides as transmembrane segments. A Web server has been developed for the discrimination and prediction, and the results are available at http://sosui.proteome.bio.tuat.ac.jp/sosuiframe0.html. An example for predicting the membrane-spanning segments of the L chain of 1PRC (photosynthetic reaction center) is shown in **Figure 5.7**. In the output, SOSUI shows the number of transmembrane segments along with the stretch of amino acid residues. Furthermore, the topology of the protein is also shown.

5.4.7   Dynamic programming

Jones et al. (1994) implemented a method, MEMSAT, for the prediction of the secondary structure and topology of integral membrane proteins based on the recognition of topological models. This method employs a set of statistical tables (log likelihoods) compiled from well-characterized membrane protein data and a novel dynamic programming algorithm to recognize membrane topology models by expectation maximization. Residues are classified as being one of the five structural states as follows: Li (inside loop), Lo (outside loop), Hi (inside helix end), Hm (helix middle), and Ho (outside helix end). Helix end caps are defined to span over four adjacent residues (one helical turn). Next the authors extracted the propensity of each amino acid for each of these five states from experimentally well-described membrane proteins. The statistical tables presented in Jones et al. (1994) show definite biases toward certain amino acid species on the inside, middle, and outside of a cellular membrane. Using these propensities, MEMSAT calculates a score relating a given sequence to a predicted topology and arrangement of membrane helices. The specific feature of MEMSAT is that it finds the optimal score through dynamic programming, i.e., an algorithm is also explored to find the optimal pairwise sequence alignment (Needleman and Wunsch, 1970; Sellers, 1974). Recently, MEMSAT has been implemented in PSIPRED protein structure prediction server and is available at http://bioinf.cs.ucl.ac.uk/psipred/psiform.html.

5.4.8   Neural networks and multiple sequence alignments

Rost et al. (1995) developed a method based on neural networks for identifying the location and topology of transmembrane α-helices. Furthermore, the method has been refined by post-processing the neural network output through a dynamic programming-like algorithm, similar to the one introduced by Jones et al. (1994). A Web server, PHDhtm, has been developed for locating transmembrane helices using the combination of various algorithms and multiple alignment information (Rost et al. 1996), which is available online at http://cubic.bioc.columbia.edu/predictprotein/.

Persson and Argos (1996) proposed a method based on multiple sequence alignment for predicting transmembrane helical segments and their topologies. They have used two sets of propensity values: one for the middle, hydrophobic portion and one for the terminal regions of the transmembrane sequence spans. Average propensity values were calculated for each position along the alignment, with the contribution from each sequence weighted according to its dissimilarity relative to the other aligned sequences. Eight-residue segments were considered as potential cores of transmembrane segments and elongated if their middle propensity values were above a given threshold. End propensity values were also considered as stop signals. Only helices with length of 15 to 29 residues were allowed, and corrections for strictly conserved charged residues were also made.

The prediction of membrane protein topology relies on residue compositional differences in the protein segments exposed at each side of the membrane. Intra/extracellular ratios were calculated for Asn, Asp, Gly, Phe, Pro, Trp, Tyr, and Val, preferably found at the outside of membranes, and for Ala, Arg, Cys and Lys, mostly occurring inside. The consensus over these 12 residues was used to predict topology. The location and topology of membrane-spanning helices are available at http://bioinfo.limbo.ifm.liu.se/tmap/.

Although multiple sequence alignments improve prediction accuracy, there are no homologues in current databases for 20% to 30% of all membrane proteins (Liu and Rost, 2001). To overcome this situation, the so-called dense alignment surface (DAS) method was developed (Cserzö et al. 1997). DAS is based on the scoring matrix originally introduced to improve alignments for G-protein coupled receptors.

It compares low-stringency dot-plots of the query protein against the background representing the universe of nonhomologous membrane proteins using the scoring matrix. DAS server is available online at http://www.sbc.su.se/~miklos/ DAS/.

5.4.9   Hidden Markov Models

Tusnady and Simon (1998) suggested a method for topology prediction of helical transmembrane proteins, HMMTOP, based on the hypothesis that the localizations of the transmembrane segments and the topology are determined by the difference in the amino acid distributions in various structural parts of these proteins rather than by specific amino acid compositions of these parts. An HMM with special architecture was developed to search the transmembrane topology corresponding to the maximum likelihood among all the possible topologies of a given protein. This model distinguishes the five structural states: inside nonmembrane region, inside TMH-cap, membrane helix, outside TMH-cap, and outside nonmembrane region. Jones et al. (1994) placed the TMH-caps inside the membrane, while in HMMTOP it has been placed outside the membrane (Tusnady and Simon, 1998). The prediction results of HMMTOP are available at http://www.enzim.hu/hmmtop/. The topology of the L chain of 1PRC is shown in **Figure 5.8** for comparison.

Krogh et al. (2001) proposed similar method using HMMs by considering a variety of constraints influencing the membrane-passing regions of proteins in one consistent methodology. They developed a method, TMHMM, which implements a cyclic model with seven states for transmembrane-helix (TMH) core, TMH-caps on the N- and C-terminal sides, nonmembrane regions on the cytoplasmic side, two nonmembrane regions on the noncytoplasmic side, and a globular domain state in the middle of each nonmembrane region. The two nonmembrane regions on the noncytoplasmic-side model short and long loops respectively, which correspond to two different membrane insertion mechanisms. The Web server is available at http://www.cbs.dtu.dk/services/TMHMM/.

5.4.10   Joint method for transmembrane helix prediction

The prediction of the secondary structure for globular proteins can be improved by combining many prediction methods (Nishikawa and Noguchi, 1991). Applying a similar average, Promponas et al. (1999) developed a method CoPreTHi, a Web-based application (http://athina.biol.uoa.gr/CoPreTHi/), which uses the results from DAS, ISREC-SAPS, PHDhtm, PRED-TMR, SOSUI, Tmpred, and TopPred2 (Promponas et al. 1999). CoPreTHi combines the results into a joint prediction histogram; residues are predicted as transmembrane if they are identified as such by at least three methods. Nilsson et al. (2000) explored consensus predictions for the membrane protein topology to derive reliability for the prediction. In particular, they used five methods (TMHMM, HMMTOP, MEMSAT, TopPred2, and PHDhtm) to evaluate a test set of 60 Escherichia coli inner membrane proteins with experimentally determined topologies. They found that the prediction performance varies strongly with the number of methods that agree, and that the topology of nearly half of all inner membrane proteins can be predicted with high reliability (>90% correct predictions) by a simple majority vote.

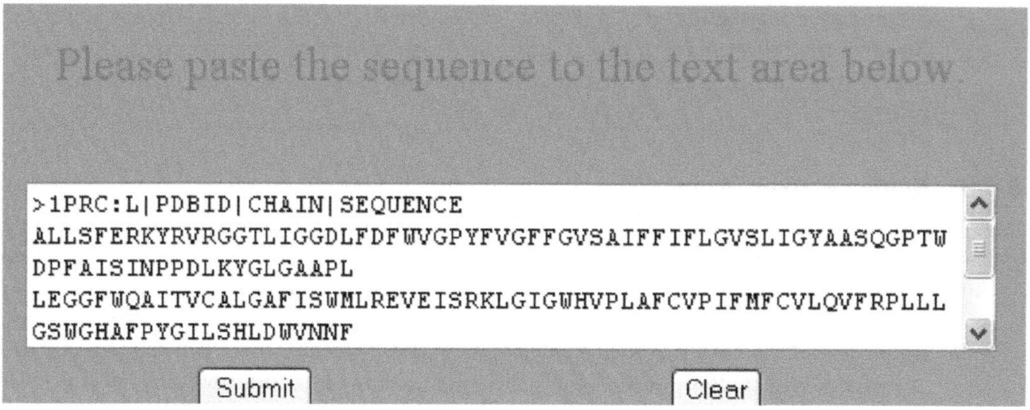

Please paste the sequence to the text area below.

```
>1PRC:L|PDBID|CHAIN|SEQUENCE
ALLSFERKYRVRGGTLIGGDLFDFWVGPYFVGFFGVSAIFFIFLGVSLIGYAASQGPTW
DPFAISINPPDLKYGLGAAPL
LEGGFWQAITVCALGAFISWMLREVEISRKLGIGWHVPLAFCVPIFMFCVLQVFRPLLL
GSWGHAFPYGILSHLDWVNNF
```

Submit   Clear

```
Protein: noname
Length:  293
N-terminus:  IN
Number of transmembrane helices: 5
Transmembrane helices: 42-66 98-122 131-154 197-221 252-276

Total entropy of the model:  17.0090
Entropy of the best path:  17.0107

The best path:

    seq  PRCLPDIDCH AINSEQENCE ALLSFERKYR VRGGTLIGGD LFDFWVGPYF    50
    pred IIIIIIIIII IIIIIIIIII IIIIIIiiii iiiiiiiiii iHHHHHHHHH

    seq  VGFFGVSAIF FIFLGVSLIG YAASQGPTWD PFAISINPPD LKYGLGAAPL   100
    pred HHHHHHHHHH HHHHHooooo oooooooooo oOoooooooo oooooooHHH

    seq  LEGGFWQAIT VCALGAFISW MLREVEISRK LGIGWHVPLA FCVPIFMFCV   150
    pred HHHHHHHHHH HHHHHHHHHH HHiiiiiiii HHHHHHHHHH HHHHHHHHHH

    seq  LQVFRPLLLG SWGHAFPYGI LSHLDWVNNF GYQYLNWHYN PGHMSSVSFL   200
    pred HHHHooooooo oooooooooO OOOOOOOOOO Ooooooooooo oooooooHHHH

    seq  FVNAMALGLH GGLILSVANP GDGDKVKTAE HENQYFRDVV GYSIGALSIH   250
    pred HHHHHHHHHH HHHHHHHHHH Hiiiiiiiii iiiiiiiiii iiiiiiiiii

    seq  RLGLFLASNI FLTGAFGTIA SGPFWTRGWP EWWGWWLDIP FWS   293
    pred iHHHHHHHHH HHHHHHHHHH HHHHHHooooo oooooooooo oOO
```

FIGURE 5.8 Prediction results obtained with HMMTOP for the L chain of 1PRC.

5.5 Discrimination of transmembrane strand proteins

Outer membrane proteins (OMPs), also known as β-barrel membrane or trans-membrane strand proteins, perform a variety of functions, such as mediating non-specific, passive transport of ions and small molecules, selectively allowing the passage of molecules such as maltose and sucrose (Schirmer et al. 1995; Forst et al. 1998; Schulz, 2000; Wimley, 2003) and are involved in voltage dependent anion channels (Mannella, 1998). These proteins contain β-strands as their membrane-spanning segments and are found in the outer membranes of bacteria, mitochondria, and chloroplasts (Schulz, 2002). A comparative analysis on the distribution of amino acid residues in α-helical and β-barrel membrane proteins shows that the membrane part of OMPs is more complex than that of transmembrane helical proteins due to the intervention of many charged and polar residues in the membrane (Gromiha et al. 1997; Gromiha, 1999). Consequently, the success rate of discriminating β-barrel membrane proteins from other proteins is significantly lower than that of α-helical membrane proteins (Hirokawa et al. 1998; Chen and Rost, 2002).

Recently, several methods have been proposed for discriminating OMPs. These methods are based on hydrophobicity, sequence alignment, neural networks, HMMs, conformational parameters, statistical methods, nearest neighbor algorithms, SVMs, and machine learning techniques. The performance of different methods has been assessed with sensitivity, specificity, accuracy, and correlation coefficient (MCC). These terms are defined as follows:

$$\text{Sensitivity (True positive rate)} = TP/(TP + FN) \tag{5.11}$$

$$\text{Specificity} = TN/(TN + FP) \tag{5.12}$$

$$\text{Accuracy} = (TP + TN)/(TP + TN + FP + FN) \tag{5.13}$$

$$MCC = \frac{TP \times TN - FP \times FN}{\sqrt{(TP + FP)(TP + FN)(TN + FP)(TN + FN)}}, \tag{5.14}$$

$$\text{False positive rate} = 1\text{-specificity} = FP/(TN + FP) \tag{5.15}$$

where TP, FP, TN, and FN refer to the number of true positives (OMPs identified as OMPs), false positives (non-OMPs identified as OMPs), true negatives (non-OMPs identified as non-OMPs), and false negatives (OMPs identified as non-OMPs), respectively. These measures can be used to assess the performance of any classification algorithms.

In addition, "receiver operator characteristics" (ROC) is widely used to assess the performance of a classification method. It is a plot connecting the sensitivity (True positive rate; Eqn. 5.11) and 1-specificity (False positive rate; Eqn. 5.15) at various thresholds (see **Figure 6.17**). A ROC curve can be interpreted numerically using the parameter, AUC (area under the curve). Currently, several discrimination programs include the ROC analysis, and it can also be done on the Web at www.jrocfit.org. Sonego et al. (2008) compiled an excellent review on ROC analysis with the applications to the classification of biological sequences and structures. It includes the principles of ROC along with creating, interpreting, and comparing ROC curves. The applications of ROC and available programs for ROC analysis as well as computing AUC have been explained in detail (Sonego et al. 2008).

5.5.1 Structure-based sequence alignment

Gnanasekaran et al. (2000) developed a structure-based sequence alignment method for discriminating β-stranded membrane proteins. In this method, selected porin structures were superimposed, and the structurally conserved transmembrane β-strand regions (SCRs) were identified. The structure-based sequence alignment was used to generate profiles for each of the SCRs. These profiles were used to search nonredundant predominantly membrane protein database and compared their ability to discriminate β-barrel membrane proteins.

5.5.2   Physicochemical and conformational parameters

Wimley (2002) analyzed the amino acid composition and architecture of β-barrel membrane proteins using known three-dimensional structures. The membrane-interacting surfaces of the β-barrels were identified with an experimentally derived, whole-residue hydrophobicity scale, and then the barrels were aligned normal to the bilayer, and the position of the bilayer midplane was determined for each protein from the hydrophobicity profile. The abundance of each amino acid, relative to the genomic abundance, was calculated for the barrel exterior and interior. The distribution of rise-per-residue values perpendicular to the bilayer plane was found to be 2.7 +/−0.25 A per residue, or about 10 +/−1 residues across the membrane. Also, nearly every known membrane-spanning β-barrel strand was found to have a short loop of seven residues or less connecting it to at least one adjacent strand. Using this information, the β-barrel membrane proteins have been discriminated from other proteins.

Liu et al. (2003) analyzed the amino acid composition in the membrane-spanning regions of 12 β-barrel membrane proteins and applied the information for discrimination, which showed 85% accuracy when tested with 241 OMPs.

5.5.3   Hidden Markov Models

Martelli et al. (2002) developed a method based on HMMs for discriminating outer membrane proteins. They have used a set of 12 OMPs for developing the model and tested it in a set of 145 well-annotated OMPs, which showed an accuracy of 84%. Bagos et al. (2004) trained the data with more number of proteins and achieved the accuracy of about 89% in a set of 133 proteins. A Web server has been developed for discriminating OMPs and is available at http://bioinformatics.biol.uoa.gr/PRED-TMBB.

5.5.4   Statistical methods

Gromiha and Suwa (2005) have systematically analyzed the amino acid composition of globular proteins from different structural classes and outer membrane proteins. It has been reported that the residues, Glu, His, Ile, Cys, Gln, Asn, and Ser, show a significant difference between globular and outer membrane proteins. On the basis of this information, statistical method has been devised for discriminating outer membrane proteins from other globular and membrane proteins. In this method, the amino acid compositions of standard sets of globular and OMPs have been computed. For a new protein, compute the composition and the difference with the amino acid compositions of precomputed globular and OMPs. The protein is predicted to be an OMP if the total deviation is lower with OMP than that with globular protein. Furthermore, the compositions of residue pairs and motifs have been used for discrimination (Gromiha et al. 2005a; Gromiha, 2005). Different Web

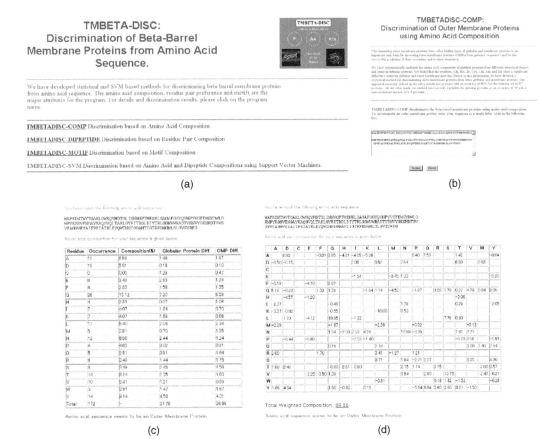

FIGURE 5.9 Discrimination of β-barrel membrane proteins: (a) different methods developed at CBRC for discrimination, (b) input page showing the query amino acid sequence, (c) results obtained with the amino acid composition, and (d) results obtained with the residue pair preference. The query sequence is identified as an outer membrane protein.

servers have been developed for discriminating OMPs using the compositions of amino acid residues, residue-pairs, and motifs. **Figure 5.9a** shows the description about these methods and are available at http://psfs.cbrc.jp/tmbetadisc/. Examples for discriminating OmpA protein using amino acid composition and residue-pair preferences have been illustrated in **Figure 5.9b–c**. **Figure 5.9b** shows the input page, and the output shows the composition along with the type of the protein (**Figure 5.9c**). In this protein, the total deviation is less for OMP than globular protein and hence it is predicted as an OMP. **Figure 5.9d** shows the output obtained with amino acid pair composition. The composition of each of the 400 residue pairs (20 × 20) has been given along with the type of the protein.

5.5.5 Nearest neighbor algorithm

Amino acid composition is one of the parameters, which can be used to identify β-barrel membrane proteins (Gromiha, 2005). Garrow et al. (2005) used the amino acid composition and proposed a modified k-nearest neighbor algorithm,

TMB-HUNT, to classify the proteins into transmembrane β-barrel (TMB) and non-TMB. This method showed an accuracy of 92.5% using weighted amino acids and evolutionary information. The discrimination results can be obtained at www. bioinformatics.leeds.ac.uk/betaBarrel.

5.5.6 Support vector machines

Park et al. (2005) developed a method based on SVMs using amino acid composition and residue pair information. The importance of all the 20 amino acid residues and 400 residue pairs was examined, and 28 parameters were selected (18 amino acid residues and 10 residue pairs) for discrimination. The approach with amino acid composition has correctly predicted the OMPs with a cross-validated accuracy of 94% in a set of 208 proteins. Furthermore, this method has successfully excluded 633 of 673 globular proteins and 191 of 206 α-helical membrane proteins. The method showed an overall accuracy of 92% for correctly picking up the OMPs from a dataset of 1087 proteins belonging to all different types of globular and membrane proteins. Furthermore, residue pair information improved the accuracy from 92% to 94%. This method is available online at http://tmbeta-svm.cbrc.jp/.

5.5.7 Radial basis function networks and PSSM profiles

Ou et al. (2008) proposed a method, TMBETADISC-RBF based on radial basis function (RBF) networks and PSSM profiles for discriminating OMPs. The main difference between RBF network and neural network is that in RBF network the hidden units perform the computations. It has a significant advantage over neural network that the first set of parameters can be determined independently of the second set and still produces accurate classifiers. This method showed an accuracy of 96% in a dataset of 206 OMPs, 667 globular proteins and 378 α-helical membrane proteins. A Web server has been developed for prediction purposes, and it is available at http://rbf. bioinfo.tw/~sachen/OMP.html. This sever can handle multiple sequences at the same time and display the results. **Figure 5.10** shows the input with multiple sequences and the results displayed in the output.

5.5.8 Comparison of machine learning techniques

Gromiha and Suwa (2006a) analyzed the performance of different machine learning methods, such as Bayes rules, logistic functions, neural networks, SVMs, and decision trees, for discriminating OMPs. The amino acid composition and amino acid properties were used as attributes, and the results obtained with amino acid properties are presented in **Table 5.3** (Gromiha and Suwa, 2006b). The results showed that there is no significant difference in the accuracy obtained with different methods. This reveals that tuning of adjustable parameters in machine learning techniques makes the better performance of a method than others.

5.6 Identification of membrane-spanning β-strand segments

Several methods have been proposed for identifying membrane-spanning β-strand segments, and most of the methods are based on physical principles and

TMBETADISC-RBF:

Discrimination of β-barrel Membrane
Proteins Using RBF Networks and PSSM Profiles

Please paste/upload the amino acid sequence in FASTA format.
20D: using amino acid composition as features
400D: using amino acid pair composition as features
420D: using both amino acid and amino acid pair composition as features
PSSM: using PSSM profiles as features

⊙ fasta data ○ From file [   ] [Browse...]

[PSSM   ▼] features are used for predicting OMPs.

```
>1061410|Genbank|Outer membrane integral membrane protein|OpcM
MDNMHNTNGLMRFAKVAAASTLLATLLAACAVGPDYQRPDAVVPAAFKEAPTLAAGEQAG
TWKTAEPSDGRTPRRMVKVFGDPVLDSLETQALAANQNLKAAAARVEEARAATRSARSQW
FPQVGAGFGPTREGLSSASQFQPQGTGPTNATLWRAQGTVSYFADLFGRVGRNVEASRAD
QAQSEALFRSVQLALQADVAQNYFELRQLDSDQDLYHRTVELREQALKLVQRRFNEGDIS
ELDVSRAKNELASAQADAVDVARPRAASEHALAILLGKAPADFAFKETPIVPVAVKIPPG
LPSALLERRPDVSAAERAMAAANARIGLAKSAYFRSSISPGRSAISVDARQFVPVVEPYF
LLGPFAGTALTLPLFDGGRRAAGVQQARAQYDEQGANYRQQVLVAFREVEDNLADLRLLD
DQIRAQEAAVNASRRAATLSRSEYQEGEVAYLDVIDSERSVLQSQLQANQLTGAQAVSTV
NLIRALGGGWGNAPAPTAVGDAASGKADVAAR
>O06873|SwissProt|Inner membrane integral membrane protein|Chemotaxis pomA protein
MDLATLLGLIGGFAFVIMAMVLGGSIGMFVDVTSILIVVGGSIFVVLMKFTMGQFFGATK
IAGKAFMFKADEPEDLIAKIVEMADAARKGGFLALEEMEINNTFMQKGIDLLVDGHDADV
VRAALKKDIALTDERHTQGTGVFRAFCDVAPAMCHIGTLVGLVAMLSNMDDFKAIGPAMA
VALLTTLYGAILSNMVFFPIADKLSLRRDQETLNRRLIMDGVLAIQDGQNPRVIDSYLKN
```

[submit] [reset]

The test file has 2 protein(s).
PSSM features are used for predicting OMPs.

---

1 >1061410|Genbank|Outer membrane integral membrane protein is Outer Membrane Protein
2 >O06873|SwissProt|Inner membrane integral membrane protein is Non-Outer Membrane Protein

---

Figure 5.10 Discrimination of OMPs using TMBETADISC-RBF.

Table 5.3 Discrimination of OMPs and non-OMPs using different machine learning approaches

	5-fold cross-validation		
Method	**Sensitivity**	**Specificity (%)**	**Accuracy**
Bayesnet	78.8	81.5	81.0
Naive Bayes	74.0	88.8	85.9
Logistic function	75.5	97.6	93.4
Neural network	81.3	97.5	94.4
Support vector machines	72.6	98.2	93.3
k-nearest neighbor	84.6	95.8	93.7
Bagging meta learning	76.9	96.4	92.6
Classification via Regression	76.4	96.3	92.5
Decision tree J4.8	75.5	94.9	91.2
NBTree	75.0	94.5	90.8
Partial decision tree	78.4	94.3	91.3

Data were taken from Gromiha and Suwa (2006b).

hydrophobicity profiles. Recently machine learning techniques have been developed for prediction.

5.6.1 Basic principles for the construction of β-barrel membrane proteins

Schultz (2000) analyzed the structures of β-barrel membrane proteins and derived 10 basic rules for understanding their folding and stability.

1. The number of β strands is even, and the N and C termini are at the periplasmic barrel end.
2. The β-strand tilt is always around 45° and corresponds to the common β-sheet twist. Only one of the two possible tilt directions is assumed, the other one is an energetically disfavored mirror image.
3. The shear number of an n-stranded barrel is positive and around $n + 2$, in agreement with the observed tilt.
4. All β strands are antiparallel and connected locally to their next neighbors along the chain, resulting in a maximum neighborhood correlation.
5. The strand connections at the periplasmic barrel end are short turns of a couple of residues named T1, T2, and so on.
6. At the external barrel end, the strand connections are usually long loops named L1, L2, and so on.
7. The β-barrel surface contacting the nonpolar membrane interior consists of aliphatic sidechains forming a nonpolar ribbon with a width of about 22 Å.
8. The aliphatic ribbon is lined by two girdles of aromatic sidechains, which have intermediate polarity and contact the two nonpolar–polar interface layers of the membrane.
9. The sequence variability of all parts of the β barrel during evolution is high when compared with soluble proteins.
10. The external loops show exceptionally high sequence variability, and they are usually mobile.

5.6.2 Turn elimination method

Paul and Rosenbusch (1985) attempted a minimal approach to predict and identify segments causing polypeptides to reverse their direction (turn identification). They tried to predict transmembrane strands by eliminating β-turns and by selecting a minimal length of six residues for a strand.

5.6.3 Hydrophobicity profiles

Vogel and Jahnig (1986) proposed a method based on the amphipathic character of β-strands for predicting the transmembrane β-strands. Furthermore, Jahnig (1990) suggested that a generalization of hydrophobicity analysis was sufficient to predict membrane-spanning amphiphilic α-helices and β-strands. However, neither all the transmembrane helices nor all the transmembrane β-strands need to be amphipathic in character, and there are several transmembrane sequences that exhibit either negligible or zero amphipathicity.

In addition, membrane strands have no long stretch of consecutive hydrophobic residues, and the overall hydrophobicity for β-barrel membrane proteins is similar to that of soluble proteins.

Welte et al. (1991) compared the hydrophilicity profiles and sequences of porin from Rhodobacter capsulatus with those of OmpF and PhoE from Escherichia coli. They determined a set of specific insertions and deletions in the alignments of these proteins, and inferred that OmpF and PhoE have similar structures in their membrane-spanning regions. Cowan et al. (1992) suggested to use the mean hydrophobicity of one side of a putative β-strand by averaging over hydrophobic moments (Eisenberg et al. 1984) of every second residue within a sliding window (Vogel and Jahnig, 1986; Schirmer and Cowan, 1993). To improve the signal to-noise-ratio, they accounted for the band of aromatic residues in flanking positions of the β-strands.

Gromiha and Ponnuswamy (1993) proposed a predictive scheme for forecasting the transmembrane strand segments in OMPs with the use of the general surrounding hydrophobicity scale developed both for the globular and membrane proteins. Two major features of the scheme are that (i) it does not solely depend on the amphipathic character of a sequence segment while identifying it as a transmembrane strand and (ii) it is capable of predicting strands in varied lengths, a facility to reflect the variation in the membrane surfaces.

5.6.4   Rule-based prediction

Gromiha et al. (1997) proposed a rule-based approach for predicting the transmembrane β-strands. In this method, several primary and secondary rules have been formulated. In the primary rules, the priority of each residue in a protein has been assigned on the basis of the preferences of residues in β-strands, surrounding hydrophobicity of amino acid residues and amphipathic character of β-strand segments. The secondary rules are formulated mainly for the search of residues with high priority for identifying the transmembrane β-strands. The method based on these rules could predict the membrane-spanning β-strands in bacterial porins successfully. This method has been used as one of the features in BioSuite, a comprehensive bioinformatics software package (The NMITLI-BioSuite team, 2007).

5.6.5   Hidden Markov Models

Martelli et al. (2002) proposed a HMM-based method, trained with a set of OMPs, and used profile information derived from PSI-BLAST for predicting the membrane-spanning segments. Bagos et al. (2004) developed a HMM method trained with a conditional maximum likelihood criterion for labeled sequences and uses as input single sequences. The prediction is performed with the N-best algorithm or dynamic programming. They have developed a Web server, PRED-TMBB, which is available online at http://bioinformatics.biol.uoa.gr/PRED-TMBB. Bigelow et al. (2004) proposed a method, ProfTMB, using profile-based HMMs for predicting transmembrane β-strand segments and made it available at http://cubic.bioc.columbia.edu/services/proftmb/.

5.6.6   Neural networks

Diederichs et al. (1998) proposed a neural network-based method to predict the topology of the bacterial β-strand proteins and to locate residues along the axes of the pores. The neural network predicts the z-coordinate of C_α atoms in a coordinate frame with the outer membrane in the xy-plane, such that low z-values indicate periplasmic turns, medium z-values indicate transmembrane β-strands, and high z-values indicate extracellular loops. Jacoboni et al. (2001) applied a method combining neural networks and dynamic programming to predict the location of membrane strands. The networks used alignment information as input and predicted whether or not a particular residue is part of a membrane strand. In the second step, the method simply finds the optimal path through the network prediction, much like the methods applied to predict membrane helical proteins (Jones et al. 1994; Rost et al. 1995). Finally, the topology is assigned based on the location of the longest loop that is taken to be exterior. They have developed a prediction server, which can be found in http://www.biocomp.unibo.it/.

Natt et al. (2004) combined both neural networks and SVMs for identifying the transmembrane segments. The neural network module uses evolutionary information, derived from multiple sequence alignments, whereas the SVM predictor uses various physicochemical properties. The user may have the option to choose one of these methods or combined them both. It has been reported that the combination of neural networks and SVMs show better accuracy, and this method, TBBpred, is available at http://www.imtech.res.in/raghava/tbbpred/.

Gromiha et al. (2004) proposed a method based on neural networks using single sequence information alone as input. While training, the network uses the central and sequence neighbors through a running window. The output results are converted into a real value for each residue, which lies between 0 and 1, and this value indicates the probability of the residue to be in transmembrane β-strand. A threshold value has been set up to identify the membrane-spanning segments, and one may change the threshold to obtain strict/loose results. Furthermore, a set of empirical rules has been introduced to refine the segments and eliminating nonplausible predictions for transmembrane β-strands. This method, TMBETA-NET, is available online at http://psfs.cbrc.jp/tmbeta-net/ (Gromiha et al. 2005b). **Figure 5.11** shows the applications of TMBETA-NET. It takes the amino acid sequence in one letter format and displays the predicted membrane-spanning β-strand segments: (i) stretch of amino acid residues in transmembrane strand, (ii) list of segments, and (iii) probability of each residue to be in β-strand.

5.6.7 Combination of methods

It has been reported that the combination of different methods improves the prediction accuracy of protein secondary structures, transmembrane helical segments, and so on. Bagos et al. (2005) combined several methods for predicting transmembrane β-strand segments and analyzed the results. They have shown that the joint method, ConBBPRED, improved the accuracy, and the results are available at http://bioinformatics.biol.uoa.gr/ConBBPRED/.

5.6.8 Radial basis function networks and PSSM profiles for predicting the number of β-strands and membrane-spanning segments

The accurate assignment of the number of β-strands in an OMP aids to predict its topology as well as to improve the accuracy of predicting the residues in membrane-spanning β-strands. Recently, Ou et al. (2009) devised a new strategy for predicting the number of β-strands in OMPs and identifying their membrane-spanning segments. They showed that the information about the total number of residues in

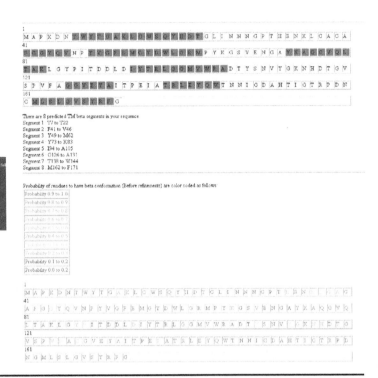

Figure 5.11 Prediction of membrane spanning β-strand segments using TMBETA-NET. Output shows the stretch of amino acid residues in strand, list of segments, and probability of reach residue to be in β-strand.

a protein is sufficient to identify the β-barrel membrane proteins with 22 and 8 strands. For example, the proteins with more than 590 residues are predicted to have 22 β-strands. The number of β-strands is 8 if the number of residues in a protein is less than 200. For proteins with the sequence length of 200 to 500 residues, RBF network and the compositions of specific amino acid residues (Ala, Asp, His, Tyr, and Val) have been used for predicting the number of β-strands (10, 12, 14, 16, and 18). This method showed an accuracy of 97% in assigning the number of β-strands in a set of 28 OMPs, which is significantly better than other prediction methods, ProfTMB (Bigelow and Rost, 2006), PRED-TMBB (Bagos et al. 2005) and TMBpro (Randall et al. 2008; http://www.igb.uci.edu/servers/psss.html).

Furthermore, PSSM profiles obtained with PSI-BLAST and nonredundant (NR) protein database have been used for predicting the membrane-spanning segments in OMPs (**Figure 5.12**). The generated PSSM profile contained the probability of occurrence of each type of amino acid residue at each position. The elements in the PSSM profile have been normalized by $1/(1 + e^{-x})$ and used a window size of 15 residues (seven residues along both sides of the central residue) to encapsulate an amino acid residue. Furthermore, the 15×20 matrix has been transformed into 15×21 vector, which has 315 elements. These elements were used as the input to RBFN classifier. This method showed an accuracy of 87%. A Web server, TMBETAPRED-RBF, has been developed for predicting the membrane-spanning β-strand segments, which is available at http://rbf.bioinfo.tw/~sachen/tmrbf.html.

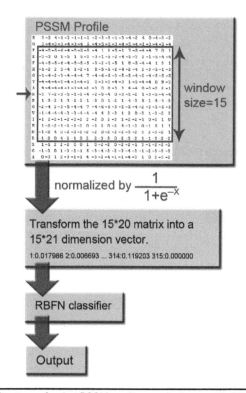

FIGURE 5.12   The architecture of using PSSM profiles as features with RBFN classifier. The PSSM profiles are generated by PSI-BLAST. A windows size of 15 was adopted to generate a 15 × 20 matrix. After normalization and adding one bit for terminal residue, this matrix transforms into a number of 315 dimensions vector and serve as inputs to the RBFN classifier. Figure was adapted from Ou et al. (2009).

5.6.9 Functional discrimination of membrane proteins

The structural discrimination of membrane proteins showed rapid progress and attained high level of accuracy. On the other hand, several methods have been proposed for discriminating membrane proteins based on their functions. Gromiha and Yabuki (2008) utilized different machine learning techniques to discriminate transporters from other α-helical and β-barrel membrane proteins and showed an accuracy of 85%. Furthermore, the method has been extended to classify them into channels/pores, electrochemical and active transporters, which showed an average accuracy of 68%. A Web server has been developed for discrimination and classification, and it is available at http://tmbeta-genome.cbrc.jp/disc-function/.

5.7 Discrimination of disordered proteins and domains

Disordered proteins are highly abundant in regulatory processes such as transcription and cell signaling. Furthermore, the disordered regions in polypeptide chains are important because such regions are essential for protein function. Hence, several methods have been proposed to identify the disordered proteins and regions.

Linding et al. (2003) presented a computational tool, DisEMBL, for the prediction of disordered/unstructured regions within a protein sequence. At the time of development, there was no clear definition for disorder, and hence they have developed parameters based on several alternative definitions based on the concept of "hot loops," i.e., coils with high temperature factors. DisEMBL is available at http://dis.embl.de. Jones and Ward (2003) developed a neural network-based method DISOPRED using PSSM for predicting disordered regions in proteins and provided source code, which can be downloaded from the URL ftp://bioinf.cs.ucl.ac.uk/pub/DISOPRED. Prilusky et al. (2005) described a program, FoldIndex, to predict intrinsically unfolded proteins based on the average residue hydrophobicity and net charge of the sequence. FoldIndex can be accessed at http://bioportal.weizmann.ac.il/fldbin/findex. Sethi et al. (2008) developed a method using sequence and profile compositions as input features and SVMs to identify the disordered proteins. They have developed a Web server, DPROT, and it is available at http://www.imtech.res.in/raghava/dprot/.

Dosztanyi et al. (2005a,b) proposed an algorithm for predicting disordered proteins and regions from amino acid sequences by estimating their total pairwise inter-residue interaction energy. It is based on the assumption that intrinsically unstructured protein sequences do not fold due to their inability to form sufficient stabilizing inter-residue interactions. A Web server was developed for prediction, and it is available at http://iupred.enzim.hu/. **Figure 5.13** shows the options

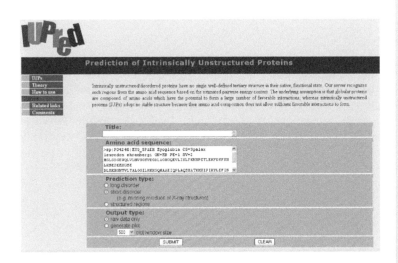

FIGURE 5.13 Prediction of disordered regions using IUPRED.

FoldUnfold: Prediction of disordered regions in protein chain

Scale   [expected number of contacts 8Å ▾]

Averaging frame [11]

Reliable frame [11]   Reliable frame is usually equal to averaging frame

Threshold [20.4]

☐ analyse regions also in those proteins which are predicted as fully disordered (average value for the whole protein is below threshold)
☐ write profile

A protein is predicted as being disordered throughout its length if the average value of the parameter over the protein is less than threshold. If the average over a protein is higher than threshold, the protein is searched for disordered regions.

A region is predicted as being disordered if the average value of the parameter over this region is less than threshold and the region is greater or equal in size t

Target sequence in FASTA format

```
>sp|P04248|MYG_SPAEH Myoglobin OS=Spalax leucodon ehrenbergi GN=MB PE=1 SV=2
MGLSDGEWQLVLNVWGKVEGDLAGHGQEVLIRLFKNHPETLEKFDKFKHLKSEDEMKGSE
DLKKHGNTVLTALGGILKKKGQHAAEIQPLAQSHATKHKIPIKYLEFISEAIIQVLQSKH
PGDFGADAQGAMSKALELFRNDIAAKYKELGFQG
```

[send]

sp|P04248|MYG_SPAEH Myoglobin OS=Spalax leucodon ehrenbergi GN=MB PE=1 SV=2

Parameters

expected number of contacts 8Å

C	M	F	I	L	V	W	Y	A	G	T	S	Q	N	E	D	H	R	K	P	X
CYS	MET	PHE	ILE	LEU	VAL	TRP	TYR	ALA	GLY	THR	SER	GLN	ASN	GLU	ASP	HIS	ARG	LYS	PRO	UNK
23.52	24.82	27.18	25.71	25.36	23.93	28.48	25.93	19.89	17.11	19.81	18.19	19.23	18.49	17.46	17.41	21.72	21.03	17.67	17.43	20.73

Averaging frame 11

Reliable frame 11

Threshold 20.40

Statistics

Composition of the protein

size	CYS	MET	PHE	ILE	LEU	VAL	TRP	TYR	ALA	GLY	THR	SER	GLN	ASN	GLU	ASP	HIS	ARG	LYS	PRO	UNK
	C	M	F	I	L	V	W	Y	A	G	T	S	Q	N	E	D	H	R	K	P	X
154	0	3	7	9	18	6	2	2	13	16	4	7	9	4	13	8	8	1	20	4	0

minimal 18.26

maximal 22.91

average 20.62

Prediction

```
disordered:  50 —— 66
disordered:  76 —— 86
disordered: 120 —— 132
```

[see profile]

FIGURE 5.14   FoldUnfold method for predicting disordered regions.

available (long and short disorder) for predicting the disordered regions and the results obtained for predicting short disordered regions. The residues with the probability of more than 0.5 are treated as disordered ones. Furthermore, they have screened the disordered proteins in genomic sequences (Csizmok et al. 2007), protein interaction networks (Dosztanyi et al. 2006), and predicted binding regions in disordered proteins (Meszaros et al. 2009).

Galzitskaya et al. (2006) proposed a parameter, mean packing density of residues to detect disordered regions in a protein sequence. It has been demonstrated that the regions with weak packing density would be responsible for the appearance of disordered regions. A Web server, FoldUnfold, has been developed to identify the disordered regions, and it is available at http://skuld.protres.ru/~mlobanov/ogu/ogu.cgi. **Figure 5.14** shows an example to predict the disordered regions in a protein sequence. Peng et al. (2006) proposed two new predictor models, VSL2-M1 and VSL2-M2, to address the length-dependency problem in prediction of intrinsic protein disorder. In both models, two specialized predictors were first built and optimized for short (≤ 30 residues) and long disordered regions (>30 residues), respectively. A meta predictor was then trained to integrate the specialized predictors into the final predictor model. The VSL2 predictors are available at http://www.ist.temple.edu/disprot/predictorVSL2.php.

Noguchi's group (Shimizu et al. 2007a; 2007b; Hirose et al. 2007) systematically analyzed the characteristic features of amino acid residues in ordered and disordered proteins and developed different methods for identifying disordered proteins and predicting short and long disordered regions. They have utilized SVMs with physicochemical properties and reduced amino acid set of PSSM as input features for prediction. A series of Web servers, POODLE-W, POODLE-S, and POODLE-L, have been developed for prediction purposes and are collectively available at http://mbs.cbrc.jp/poodle/.

Recently, Schlessinger et al. (2009) combined different approaches based on statistical potentials, hydrophobicity/charge-based methods, and other simple concepts, and developed a META-Disorder (MD) prediction for identifying the disordered proteins and regions. The prediction results can be obtained at http://www.rostlab.org/services/md/index.php.

5.8 Solvent accessibility

Solvent accessibility or accessible surface area (ASA) is one of the important factors for the structure and function of proteins. As the information about ASA has a wide range of biological applications, the prediction of ASA from amino acid sequences has been a long-standing goal in computational biology. Several methods have been put forward to assign the location of residues as buried, partially buried, and exposed based on specific thresholds of ASA; and recently, methods have been developed for predicting the real value of ASA. These methods are based on neural networks (Rost and Sander, 1994b; Fariselli and Casadio, 2001; Cuff and Barton, 2000; Ahmad and Gromiha, 2002; Pollastri et al. 2002b; Ahmad et al. 2003a,b), Bayesian statistics (Thompson and Goldstein, 1996), multiple sequence alignments (Pascarella et al. 1998), logistic function (Mucchielli-Giori et al. 1999), sequence environment (Carugo, 2000; Wang et al. 2004), information theory (Naderi-Manesh et al. 2001), physical-chemical properties (Li and Pan, 2001), SVMs (Yuan et al. 2002; Kim and Park, 2004), and probability profiles (Gianese et al. 2003).

5.8.1 Category prediction: neural network

Rost and Sander (1994b) presented an analysis to project the three-dimensional structure of a protein from its amino acid sequence by means of relative solvent accessibility. They have developed a neural network system for predicting the ASA of each residue based on several states (2, 3, and 10) using evolutionary profiles of amino acid substitutions derived from multiple sequence alignments. The predicted accuracy depends on the number of states, and the correlation between predicted and observed relative ASA is reported to be 0.54 for the two-state prediction with a threshold of 16%. In terms of the three-state description of relative ASA (buried, partially buried, and exposed), the fraction of correctly predicted residue states is about 58%. The best prediction of 86% is obtained for completely buried residues (0% relative accessibility). They have developed a Web server, PHDacc, for assigning the location of residues based on solvent accessibility. It is available at http://cubic.bioc.columbia.edu/predictprotein/. Cuff and Barton (2000) used neural networks along with PSIBLAST and HMMER profiles to improve the accuracy of solvent accessibility prediction. They have tested their method with several datasets and cross-validation procedures. It has been claimed that the method, Jnet (Cuff and Barton, 2000), predicts the ASA of residues more accurately than PHDacc (Rost and Sander, 1994b). Jnet is available at http://www.compbio.dundee.ac.uk/.

5.8.2 Multiple sequence alignment

Pascarella et al. (1998) proposed an easy and uncomplicated method to predict the solvent accessibility state of a site in a multiple protein sequence alignment. This approach is based on amino acid exchange and compositional preference matrices for each of the three accessibility states: buried, exposed, and intermediate. They reported that this technique achieves the same accuracy as much more complex methods. Hence, it has the advantages of computational affordability, facile updating, and easily understood residue substitution patterns, which are useful to biochemists involved in protein engineering, design, and structural prediction. Five years later, the same group (Gianase et al. 2003) proposed a method based on probability profiles calculated on an amino acid sequence centered on the residue whose accessibility has to be predicted. The prediction accuracy is in the range of 71% to 90% for two-state and 58% to 67% for three-state predictions.

5.8.3 Machine learning

Mucchielli-Giorgi et al. (1999) proposed a method, PredAcc, for predicting ASA in four states with machine learning based on an improved logistic function and starting from a single sequence. This method is reported to have an accuracy in the range of 70% to 86% in two-state prediction. PredAcc is available at http://bioserv.rpbs.jussieu.fr/RPBS/cgi-bin/Ressource.cgi?chzn_lg=an&chzn_rsrc=PredAcc. Richardson and Barlow (1999) suggested a simple method for predicting residue solvent accessibility so that it should be used as a baseline for judging other sophisticated approaches. This method does not take into account the local sequence information, and the predictions are solely based on the exposure category in which a residue is often found. The accuracy of this simple method is few percent less than that of other methods, such as Bayesian statistics and neural network.

5.8.4 Knowledge-based prediction

Carugo (2000) proposed a prediction method, which is based on the comparison of the observed and the average values of the solvent-accessible area. This algorithm showed that the prediction accuracy is significantly improved by considering the residue types preceding and/or following the residue whose accessibility must be predicted. In contrast, the separate treatment of different secondary structural types does not improve the quality of the prediction. Furthermore, this method better predicts the residue accessibility of small proteins than that in larger ones.

5.8.5 Information theory

Naderi-Manesh et al. (2001) introduced a simple method based on information theory to predict the solvent accessibility of residues in various states defined by their different thresholds. Prediction is achieved by the application of information obtained from a single amino acid position or pair-information for a window of 17 amino acids around the desired residue. It has been shown that the results obtained by pair-wise information values are better than the results from single amino acids. This reinforces the effect of the local environment on the accessibility of residues.

5.8.6 Support vector machine

Yuan et al. (2002) trained an SVM learning system to predict protein solvent accessibility from the primary structure. They have examined different kernel functions and sliding window sizes to explore the effect of prediction performance. Using a cutoff threshold of 15% that splits the dataset evenly (an equal number of exposed and buried residues), this method was able to achieve a prediction accuracy of 70.1% for the single sequence input and 73.9% for the multiple alignment sequence input, respectively. Kim and Park (2004) combined the algorithm of SVMs with the PSSM generated from PSI-BLAST to achieve better prediction accuracy of the relative solvent accessibility. They have introduced a three-dimensional local descriptor that contains information about the expected remote contacts by both the long-range interaction matrix and neighbor sequences. The feature weights to kernels in SVMs have also been applied in order to consider the degree of significance that depends on the distance from the specific amino acid. This method predicted the relative solvent accessibility in the range of 78% to 88% based on a two-state model.

5.8.7 NETASA

Ahmad and Gromiha (2002) implemented a server, NETASA, for predicting the solvent accessibility of residues using an optimized neural network algorithm. In this method, a feed forward neural network, consisting of input, output, and hidden layer has been designed for ASA training. The input layer consists of 17 units of 21 bit binary vectors, when the prediction is made for the central (9th) residue. Each

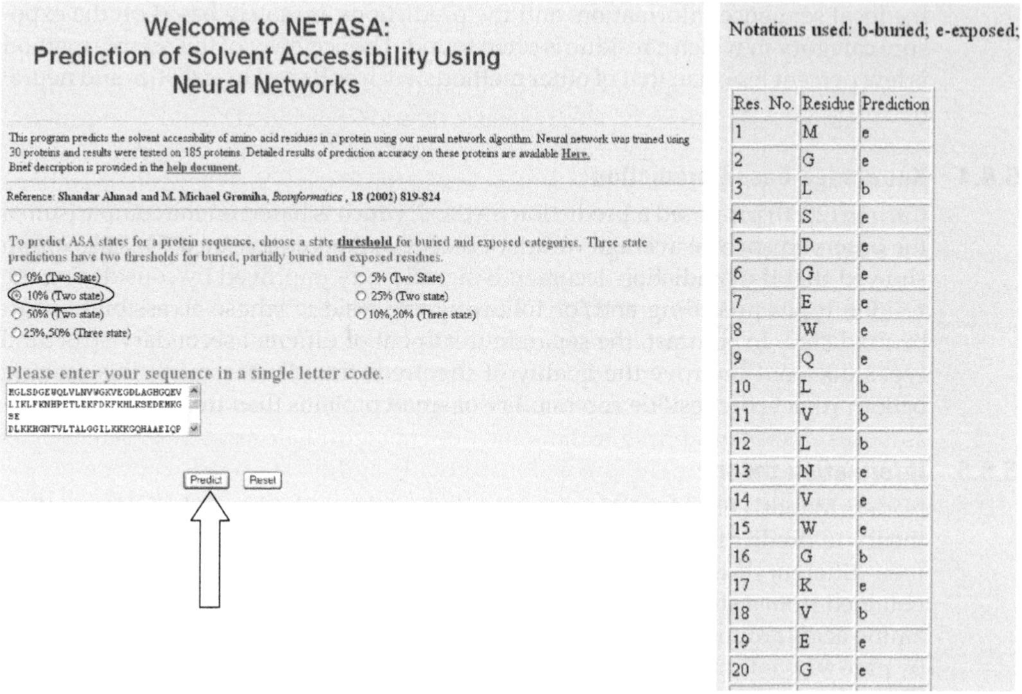

Res. No.	Residue	Prediction
1	M	e
2	G	e
3	L	b
4	S	e
5	D	e
6	G	e
7	E	e
8	W	e
9	Q	e
10	L	b
11	V	b
12	L	b
13	N	e
14	V	e
15	W	e
16	G	b
17	K	e
18	V	b
19	E	e
20	G	e

FIGURE 5.15   Prediction of ASA states using NETASA.

of these 21 bit vectors represents the residue at the location being encoded. All 17 units represent 8 neighbors on both sides and the residue for which prediction has to be made.

Hidden and output layers consist of the same number of units each as the number of accessibility states, n, desired in output (e.g., two bits each in the case of the two-state classification). Solvent accessibility is encoded by an n bit binary vector in which all the coding bits are zero except one which corresponds to one of the accessibility states.

The performance of NETASA was tested with several thresholds in two- and three-state classification systems. It predicts the ASA up to an accuracy of 90% for a 0% threshold of two-state predictions. Furthermore, the average accuracy lies between 70% and 90% for the two-state predictions at different thresholds. In three-state predictions, the average accuracy lies in the range of 55% to 65%. NETASA is available at http://www.netasa.org/netasa/. **Figure 5.15** shows the two-sate prediction with the threshold of 10% using NETASA. Furthermore, Adamczak et al. (2004) reported that NETASA could predict the ASA with similar accuracy for different sets of proteins.

5.8.8   Real Value ASA Prediction

In the earlier section, the methods developed for predicting ASA of residues in several states (e.g., buried, partially buried, and exposed) based on specific thresholds of arbitrary choice have been described. However, it is not uncommon for mutually contradictory predictions to be made when the same method and model are applied to more than one threshold; for example, while a residue may be predicted to be exposed based on a 5% threshold, the same residue may be predicted to be buried based on a 25% threshold. Furthermore, the very act of a state classification of accessible areas means that a subsequently developed model relies upon less information than is actually available from the structural data. Hence, Ahmad et al. (2003a,b) developed a novel method for predicting the real value of ASA instead of "state" prediction. Although the values of ASA obtained from different computational methods are not very accurate, the prediction of the real value provides a lot of additional information instead of specifying them to be in an arbitrary state. Indeed, predicting the real value ASA with 25% error is more informative than 100% correct binary classification.

5.8.9 Neural network

Ahmad et al. (2003a) developed a neural network, which consists of one input layer with 147 bits (21 bits for each residue position, accommodating 3 neighbors on either side), one hidden layer and one output layer, each with two units. The inclusion of neighboring residues provides additional information as residue–residue contacts are important for the structure and stability of globular proteins (Gromiha and Selvaraj, 2004). In the final layer, the difference between these activation signals received by the two units is normalized by a sigmoidal function, which in turn is transformed into a percentage scale, providing the final prediction of ASA. This method predicted the real value ASA of 215 high-resolution structures of nonhomologous proteins with a mean absolute error (MAE; Eqn. 6.18) of 17.6%, 18.0%, and 18.0%, respectively, for the training, test, and validation sets. The correlations between experimental and predicted values are 0.52, 0.50, and 0.50, respectively.

Furthermore, a Web server, RVP-net, has been developed for predicting the real value ASA from amino acid sequence. In the output, the numerical values of ASA were provided along with the result in graphics (Ahmad et al. 2003b). The Web server is available at http://www.netasa.org/rvp-net/. **Figure 5.16a** shows an example to input the sequence, and the output results are displayed in **Figure 5.16b**.

Followed by RVP-net on the real-value prediction of ASA, several researchers stressed the importance of it and moved their focus on predicting the real value ASA. Garg et al. (2005) developed a neural network-based method for predicting the real value of solvent accessibility from the sequence using evolutionary information in the form of multiple sequence alignment. In this method, two feedforward networks with a single hidden layer have been trained with standard back-propagation as a learning algorithm. The Pearson's correlation coefficient increases from 0.53 to 0.63, and mean absolute error decreases from 18.2% to 16% when multiple-sequence alignment obtained from PSI-BLAST is used as input instead of a single sequence. The performance of the method further improves from a correlation coefficient of 0.63 to 0.67 when secondary structure information predicted by PSIPRED is incorporated in the prediction. The method consists of two

Welcome to RVP-NET:
Real Value Prediction of Solvent
Accessibility

This program predicts the Real Values of Solvent Accessibility in a protein using our neural network algorithm. In this method, single residue information of neighbours is used for making prediction of Solvent accessibility. Algorithmic details of real value predictions may be found Here. Brief description of this server may be found Here.
Reference

Shandar Ahmad, M. Michael Gromiha and Akinori Sarai *Bioinformatics 2003 19: 1849-1851*

Note: RVP-Net version 2.0 was released on June 26, 2003, which provides graphical outputs to RVP-Net predictions.
To keep this service faster for text-only output, graphical outputs for this version are being stopped w.e.f. Sept 24, 2004.
Please use Version 2.0 to get graphical predictions.

April 27, 2005
We apologize for some errant behavior and erroneous results from this server during last week or so. We have now corrected the problem

To predict ASA for a protein, Enter your sequence below in the text box or upload a file by clicking the "Browse" button. Plain text sequence or FASTA, SwissProt and PIR formats are accepted for the input. Lower and upper cases are accepted for residue notations in single latter code. Letters not representing any residue (e.g. X and Z) and other symbols will be treated as unknown residues.

File Upload: [   ] [Browse]
Or paste your sequence here:
MNIFEMLRIDRGLRLKIYKDTEGYYTIGIG
NLLTKSPSLNAAKHELDKAIGRNCNGVITK
DEAEKLFNQDVDAAVRGILR
NAKLKPVYDSLDAVRRCALINHVFQMGETG

[Predict] [Reset]

R. No.	Res	ASA (%)	ASA (A^2)	Category
1	M	8.6	17.2	B
2	N	47.9	70.1	E
3	I	2.7	5.1	B
4	F	3.7	7.5	B
5	E	33.2	58.1	E
6	M	1.8	3.6	B
7	L	3.2	5.9	B
8	R	32.8	75.0	E
9	I	5.1	9.4	B
10	D	37.2	53.7	E
11	E	58.7	102.6	E
12	G	16.6	13.1	E
13	L	5.7	10.3	B
14	R	44.8	102.6	E
15	L	4.8	8.8	B
16	K	37.2	76.6	E
17	I	6.5	12.0	B
18	Y	11.0	23.4	B
19	K	62.7	129.1	E
20	D	32.8	47.3	E
21	T	27.4	38.1	E
22	E	56.2	98.1	E
23	G	23.9	18.8	E
24	Y	7.4	15.8	B
25	Y	11.6	24.8	B
26	T	12.3	17.1	B
27	I	1.9	3.5	B

Predicted ASA values (Percentage) are given below.

(a)   (b)

FIGURE 5.16   Prediction of the real value solvent accessibility: (a) input page and (b) results obtained with RVP-NET. It shows the residue number and name, ASA in % and $Å^2$ along with the category based on a specific threshold.

steps: (1) In the first step, a sequence-to-structure network is trained with the multiple alignment profiles in the form of PSI-BLAST-generated PSSMs, and (2) in the second step, the output obtained from the first network and PSIPRED-predicted secondary structure information is used as an input to the second structure-to-structure network. On the basis of this, a server, SARpred, has been developed and is available at http://www.imtech.res.in/raghava/sarpred/.

5.8.10 Nonlinear regression and neural network

Adamczak et al. (2004) developed a continuous approximation of the real value ASA using nonlinear regression, with several feed forward and recurrent neural networks, which are then combined into a consensus predictor. Furthermore, the effects of evolutionary profiles have been included with respect to the growth of sequence databases. This method performs well for predicting the real value ASA with an absolute error of 16%. The improvement in accuracy over RVP-net method is due to the additional information obtained from sequence alignments. It is available at http://sable.cchmc.org.

5.8.11 Support vector machines for real-value prediction

Yuan and Huang (2004) developed a support vector regression (SVR) approach to predict the real value ASAs of proteins from their primary structures. They have reported the real values of ASA in squared angstroms instead of the relative solvent accessibility. This method could predict the real value ASA at an average error of 26.0 Å^2. The correlation coefficient between the predicted and observed ASAs is 0.66. It has been mentioned that Cys is the best-predicted residue with a mean absolute error of 13.8 Å^2, while Arg is the least predicted amino acid with a mean absolute error of 42.7 Å^2. Recently, Nguyen and Rajapakse (2006) proposed a two-stage SVM approach along with PSI-BLAST profiles to predict the real value of solvent accessibility and reported an improvement of 3% over other methods. A Web server has been set up for predicting real value ASA, and it is available at http://birc.ntu.edu.sg/~pas0186457/asa.html.

5.8.12 Look-up tables

Wang et al. (2004) followed the method of Richardson and Barlow (1999) and Carugo (2000) in predicting the real value ASA. These authors calculated the ASA of each residue or for patterns of two (or more) residues and fixed it as buried or exposed based on specific threshold. The ASA state of an unknown residue can be assigned from the library of a known state. The libraries of ASA have been developed for each di-, tri- and pentapeptides for a larger set of proteins. These dictionaries serve as a look-up table for making subsequent predictions of solvent accessibility of amino-acid residues (Wang et al. 2004). The predictions made in such a way are very close to more sophisticated methods of solvent accessibility prediction. Furthermore, one can use this method as a baseline for testing the predictive ability of any new method for real value ASA.

5.9   Inter-residue contact prediction

Gobel et al. (1994) proposed a simple and general method to analyze correlations in mutational behavior between different positions in a multiple sequence alignment. They have then used these correlations to predict contact maps of protein structures. Furthermore, a neural network based prediction has been proposed for predicting the inter-residue contacts in proteins (Fariselli et al. 2001). In this procedure, two residues in a protein are contact with each other if the distance between the C$_\beta$ atoms is within 8 Å, which are separated by more than seven residues. The program CORNET has been developed, including the evolutionary information in the form of sequence profile, sequence conservation, correlated mutations, and predicted secondary structures. The predictor was trained and cross-validated on a dataset of 173 nonhomologous proteins. It has been reported that the performance of this method is poor for all-α proteins, and the accuracy is high for all-β and mixed proteins. CORNET is available at http://gpcr.biocomp.unibo.it/cgi/predictors/cornet/pred_cmapcgi.cgi.

Pollastri and Baldi (2002) developed a set of flexible machine learning architectures for the prediction of contact maps. The architectures can be viewed as recurrent neural network implementations of a class of Bayesian networks. This method could correctly predict 60% of contacts at a distance cutoff of 8 Å and 45% of distant contacts at 10 Å, for proteins of length up to 300. The Web server CMAPpro is available at http://www.ics.uci.edu/~baldig/.

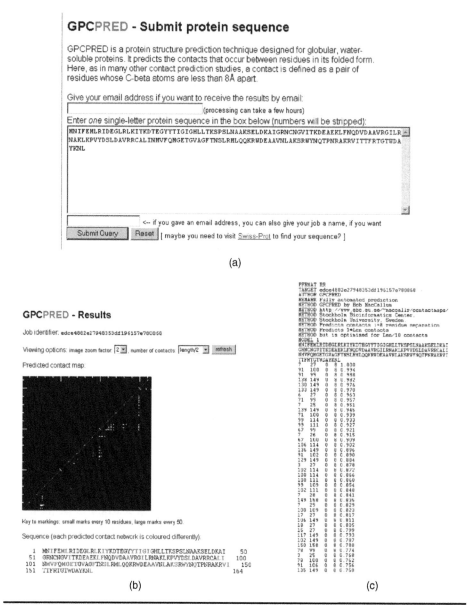

FIGURE 5.17 Prediction of residue contacts in 2LZM using the GPCPRED server: (a) input pages showing the amino acid sequence, (b) the contact map, and (c) residues that are in contact within the distance of 8 Å and the probability of being the residues in contact with each other.

Shao and Bystroff (2003) presented a method, HMMSTR-CM, for protein contact map predictions. Contact potentials were calculated by using HMMSTR, an HMM for local sequence structure correlations. Targets were aligned against protein templates using a Bayesian method, and contact maps were generated by using these alignments. Contact potentials then were used to evaluate these templates.

An ab initio method based on the target contact potentials using a rule-based strategy to model the protein-folding pathway was developed. Fold recognition and ab initio methods were combined to produce accurate, protein-like contact maps.

MacCallum (2004) developed a method based on genetic programming to identify the contacting residue pairs from local sequence patterns extracted with the help of self-organizing maps. This method could predict with the accuracy of 27% on a validation set of 156 domains up to 400 residues in length, where contacts are separated by at least 8 residues. A Web server (GPCPRED) has been developed for predicting the contacts and is available at http://www.sbc.su.se/~maccallr/contactmaps. An example is shown in **Figure 5.17**. It takes the amino acid sequence as input (**Figure 5.17a**) and displays the contacts in the output. **Figure 5.17b** shows the contact map, and **Figure 5.17c** shows the residues that are in contact within the distance of 8 Å and the probability of being the residues in contact with each other.

Hamilton et al. (2004) described a method using neural networks to predict residue contact pairs in a protein. The main inputs to the neural network are a set of 25 measures of correlated mutation between all pairs of residues in two "windows" of size 5 centered on the residues of interest. This method could predict the residue contacts with an accuracy of 21.7% in a test set of 1033 protein structures. The server, PoCM, is available at http://foo.acmc.uq.edu.au/~nick/Protein/contact_casp6.html.

Punta and Rost (2005) developed a contact prediction method, PROFcon, that combines information from alignments, predictions of secondary structure and solvent accessibility, region between two residues, and the average properties of the entire protein. This method performs well in α/β class of proteins, and the accuracy is irrespective of the protein size. A Web server has been developed to predict the contacts, which takes the input sequence and give the preferred contacts between residues at the distance of 8 Å. It is available at http://www.predictprotein.org/submit_profcon.html.

Furthermore, Grana et al. (2005) evaluated the performances of different methods that predicted the long-range contacts for the target proteins given in CASP6. They reported that the new methods improved the accuracy compared with previous predictions and suggested that further developments are necessary for accurate prediction.

5.10   Protein tertiary structure prediction

Predicting protein three-dimensional structure from its amino acid sequence, termed as protein folding problem, is one of the long-standing goals in bioinformatics and computational biology. Several methods have been proposed to address this problem, and each method has its own merits and demerits. A committee has been set up to evaluate the performance of several structure prediction algorithms using blind tests on several proteins. This Critical Assessment of Techniques for Protein Structure Prediction (CASP) provides a forum for the testing and comparison of structure prediction methods. It operates by soliciting from interested groups to evaluate structure predictions of target molecules whose structure was known but had not been made public at the time of assessment. The prediction methods have been categorized into comparative modeling, fold recognition, and new folds.

5.10.1 Comparative modeling

When the sequence of the target structure is clearly related to that of one or more structures, the structures will also be similar. Thus, an approximate model can be created simply by copying related regions of polypeptide from the parent structures and changing the side-chains where necessary. A total of 37 target domains were considered by the assessors to be in the comparative modeling category in CASP5. These domains were divided into two finer categories: 22, which could be related to known structures using a simple BLAST search (high-sequence identity) and 15, where a relationship to a known structure could be identified by using moderately sophisticated PSI-BLAST searches (low-sequence identity). Models for the high-sequence identity set were analyzed in more detail than the rest, considering the accuracy of side-chains, the construction of regions not present in available template structures, and whether the overall backbone accuracy is higher than that obtained by simply copying the best template. The boundary between the comparative modeling category and the fold recognition has become blurred, because continually improving sequence comparison techniques have made it possible to reliably identify folds where the sequence relationship would once have been regarded as being in the twilight zone.

Sali and Blundell (1993) developed MODELLER, a comparative protein modeling method designed to find the most probable structure for a sequence given its alignment with related structures. The three-dimensional model is obtained by optimally satisfying spatial restraints derived from the alignment and expressed as probability density functions (pdfs) for the features restrained. For example, the probabilities for main-chain conformations of a modeled residue may be restrained by its residue type, main-chain conformation of an equivalent residue in a related protein, and the local similarity between the two sequences. Several such pdfs are obtained from the correlations between structural features in 17 families of homologous proteins, which have been aligned on the basis of their 3D structures. The pdfs restrain $C_\alpha - C_\alpha$ distances, main-chain N—O distances, and main-chain and side-chain dihedral angles. A smoothing procedure is used in the derivation of these relationships to minimize the problem of a sparse database. The 3D model of a protein is obtained by optimization of the molecular pdf such that the model violates the input restraints as little as possible. The molecular pdf is derived as a combination of pdfs restraining individual spatial features of the whole molecule. The optimization procedure is a variable target function method that applies the conjugate gradients algorithm to positions of all nonhydrogen atoms. The method is automated and it is available at http://salilab.org/modeller/modeller.html.

Kosinski et al. (2003) applied a multistep protocol to predict the structures of all targets during CASP5: (1) Diverse fold-recognition (FR) methods to generate initial target-template alignments, which were converted into preliminary full-atom models by comparative modeling. All preliminary models were evaluated (scored) by VERIFY3D to identify well and poorly folded fragments.

(2) Preliminary models with similar 3D folds were superimposed, poorly scoring regions were deleted, and the "average model" structure was created by merging the remaining segments. All template structures reported by FR were superimposed, and a composite multiple-structure template was created from the most conserved fragments. (3) The average model was superimposed onto the composite template, and the structure-based target-template alignment was inferred. This alignment was used to build a new (intermediate) comparative model of the target, again scored with VERIFY3D. (4) For all poorly scoring regions, series of alternative alignments were generated by progressively shifting the "unfit" sequence fragment in either direction. Here, the additional information such as secondary structure, placement of insertions and deletions in loops, conservation of putative catalytic residues, and the necessity to obtain a compact, well-folded structure has been considered. For all alternative alignments, new models were built and evaluated. (5) All models were superimposed, and the "FRankenstein's monster" (FR, fold recognition) model was built from best-scoring segments. The final model was obtained after limited energy minimization to remove steric clashes between sidechains from different fragments. The novelty of this approach is in the focus on "vertical" recombination of structure fragments, typical for the ab initio field, rather than "horizontal" sequence alignment typical for comparative modeling. The automatic version of the multistep protocol is being developed as a meta-server; the prototype is freely available at http://genesilico.pl/meta/.

Sasson and Fischer (2003) generated full atom models for all CASP5 targets by using the fully automated 3D-SHOTGUN fold recognition meta-predictors (Fischer, 2003). The 3D-SHOTGUN meta-predictors assemble hybrid 3D models by combining structural information of a number of independently generated, fold recognition models. They applied a very simple approach using the multiple parent option of the Modeller program (Sali and Blundell, 1993). The input to Modeller was different combinations of the unrefined 3D-SHOTGUN models and the sequence-template alignments used by 3D-SHOTGUN's assembly step.

5.10.2 Fold recognition

Increasingly, new structures deposited in the PDB turn out to have folds that have been seen before, even though there is no obvious sequence relationship between the related structures. Thus, methods of identifying folds from sequence information continue to grow in importance. There are two main questions to be asked: How successful are the different methods at identifying fold relationships, and when successful, what is the quality of the models produced? Techniques for fold recognition include advanced sequence comparison methods, comparison of predicted secondary structure strings with those of known folds, tests of the compatibility of sequences with three-dimensional folds (threading), and the use of human expert knowledge. Evaluation of the quality of the models produced has common components with comparative modeling, specifically alignment accuracy, and with new fold methods, specifically recognizing correct architecture, even in cases where the topology is incorrect.

Jones (1999b) described a fold recognition method, GenTHREADER, which uses a traditional sequence alignment algorithm to generate alignments, which are then evaluated by a method derived from threading techniques. Threading is an approach to fold recognition that used a detailed 3D representation of protein structure. It is to physically "thread" a sequence of amino acid side chains onto a backbone structure (a fold) and to evaluate this proposed 3D structure using a set of pair potentials and (importantly) a separate solvation potential. As a final step, each threaded model is evaluated by a neural network in order to produce a single measure of confidence in the proposed prediction. The speed of the method, along with its sensitivity and very low false-positive rate, makes it ideal for automatically predicting the structure of all the proteins in a translated bacterial genome (proteome). The program is freely available on the Web at http://bioinf.cs.ucl.ac.uk/psipred/psiform.html.

Ginalski and Rychlewski (2003) modeled proteins using a combination of consensus alignment strategy and 3D assessment. A large number and a broad variety of prediction targets, with sequence identity between each modeled domain and the related known structure, ranging from 6% to 49%, represented all difficulty levels in comparative modeling and fold recognition. The critical steps in modeling, selection of template(s), and generation of sequence-to-structure alignment were based on the results of secondary structure prediction and tertiary fold recognition carried out using the Meta Server coupled with the 3D-Jury system. The main idea behind the modeling procedure was to select the most common alignment variants provided by individual servers, as well as to generate several alternatives for questionable regions and to evaluate them in 3D by building corresponding molecular models. Analysis of fold-specific features and sequence conservation patterns for the target family was also widely used at this stage. For both CM and FR targets, remote homologs of the known structure were clearly recognized by the 3D-Jury system.

Petrey et al. (2003) proposed a method for structure prediction, and the central feature of their structure prediction strategy involved the ability to generate good sequence-to-structure alignments and to quickly transform them into models that could be evaluated both with energy-based methods and manually. They have used the following programs: (a) HMAP (Hybrid Multidimensional Alignment Profile)—a profile-to-profile alignment method that is derived from sequence-enhanced multiple structure alignments in core regions and sequence motifs in nonstructurally conserved regions. (b) NEST—a fast model building program that applies an "artificial evolution" algorithm to construct a model from a given template and alignment. (c) GRASP2—a new structure and alignment visualization program incorporating multiple structure superposition and domain database scanning modules. These methods were combined with model evaluation based on all atom and simplified physical-chemical energy functions.

Tomii et al. (2005) developed a method, FORTE, based on profile–profile comparison to make use of known structural information for protein structure prediction. The following prediction strategy has been adopted for the targets in CASP6: (i) Derived target-template alignments from several versions of profile–profile comparisons, (ii) Constructed and exhaustively evaluated 3D models based on those alignments and (iii) Selected proper model(s). All generated models were evaluated based on a structural quality score calculated by both Verify3D and Prosa2003 programs. FORTE takes the amino acid sequence as input and provides the details of the structure as output through e-mail. FORTE is available at http://www.cbrc.jp/htbin/forte-cgi/forte_form.pl.

5.10.3 New fold methods

In new fold methods, a wide range of knowledge-based techniques are used: well-established secondary structure prediction tools, sequence-based identification of sets of possible conformations for short fragments of chain, methods that assemble three-dimensional folds from candidate fragments, predicted secondary structure, prediction of contacting residues in the structure, "mini-threading" methods that identify supersecondary structure motifs and full-domain fold recognition methods that may establish an approximate or partial topology. These approaches are sometimes combined with numerical search methods such as molecular dynamics, Monte Carlo, and genetic algorithms. There are few pure ab initio methods, usually based on some form of numerical simulation techniques together with more traditional "empirical potentials."

Bradley et al. (2003) described predictions of the structures using Rosetta. The Rosetta fragment insertion protocol was used to generate models for entire target domains without a detectable sequence similarity to a protein of the known structure and to build long loop insertions (and N-and C-terminal extensions) in cases where a structural template was available. In particular, de novo predictions failed for large proteins that were incorrectly parsed into domains and for topologically complex (high contact order) proteins with swapping of segments between domains. However, for the remaining targets, at least one of the five submitted models had a long fragment with significant similarity to the native structure. Recently, ROSETTA has been applied to model the structures of potassium channels (Yarov-Yarovov et al. 2006). Furthermore, a fully automated prediction server, Robetta, has been developed that uses the Rosetta fragment-insertion method (Chivian et al. 2003). It combines template-based and de novo structure prediction methods in an attempt to produce high-quality models that cover every residue of a submitted sequence. The first step in the procedure is the automatic detection of the locations of domains and the selection of the appropriate modeling protocol for each domain. For domains matched to a homolog with an experimentally characterized structure by PSI-BLAST or Pcons2, Robetta uses a new alignment method, called K*Sync, to align the query sequence onto the parent structure. It then models the variable regions by allowing them to explore conformational space with fragments in fashion similar to the de novo protocol, but in the context of the template. When no structural homolog is available, domains are modeled with the Rosetta de novo protocol, which allows the full length of the domain to explore conformational space via fragment-insertion, producing a large decoy ensemble from which the final models are selected. ROBETTA is available at http://robetta.bakerlab.org/.

Skolnick et al. (2003) have applied the TOUCHSTONE structure prediction algorithm that spans the range from homology modeling to ab initio folding to all protein targets. Using the threading algorithm PROSPECTOR that does not utilize input from metaservers, one threads against a representative set of PDB templates. If a template is significantly hit, generalized comparative modeling designed to span the range from closely to distantly related proteins from the template is done.

This involves freezing the aligned regions and relaxing the remaining structure to accommodate insertions or deletions with respect to the template. For all targets, consensus predicted side chain contacts from at least weakly threading templates are pooled and incorporated into ab initio folding. Often, TOUCHSTONE performs well in the CM to FR categories, with PROSPECTOR showing significant ability to identify analogous templates. When ab initio folding is done, frequently the best models are closer to the native state than the initial template.

Jones and McGuffin (2003) developed a method, FRAGFOLD, for protein structure prediction, which is based on the assembly of supersecondary structural fragments taken from highly resolved protein structures using a simulated annealing algorithm. In this method, the heart of the objective function is a set of pairwise potentials of mean force, determined by a statistical analysis of highly resolved protein X-ray crystal structures, and the application of the inverse Boltzmann equation to convert observed frequencies of residue pair interactions to free energy changes. In addition to the pair potentials, a solvation potential is also employed. These potentials are identical to those currently in use by the latest versions of their threading programs.

Fang and Shortle (2003) developed a prediction strategy on the assumption that local side-chain/backbone interactions are the principal determinants of protein structure at low resolution. The implementation of this assumption made an extensive use of a scoring function based on the propensities of the 20 amino acids for 137 different subregions of the Ramachandran plot, allowing estimation of the quality of fit between a sequence segment and a known conformation. New folds were predicted in three steps: prediction of secondary structure, threading to isolate fragments of protein structures corresponding to one turn plus flanking helices/strands, and recombination of overlapping fragments. The most important step in this fragment ensemble approach, the isolation of turn fragments, employed two to six sequence homologues when available, with clustering of the best scoring fragments to recover the most common turn arrangement. Recombinants formed between three to eight turn fragments, with crossovers confined to helix/strand segments, were selected for compactness plus low energy as estimated by empirical amino acid pair potentials, and the most common overall topology identified by visual inspection. Because significant amounts of steric overlap were permitted during the recombination step, the final model was manually adjusted to reduce the overlap and to enhance protein-like structural features.

5.10.4 Ab initio protein structure prediction

Skolnick and colleagues (Skolnick and Kihara, 2001; Skolnick et al. 2001) have developed a program, PROSPECTOR (PROtein Structure Predictor Employing Combined Threading to Optimize Results), for identifying templates of protein structures, side-group contact prediction and for the prediction of short-range and intermediate-range distances between the side groups along the polypeptide chain. It is a threading approach that uses sequence profiles to generate an initial probe-template alignment and then uses this "partly thawed" alignment in the evaluation of pair interactions. Two types of sequence profiles are used: the close set, composed of sequences in which sequence identity lies between 35% and 90%, and the distant set, composed of sequences with a FASTA e-score less than 10. Thus, a total of four scoring functions are used in a hierarchical method: The close (distant) sequence profiles screen a structural database to provide an initial alignment of the probe sequence in each of the templates. The same database is then screened with a scoring function composed of sequence plus secondary structure plus pair interaction profiles. This combined hierarchical threading method is called PROSPECTOR1. Next, the set of the top 20 scoring sequences (four scoring functions times the top five structures) is used to construct a protein-specific pair potential based on consensus side-chain contacts occurring in 25% of the structures. In subsequent threading iterations, this protein-specific pair potential, when combined in a composite manner, is found to be more sensitive in identifying the correct pairs than when the original statistical potential is used, and it increases the number of recognized structures for the combined scoring functions, termed PROSPECTOR2. Overall, these studies show that the use of pair interactions as assessed by the improved Z-score enhances the specificity of probe-template matches. Thus, when the hierarchy of scoring functions is combined, the ability to identify correct probe-template pairs is significantly enhanced. Finally, a Web server has been established, which is available at http://128.205.242.1/current_buffalo/skolnick/prospector.html.

Jayaram et al. (2006) described an energy-based computer software for narrowing down the search space of tertiary structures of small globular proteins. The protocol comprises eight different computational modules that combine physics-based potentials with biophysical filters starting from sequence and secondary structure information. The protocol has been Web enabled and is accessible at http://www.scfbio-iitd.res.in/bhageerath.

5.10.5 Modeling on the Web

Most of the three-dimensional structure prediction algorithms require much computational time, and hence the developers provided the executables to download and run locally. However, few methods can be directly used online. SWISS-MODEL is an automated comparative protein modeling server, and one can get the structure for a given amino acid sequence directly (Schwede et al. 2003). SWISS-MODEL provides several levels of user interaction through its WWW interface: In the "first approach mode," only an amino acid sequence of a protein is submitted to build a 3D model. Template selection, alignment, and model building are done completely automated by the server. In the "alignment mode," the modeling process is based on a user-defined target-template alignment. Complex modeling tasks can be handled with the "project mode" using DeepView (Swiss-PdbViewer), an integrated sequence-to-structure workbench. SWISS-MODEL server is available at http://swissmodel.expasy.org.

Figure 5.18 shows an example to obtain the 3D structure for a given amino acid sequence using SWISS-MODEL. It takes the amino acid sequence as input (**Figure 5.18a**) and displays the information on model details, alignment, tertiary structure, template selection, etc. **Figure 5.18b** shows the alignment with the automatically selected template PDB (1myhA). The predicted three-dimensional structure can be viewed through Swiss-viewer (**Figure 5.18c**). The detailed information on 3D structure and atomic coordinates in PDB format is shown in **Figure 5.18d**.

SWISS MODEL WORKSPACE

[Workspace] [Repository] [Modelling] [Tools]

SwissModel Automatic Modelling Mode

Email:

Project Title:

Provide a protein sequence or a UniProt AC Code:

MGLSDGEWQLVLNVWGKVEGDLAGHGQEVLIKLFKNHPETLEKFDKFKHLKSEDEMKGSE
DLKKHGNTVLTALGGILKKKGQHAAEIQPLAQSHATKHKIPIKYLEFISEAIIQVLQSKH
PGDFGADAQGAMSKALELFRNDIAAKYKELGFQG

Submit Modelling Request

Options:

Use a specific template:

(a)

Workunit: P000331 Title:

1   154

Go to: [Template Selection] [Alignment] [Modelling Log] [Evaluation]

Model Details: ⦿ Segment 1

Model info:
modelled residue range: 2 to 154
based on template 1myhA (1.90 Å)
Sequence Identity [%]: 92.157
Evalue: 9.74e-71

display model: as pdb - as DeepView project
download model: as pdb - as Deepview project - as text

Alignment [top]

```
TARGET    2     GLSDGEWQ LVLNVWGKVE GDLAGHGQEV LIKLFKNHPE TLEKFDKFKH
1myhA     1     glsdgewq lvlnvwgkve advaghgqev liritkghpe tlekfdrtkh

TARGET          hhhhh hhhhhhhhhh     hhhhhhh hhhhhh  h hhh
1myhA           hhhhh hhhhhhhhhh     hhhhhhh hhhhhh  h hhh
TARGET    50    LKSEDEMKGS EDLKKHGNTV LTALGGILKK KGQHAAEIQP LAQSHATKHK
1myhA     49    lksedemkas edlkkhgntv ltalggilkk kghheaeltp laqshatkhk

TARGET          hhhhh    hhhhhhhh hhhhhhhh       hhhhhh hhhhhhhh
1myhA           hhhhh    hhhhhhhh hhhhhhhh       hhhhhh hhhhhhhh
TARGET    100   IPIKYLEFIS EAIIQVLQSK HPGDFGADAQ GAMSKALELF RNDIAAKYKE
1myhA     99    ipvkylefis eaiiqvlqsk hpgdfgadaq gamskalelf rndmaakyke

TARGET          hhhhhhh hhhhhhhhhh      hhhh hhhhhhhhhh hhhhhhhhhh
1myhA           hhhhhhh hhhhhhhhhh      hhhh hhhhhhhhhh hhhhhhhhhh
TARGET    150   LGFQG
1myhA     149   lgfqg-

TARGET          h
1myhA
```

(b)

```
HEADER    SWISS-MODEL SERVER (http://swissmodel.expasy.org)
TITLE  SWISS-MODEL
EXPDTA    THEORETICAL MODEL (SWISS-MODEL SERVER)
AUTHOR    SWISS-MODEL SERVER (SEE REFERENCE IN JRNL Records)
REVDAT  1   Fri May 29 02:48:35 2009
JRNL       AUTH   T.SCHWEDE,J.KOPP,N.GUEX,M.C.PEITSCH
JRNL       TITL   SWISS_MODEL: AN AUTOMATED PROTEIN HOMOLOGY-MODELING
JRNL       TITL 2 SERVER
JRNL       REF    NUCLEIC ACIDS RESEARCH        V.  31  3381 2003
JRNL       REFN   ASTM NARHAD  UK ISSN 0305-1048
REMARK   1 -----------------------------------------------------------------
REMARK   1 DISCLAIMER
REMARK   1
REMARK   1 The SWISS-MODEL SERVER produces theoretical models for proteins. The
REMARK   1 results of any theoretical modelling procedure is NON-EXPERIMENTAL
REMARK   1 and MUST be considered with care. These models may contain significant
REMARK   1 errors. This  is  especially true  for  automated modeling since there
REMARK   1 is no human intervention during model building.
REMARK   1 Please read the header section and the logfile carefully to  know what
REMARK   1 templates and alignments were used during the model building  process.
REMARK   1 All information by the SWISS-MODEL SERVER is provided "AS-IS", without
REMARK   1 any warranty, expressed or implied.
REMARK   1
REMARK   1 -----------------------------------------------------------------
REMARK 999
REMARK 999 SEQUENCE
REMARK 999 SPAC SWS  SPAC      1 -     1 NOT IN ATOMS LIST
DBREF  SPAC        1   153 SWS  SPAC       2    154
IPREF  SPAC  SPAC 2 154
CRYST1    1.000    1.000    1.000  90.00  90.00  90.00 P 1          1
ORIGX1      1.000000 0.000000 0.000000        0.00000
ORIGX2      0.000000 1.000000 0.000000        0.00000
ORIGX3      0.000000 0.000000 1.000000        0.00000
SCALE1      1.000000 0.000000 0.000000        0.00000
SCALE2      0.000000 1.000000 0.000000        0.00000
SCALE3      0.000000 0.000000 1.000000        0.00000
ATOM      1  N   GLY     2      49.627 -10.303  26.376  1.00 50.00
ATOM      2  CA  GLY     2      49.507 -10.855  27.744  1.00 50.00
ATOM      3  C   GLY     2      50.884 -10.870  28.427  1.00 50.00
ATOM      4  O   GLY     2      51.440  -9.823  28.747  1.00 50.00
ATOM      5  N   LEU     3      51.440 -12.070  28.554  1.00 50.00
ATOM      6  CA  LEU     3      52.872 -12.244  28.841  1.00 50.00
ATOM      7  C   LEU     3      53.467 -12.997  27.654  1.00 50.00
ATOM      8  O   LEU     3      52.877 -13.956  27.161  1.00 50.00
ATOM      9  CB  LEU     3      53.115 -13.050  30.124  1.00 50.00
ATOM     10  CG  LEU     3      52.943 -12.263  31.430  1.00 50.00
ATOM     11  CD1 LEU     3      51.492 -11.864  31.712  1.00 50.00
ATOM     12  CD2 LEU     3      53.483 -13.095  32.592  1.00 50.00
ATOM     13  N   SER     4      54.587 -12.487  27.153  1.00 50.00
ATOM     14  CA  SER     4      55.305 -13.182  26.067  1.00 50.00
ATOM     15  C   SER     4      55.826 -14.525  26.608  1.00 50.00
ATOM     16  O   SER     4      55.883 -14.737  27.821  1.00 50.00
ATOM     17  CB  SER     4      56.446 -12.321  25.513  1.00 50.00
ATOM     18  OG  SER     4      57.525 -12.186  26.449  1.00 50.00
ATOM     19  N   ASP     5      56.324 -15.370  25.707  1.00 50.00
```

(c)

(d)

FIGURE 5.18 Steps to construct a 3D model using SWISS-MODEL: (a) input amino acid sequence, (b) alignment details with an automatically selected template, (c) predicted 3D structure, and (d) structural coordinates and other information in the PDB format.

5.11 Exercises

1. Discriminate different structural classes of proteins using amino acid composition and correlation coefficient.
   Hint: Develop datasets for different structural classes. Compute amino acid composition for each protein and average for each class (**Section 2.3.2**). Compute correlation coefficient for each protein with the average of four structural classes. The structural class is the one with the highest correlation coefficient.

2. Discriminate different structural classes of proteins using the residue pair preference and the correlation coefficient.
   Hint: Repeat question 1 by replacing amino acid composition into residue pair preference (**Section 2.3.3**).

3. Predict the secondary structure content and class for the following sequence:

   ```
   > SEQUENCE1
   MNIFEMLRIDEGLRLKIYKDTEGYYTIGIGHLLTKSPSLNAAKSELDKAIGRNCNGVITK
   DEAEKLFNQDVDAAVRGILRNAKLKPVYDSLDAVRRCALINMVFQMGETGVAGFTNSLRM
   LQQKRWDEAAVNLAKSRWYNQTPNRAKRVITTFRTGTWDAYKNL
   ```

   Hint: Use SSCP server.

4. Predict the secondary structure of the above sequence using the Chou–Fasman method.
   Hint: Input the sequence at the server discussed in 5.3.2.

5. Comment on the secondary structure content predicted directly and from the results obtained with secondary structure prediction methods.
   Hint: Use different methods used to predict the content (**Section 5.2**) and predicted secondary structure (**Section 5.3**).

6. Compare the predicted secondary structure for the following sequence using different methods.

   ```
   >sp|P04248|
   MGLSDGEWQLVLNVWGKVEGDLAGHGQEVLIKLFKNHPETLEKFDKFKHLKSEDEMKGSE
   DLKKHGNTVLTALGGILKKKGQHAAEIQPLAQSHATKHKIPIKYLEFISEAIIQVLQSKH
   PGDFGADAQGAMSKALELFRNDIAAKYKELGFQG
   ```

   Hint: Use different online methods discussed in **Section 5.3**.

7. Predict the transmembrane helices in bacteriorhodopsin using SOSUI, HMM-TOP, and TMHMM. Compare the results.
   Hint: Use the servers discussed in **Section 5.4**.

8. Evaluate the performance of different methods to discriminate the following sequences into β-barrel membrane proteins.

   ```
   >1061410|Genbank|
   MDNMHNTNGLMRFAKVAAASTLLATLLAACAVGPDYQRPDAVVPAAFKEAPTLAAGEQAG
   TWKTAEPSDGRTPRRMVKVFGDPVLDSLETQALAANQNLKAAAARVEEARAATRSARSQW
   FPQVGAGFGPTREGLSSASQFQPQGTGPTNATLWRAQGTVSYEADLFGRVGRNVEASRAD
   QAQSEALFRSVQLALQADVAQNYFELRQLDSDQDLYHRTVELREQALKLVQRRFNEGDIS
   ELDVSRAKNELASAQADAVDVARRRAASEHALAILLGKAPADFAFKETPIVPVAVKIPPG
   ```

```
LPSALLERRPDVSAAERAMAAANARIGLAKSAYFRSSISPGRSAISVDARQPVPVVEPYF
LLGPFAGTALTLPLFDGGRRAAGVQQARAQYDEQGANYRQQVLVAFREVEDNLADLRLLD
DQIRAQEAAVNASRRAATLSRSEYQEGEVAYLDVIDSERSVLQSQLQANQLTGAQAVSTV
NLIRALGGGWGNAPAPTAVGDAASGKADVAAR
>006873|
MDLATLLGLIGGFAFVIMAMVLGGSIGMFVDVTSILIVVGGSIFVVLMKFTMGQFFGATK
IAGKAFMFKADEPEDLIAKIVEMADAARKGGFLALEEMEINNTFMQKGIDLLVDGHDADV
VRAALKKDIALTDERHTQGTGVFRAFGDVAPAMGMIGTLVGLVAMLSNMDDPKAIGPAMA
VALLTTLYGAILSNMVFFPIADKLSLRRDQETLNRRLIMDGVLAIQDGQNPRVIDSYLKN
YLNEGKRALEIDE
```

Hint: Use the methods described in **Section 5.5.**

9. For the β-barrel membrane proteins, predict the segments using different methods, and compare the results.

Hint: Use the methods described in **Section 5.6.**

10. Predict the disordered regions in the following protein using different methods, and compare the results.

```
>sp|P04248|
MGLSDGEWQLVLNVWGKVEGDLAGHGQEVLIKLFKNHPETLEKFDKFKHLKSEDEMKGSE
DLKKHGNTVLTALGGILKKKGQHAAEIQPLAQSHATKHKIPIKYLEFISEAIIQVLQSKH
PGDFGADAQGAMSKALELFRNDIAAKYKELGFQG
```

Hint: Use the methods described in **Section 5.7.**

11. Predict solvent accessibility of the following sequence using NETASA and RVP-net, compare the results.

```
>sp|P04248|
MGLSDGEWQLVLNVWGKVEGDLAGHGQEVLIKLFKNHPETLEKFDKFKHLKSEDEMKGSE
DLKKHGNTVLTALGGILKKKGQHAAEIQPLAQSHATKHKIPIKYLEFISEAIIQVLQSKH
PGDFGADAQGAMSKALELFRNDIAAKYKELGFQG
```

Hint: Use the servers NETASA and RVP-net.

12. Compare the state-wise solvent accessibility predictors for the same sequence.
Hint: Use the methods described in **Sections 5.8.1 to 5.8.7.**

13. Compare the performance of real-value solvent accessibility predictors for the same sequence.
Hint: Use the methods described in **Sections 5.8.8 to 5.8.12.**

14. Discuss the performance of contact prediction methods for 2LZM and compare with experimental data.
Hint: Use the methods described in **Section 5.9.** For the experimental data, take the PDB coordinates and compute the contacting residues with specific cutoff distance.

15. Build a three-dimensional model structure of the following sequence using SWISS-MODEL.

```
>sp|P59799|
MPKVIVANINAEFEGIENETIMQILYRNGIEIDSACGGHGQCTSCKVLIISGSENLYPAE
FEEKDTLEENGMDPETERLSCQAKLNGKGDVVIYLP
```

Hint: Follow the steps described in **Section 5.10.5.**

16. Compare the changes in 3D structure due to the substitution of amino acid residues for the following sequences.

```
>  |SEQUENCE2
ADQLTEEQIAEFKEAFSLFDKDGDGTITTKELGTVMRSLGQNPTEAELQDMINEVDADGN
GTIDFPEFLTMMARKMKDTDSEEEIREAFRVFDKDGNGFISAAELRHVMTNLGEKLTDEE
VDEMIREADIDGDGQVNYEEFVTMMTSK
>  |SEQUENCE3
ADQLTEEQIAEFKEAFSLFDKDGDGTITTKELGTVMRSLGQNPTEAELQDMINEVDADGN
GTGDFPEFLTMMARKMKDTDSEEEIREAFRVFDKDGNGFISAAELRHVMTNLGEKLTDEE
VDEMIREADIDGDGQVNYEEFVTMMTSK
```

Hint: Ile 63 is replaced with Gly. Use SWISS-MODEL to get the PDB for new sequence, and identify the changes.

17. Compute the SOV measure for the following assignment.

```
>Observed
CCCHHHCCHHHCEEEEEEECCCCHHHHHHHC
>Ppredicted
CCHHHHCCCHHHCCCEEECCCCCCCEEEEEE
>Amino-acid sequence
MQTRSIGVQWERTYASDFGHKLPACVMAVAA
```

Hint: Use SOV server.

References

Adamczak R, Porollo A, Meller J. Accurate prediction of solvent accessibility using neural networks-based regression. Proteins. 2004;56:753–767.

Ahmad S, Gromiha MM. NETASA: neural network based prediction of solvent accessibility. Bioinformatics. 2002;18:819–824.

Ahmad S, Gromiha MM, Sarai A. Real-value prediction of solvent accessibility from amino acid sequence. Proteins 2003a;50:629–635.

Ahmad S, Gromiha MM, Sarai A. RVP-net: online prediction of real valued accessible surface area of proteins from single sequences. Bioinformatics. 2003b;19:1849–1851.

Altschul SF, Koonin EV. Iterated profile searches with PSI-BLAST–a tool for discovery in protein databases. Trends Biochem Sci. 1998;23(11):444–447.

Altschul SF, Madden TL, Schaffer AA, Zhang J, Zhang Z, Miller W, Lipman DJ. Gapped BLAST and PSI-BLAST: a new generation of protein database search programs. Nucleic Acids Res. 1997;25(17):3389–3402.

Bagos PG, Liakopoulos TD, Spyropoulos IC, Hamodrakas SJ. A Hidden Markov Model method, capable of predicting and discriminating b-barrel outer membrane proteins. BMC Bioinformatics. 2004;5:29.

Bagos PG, Liakopoulos TD, Hamodrakas SJ. Evaluation of methods for predicting the topology of beta-barrel outer membrane proteins and a consensus prediction method. BMC Bioinformatics. 2005;6:7.

Barton GJ. Protein sequence alignment and database scanning. In: Sternberg MJE, ed. *Protein Structure Prediction: A Practical Approach.* Oxford, UK: IRL Press; 1996: 31–63.

Bigelow H, Rost B. PROFtmb: a web server for predicting bacterial transmembrane beta barrel proteins. Nucleic Acids Res. 2006;34(Web Server issue):W186–W188.

Bigelow HR, Petrey DS, Liu J, Przybylski D, Rost B. Predicting transmembrane beta-barrels in proteomes. Nucleic Acids Res. 2004;32(8):2566–2577.

Bradley P, Chivian D, Meiler J, Misura KM, Rohl CA, Schief WR, Wedemeyer WJ, Schueler-Furman O, Murphy P, Schonbrun J, Strauss CE, Baker D. Rosetta predictions in CASP5: successes, failures, and prospects for complete automation. Proteins. 2003;53(Suppl 6):457–468.

Bu WS, Feng ZP, Zhang Z, Zhang CT. Prediction of protein (domain) structural classes based on amino-acid index. Eur J Biochem. 1999;266(3):1043–1049.

Cai YD, Feng KY, Lu WC, Chou KC. Using LogitBoost classifier to predict protein structural classes. J Theor Biol. 2006;238(1):172–176.

Cai YD, Liu XJ, Chou KC. Prediction of protein secondary structure content by artificial neural network. J Comput Chem. 2003;24(6):727–731.

Carugo O. Predicting residue solvent accessibility from protein sequence by considering the sequence environment. Protein Eng. 2000;13:607–609.

Chandonia JM, Karplus M. The importance of larger data sets for protein secondary structure prediction with neural networks. Protein Sci. 1996;5(4):768–774.

Chen C, Tian YX, Zou XY, Cai PX, Mo JY. Using pseudo-amino acid composition and support vector machine to predict protein structural class. J Theor Biol. 2006;243(3):444–448.

Chen CP, Rost B. State-of-the-art in membrane protein prediction. Appl Bioinformatics. 2002;1:21–35.

Chivian D, Kim DE, Malmstrom L, Bradley P, Robertson T, Murphy P, Strauss CE, Bonneau R, Rohl CA, Baker D. Automated prediction of CASP-5 structures using the Robetta server. Proteins. 2003;53(Suppl 6):524–533.

Chou KC. Using pair-coupled amino acid composition to predict protein secondary structure content. J Protein Chem. 1999;18(4):473–480.

Chou KC. Prediction of protein structural classes and subcellular locations. Curr Protein Pept Sci. 2000;1(2):171–208.

Chou KC. Progress in protein structural class prediction and its impact to bioinformatics and proteomics. Curr Protein Pept Sci. 2005;6(5):423–436.

Chou KC, Cai YD. Predicting protein structural class by functional domain composition. Biochem Biophys Res Commun. 2004;321(4):1007–1009.

Chou KC, Elrod DW. Protein subcellular location prediction. Protein Eng. 1999;12(2):107–118.

Chou KC, Maggiora GM. Domain structural class prediction. Protein Eng. 1998;11(7):523–538.

Chou KC, Zhang CT. A correlation-coefficient method to predicting protein-structural classes from amino acid compositions. Eur J Biochem. 1992;207(2):429–433.

Chou PY, Fasman GD. Prediction of protein conformation. Biochemistry. 1974;13(2):222–245.

Cid H, Bunster M, Canales M, Gazitua F. Hydrophobicity and structural classes in proteins. Protein Eng. 1992;5(5):373–375.

Claverie J-M, Daulmiere C. Smoothing profiles with sliding windows: better to wear a hat! CABIOS. 1991;7:113–115.

Cole C, Barber JD, Barton GJ. The Jpred 3 secondary structure prediction server. Nucleic Acids Res. 2008;36(Web Server issue):W197–W201.

Cowan SW, Schirmer T, Rummel G, Steiert M, Ghosh R, Pauptit RA, Jansonius JN, Rosenbusch JP. Crystal structures explain functional properties of two E. coli porins. Nature. 1992;358(6389):727–733.

Cristianini N, Shawe-Taylor J. An Introduction to Support Vector Machines and other kernel-based learning methods. Cambridge, UK: Cambridge University Press; 2000.

Cserzö M, Wallin E, Simon I, von Heijne G, Elofsson A. Prediction of transmembrane α-helices in prokaryotic membrane proteins: the dense alignment surface method. Protein Eng, 1997;10:673–676.

Csizmók V, Dosztányi Z, Simon I, Tompa P. Towards proteomic approaches for the identification of structural disorder. Curr Protein Pept Sci. 2007;8(2):173–179.

Cuff JA, Barton GJ. Application of multiple sequence alignment profiles to improve protein secondary structure prediction. Proteins. 2000;40:502–511.

Deber CM, Wang C, Liu LP, Prior AS, Agrawal S, Muskat BL, Cuticchia AJ. TM Finder: a prediction program for transmembrane protein segments using a combination of hydrophobicity and nonpolar phase helicity scales. Protein Sci. 2001;10:212–219.

Diederichs K, Freigang J, Umhau S, Zeth K, Breed J. Prediction by a neural network of outer membrane beta-strand protein topology. Protein Sci. 1998;7(11):2413–2420.

Dong L, Yuan Y, Cai Y. Using bagging classifier to predict protein domain structural class. J Biomol Struct Dyn. 2006;24(3):239–242.

Dosztányi Z, Chen J, Dunker AK, Simon I, Tompa P. Disorder and sequence repeats in hub proteins and their implications for network evolution. J Proteome Res. 2006;5(11):2985–2995.

Dosztányi Z, Csizmok V, Tompa P, Simon I. IUPred: web server for the prediction of intrinsically unstructured regions of proteins based on estimated energy content. Bioinformatics. 2005a;21(16):3433–3434.

Dosztányi Z, Csizmók V, Tompa P, Simon I. The pairwise energy content estimated from amino acid composition discriminates between folded and intrinsically unstructured proteins. J Mol Biol. 2005b;347(4):827–839.

Eisenberg D, Schwartz E, Komaromy M, Wall R. Analysis of membrane and surface protein sequences with the hydrophobic moment plot. J Mol Biol. 1984;179:125–142.

Eisenberg D, Weiss RM, Terwilliger TC. The helical hydrophobic moment: a measure of the amphiphilicity of a helix. Nature. 1982;299:371–374.

Eisenhaber F, Imperiale F, Argos P, Frommel C. Prediction of secondary structural content of proteins from their amino acid composition alone. I. New analytic vector decomposition methods. Proteins. 1996;25(2):157–168.

Engelman DM, Steitz TA, Goldman A. Identifying nonpolar transbilayer helices in amino acid sequences of membrane proteins. Annu Rev Biophys Biophys Chem. 1986;15:321–353.

Fang Q, Shortle D. Prediction of protein structure by emphasizing local side-chain/backbone interactions in ensembles of turn fragments. Proteins. 2003;53 (Suppl 6): 486–490.

Fariselli P, Casadio R. RCNPRED: prediction of the residue co-ordination numbers in proteins. Bioinformatics. 2001;17:202–204.

Fariselli P, Olmea O, Valencia A, Casadio R. Prediction of contact maps with neural networks and correlated mutations. Protein Eng. 2001;14(11):835–843.

Fasman GD. The development of the prediction of protein structure. In Fasman GD, ed. Prediction of Protein Structure and the Principles of Protein Conformation. New York, London: Plenum Press; 1989:193–303.

Fischer D. 3D-SHOTGUN: a novel, cooperative, fold-recognition meta-predictor. Proteins. 2003;51(3):434–441.

Forst D, Welte W, Wacker T, Diederichs K. Structure of the sucrose-specific porin ScrY from Salmonella typhimurium and its complex with sucrose. Nature Struct Biol. 1998;5:37–46.

Galzitskaya OV, Garbuzynskiy SO, Lobanov MY. FoldUnfold: web server for the prediction of disordered regions in protein chain Bioinformatics. 2006;22(23):2948–2949.

Garg A, Kaur H, Raghava GP. Real value prediction of solvent accessibility in proteins using multiple sequence alignment and secondary structure. Proteins. 2005;61(2): 318–324.

Garnier J, Osguthorpe DJ, Robson B. Analysis of the accuracy and implications of simple methods for predicting the secondary structure of globular proteins. J Mol Biol. 1978;120(1):97–120.

Garrow AG, Agnew A, Westhead DR. TMB-Hunt: a web server to screen sequence sets for transmembrane beta-barrel proteins. Nucleic Acids Res. 2005;33:W188–W192.

Gianese G, Bossa F, Pascarella S. Improvement in prediction of solvent accessibility by probability profiles. Protein Eng. 2003;16:987–992.

Gibrat JF, Garnier J, Robson B. Further developments of protein secondary structure prediction using information theory. New parameters and consideration of residue pairs. J Mol Biol. 1987;198(3):425–443.

Ginalski K, Rychlewski L. Protein structure prediction of CASP5 comparative modeling and fold recognition targets using consensus alignment approach and 3D assessment. Proteins. 2003;53 (Suppl 6):410–417.

Gnanasekaran TV, Peri S, Arockiasamy A, Krishnaswamy S. Profiles from structure based sequence alignment of porins can identify beta stranded integral membrane proteins. Bioinformatics. 2000;16:839–842.

Gobel U, Sander C, Schneider R, Valencia A. Correlated mutations and residue contacts in proteins. Proteins. 1994;18(4):309–317.

Grana O, Baker D, MacCallum RM, Meiler J, Punta M, Rost B, Tress ML, Valencia A. CASP6 assessment of contact prediction. Proteins. 2005;61 (Suppl 7):214–224.

Gromiha MM. A simple method for predicting transmembrane alpha helices with better accuracy. Protein Eng. 1999;12(7):557–561.

Gromiha MM. Motifs in outer membrane protein sequences: applications for discrimination. Biophys Chem. 2005;117:65–71.

Gromiha MM, Ponnuswamy PK. Prediction of transmembrane beta strands from hydrophobic characteristics of proteins. Int J Peptide Protein Res. 1993;42:420–431.

Gromiha MM, Ponnuswamy PK. Prediction of protein secondary structures from their hydrophobic characteristics. Int J Pept Protein Res. 1995;45(3):225–240.

Gromiha MM, Selvaraj S. Protein secondary structure prediction in different structural classes. Protein Eng. 1998;11(4):249–251.

Gromiha MM, Selvaraj S. Inter-residue interactions in protein folding and stability. Prog Biophys Mol Biol. 2004;86(2):235–277.

Gromiha MM, Suwa M. A simple statistical method for discriminating outer membrane proteins with better accuracy. Bioinformatics. 2005;21(7):961–968.

Gromiha MM, Suwa M. Discrimination of outer membrane proteins using machine learning algorithms. PROTEINS: Struct Funct Bioinf. 2006a;63:1031–1037.

Gromiha MM, Suwa M. Influence of amino acid properties for discriminating outer membrane proteins at better accuracy. Biochim Biophys Acta. 2006b;1764:1493–1497.

Gromiha MM, Yabuki Y. Functional discrimination of membrane proteins using machine learning techniques. BMC Bioinformatics. 2008;9:135.

Gromiha MM, Majumdar R, Ponnuswamy PK. Identification of membrane spanning beta strands in bacterial porins. Protein Eng. 1997;10(5):497–500.

Gromiha MM, Ahmad S, Suwa M. Neural network-based prediction of transmembrane beta-strand segments in outer membrane proteins. J Comp Chem. 2004;25:762–767.

Gromiha MM, Ahmad S, Suwa M. Application of residue distribution along the sequence for discriminating outer membrane proteins. Comp Biol Chem. 2005a;29: 135–142.

Gromiha, MM. Ahmad S, Suwa M. TMBETA-NET: discrimination and prediction of membrane spanning beta-strands in outer membrane proteins. Nucleic Acids Res. 2005b;33:W164–W167.

Hamilton N, Burrage K, Ragan MA, Huber T. Protein contact prediction using patterns of correlation. Proteins. 2004;56(4):679–684.

Hartmann E, Rapoport TA, Lodish HF. Predicting the orientation of eukaryotic membrane spanning proteins. Proc Natl Acad Sci USA, 1989;86:5786–5790.

Hirokawa T, Boon-Chieng S, Mitaku S. SOSUI: classification and secondary structure prediction system for membrane proteins. Bioinformatics.1998;14:378–379.

Hirose S, Shimizu K, Kanai S, Kuroda Y, Noguchi T. POODLE-L: a two-level SVM prediction system for reliably predicting long disordered regions. Bioinformatics. 2007;23(16):2046–2053.

Hofmann K, Stoffel W. TMBASE—a database of membrane spanning protein segments. Biol Chem. 1993;374:166.

Hua S, Sun Z. A novel method of protein secondary structure prediction with high segment overlap measure: support vector machine approach. J Mol Biol. 2001;308(2): 397–407.

Jacoboni I, Martelli PL, Fariselli P, De Pinto V, Casadio R. Prediction of the transmembrane regions of beta-barrel membrane proteins with a neural network-based predictor. Protein Sci. 2001;10(4):779–787.

Jahnig F. Structure predictions of membrane proteins are not that bad. Trends Biochem Sci. 1990;15(3):93–95.

Jayaram B, Bhushan K, Shenoy SR, Narang P, Bose S, Agrawal P, Sahu D, Pandey V. Bhageerath: an energy based web enabled computer software suite for limiting the search space of tertiary structures of small globular proteins. Nucleic Acids Res. 2006;34(21):6195–6204.

Jayasinghe S, Hristova K, White SH. Energetics, stability, and prediction of transmembrane helices. J Mol Biol, 2001; 312:927–934.

Jin L, Fang W, Tang H. Prediction of protein structural classes by a new measure of information discrepancy. Comput Biol Chem. 2003;27(3):373–380.

Joachims T. Making Large-Scale SVM Learning Practical. Advances in Kernel Methods—Support Vector Learning. Schölkopf B, Burges C, Smola A, eds. MIT Press; 1999.

Jones DT, McGuffin LJ. Assembling novel protein folds from super-secondary structural fragments. Proteins. 2003;53 (Suppl 6):480–485.

Jones DT, Taylor WR, Thornton JM. A model recognition approach to the prediction of all-helical membrane protein structure and topology. Biochem. 1994;33:3038–3049.

Jones DT, Ward JJ. Prediction of disordered regions in proteins from position specific score matrices. Proteins. 2003;53 (Suppl 6):573–578.

Jones DT. GenTHREADER: an efficient and reliable protein fold recognition method for genomic sequences. J Mol Biol. 1999a;287(4):797–815.

Jones DT. Protein secondary structure prediction based on position-specific scoring matrices. J Mol Biol. 1999b;292(2):195–202.

Juretic D, Lee B, Trinajstic N, Williams RW. Conformational preference functions for predicting helices in membrane proteins. Biopolymers. 1993;33:255–273.

Juretic D, Zucic D, Lucic B, Trinajstic N. Preference functions for prediction of membrane-buried helices in integral membrane proteins. Comput Chem. 1998;22: 279–294.

Kaur H, Raghava GP. A neural network method for prediction of beta-turn types in proteins using evolutionary information. Bioinformatics. 2004;20(16):2751–2758.

Kaur H, Raghava GP. A neural-network based method for prediction of gamma-turns in proteins from multiple sequence alignment. Protein Sci. 2003;12(5):923–929.

Kim H, Park H. Prediction of protein relative solvent accessibility with support vector machines and long-range interaction 3D local descriptor. Proteins. 2004;54:557–562.

Kim H, Park H. Protein secondary structure prediction based on an improved support vector machines approach. Protein Eng. 2003;16(8):553–560.

Klein P, Kanehisa M, De Lisi C. The detection and classification of membrane-spanning proteins. Biochim Biophys Acta. 1985;815:468–476.

Klein P. Prediction of protein structural class by discriminant analysis. Biochim Biophys Acta. 1986;874(2):205–215.

Kosinski J, Cymerman IA, Feder M, Kurowski MA, Sasin JM, Bujnicki JM. A "FRanken-stein's monster" approach to comparative modeling: merging the finest fragments of fold-recognition models and iterative model refinement aided by 3D structure evaluation. Proteins. 2003;53(Suppl 6):369–379.

Krogh A, Larsson B, von Heijne G, Sonnhammer EL. Predicting transmembrane protein topology with a hidden Markov model: application to complete genomes. J Mol Biol. 2001;305:567–80.

Kumarevel TS, Gromiha MM, Ponnuswamy MN. Structural class prediction: an application of residue distribution along the sequence. Biophys Chem. 2000;88(1–3):81–101.

Kurgan L, Cios K, Chen K. SCPRED: accurate prediction of protein structural class for sequences of twilight-zone similarity with predicting sequences. BMC Bioinformatics. 2008;9:226.

Kyte J, Doolittle RF. A simple method for displaying the hydropathic character of a protein. J Mol Biol. 1982;157(1):105–132.

Levin JM, Robson B, Garnier J. An algorithm for secondary structure determination in proteins based on sequence similarity. FEBS Lett. 1986;205(2):303–308.

Li X, Pan XM. New method for accurate prediction of solvent accessibility from protein sequence. Proteins. 2001;42:1–5.

Lin K, Simossis VA, Taylor WR, Heringa J. A simple and fast secondary structure prediction method using hidden neural networks. Bioinformatics. 2005;21(2):152–159.

Lin Z, Pan XM. Accurate prediction of protein secondary structural content. J Protein Chem. 2001;20(3):217–220.

Linding R, Jensen LJ, Diella F, Bork P, Gibson TJ, Russell RB. Protein disorder prediction: implications for structural proteomics. Structure. 2003;11(11):1453–1459.

Liu J, Rost B. Comparing function and structure between entire proteomes. Protein Sci. 2001;10:1970–1979.

Liu Q, Zhu Y, Wang B, Li Y. Identification of b-barrel membrane proteins based on amino acid composition properties and predicted secondary structure. Comput Biol Chem. 2003;27:355–361.

Luo RY, Feng ZP, Liu JK. Prediction of protein structural class by amino acid and polypeptide composition. Eur J Biochem. 2002;269(17):4219–4225.

MacCallum RM. Striped sheets and protein contact prediction. Bioinformatics. 2004 Aug 4;20:I224–I231.

Mannella CA. Conformational changes in the mitochondrial channel protein, VDAC, and their functional implications. J Struct Biol. 1998;121:207–218.

Martelli PL, Fariselli P, Krogh A, Casadio R. A sequence-profile-based HMM for predicting and discriminating β barrel membrane proteins. Bioinformatics. 2002;18: S46–S53.

McGregor MJ, Flores TP, Sternberg MJ. Prediction of beta-turns in proteins using neural networks. Protein Eng. 1989;2(7):521–526.

McGuffin LJ, Bryson K, Jones DT. The PSIPRED protein structure prediction server. Bioinformatics. 2000;16(4):404–405.

Mehta PK, Heringa J, Argos P. A simple and fast approach to prediction of protein secondary structure from multiply aligned sequences with accuracy above 70%. Protein Sci. 1995;4(12):2517–2525.

Mészáros B, Simon I, Dosztányi Z. Prediction of protein binding regions in disordered proteins. PLoS Comput Biol. 2009;5(5):e1000376.

Mucchielli-Giorgi MH, Hazout S, Tuffery P. PredAcc: prediction of solvent accessibility. Bioinformatics. 1999;15(2):176–177.

Mugilan SA, Veluraja K. Generation of deviation parameters for amino acid singlets, doublets and triplets from three-dimentional structures of proteins and its implications for secondary structure prediction from amino acid sequences. J Biosci. 2000; 25(1):81–91.

Muskal SM, Kim SH. Predicting protein secondary structure content. A tandem neural network approach. J Mol Biol. 1992;225(3):713–727.

Naderi-Manesh H, Sadeghi M, Arab S, Moosavi Movahedi AA. Prediction of protein surface accessibility with information theory. Proteins. 2001;42(4):452–459.

Nakai K, Kanehisa M. A knowledge base for predicting protein localization sites in eukaryotic cells. Genomics. 1992;14:897–911.

Nakashima H, Nishikawa K, Ooi T. The folding type of a protein is relevant to the amino acid composition. J Biochem (Tokyo). 1986;99(1):153–162.

Natt NK, Kaur H, Raghava GP: Prediction of transmembrane regions of beta-barrel proteins using ANN- and SVM-based methods. Proteins 2004;56(1):11–18.

Needleman SB, Wunsch CD. A general method applicable to the search for similarities in the amino acid sequence of two proteins. J Mol Biol. 1970;48:443–453.

Nguyen MN, Rajapakse JC. Two-stage support vector regression approach for predicting accessible surface areas of amino acids. Proteins. 2006;63(3):542–550.

Nilsson J, Persson B, von Heijne G. Consensus prediction of membrane protein topology. FEBS Lett. 2000;486:267–269.

Nishikawa K, Noguchi T. Predicting protein secondary structure based on amino acid sequence. Methods Enzymol. 1991;202:31–44.

Nishikawa K, Ooi T. Amino acid sequence homology applied to the prediction of protein secondary structures, and joint prediction with existing methods. Biochim Biophys Acta. 1986;871(1):45–54.

Niu B, Cai YD, Lu WC, Li GZ, Chou KC. Predicting protein structural class with AdaBoost Learner. Protein Pept Lett. 2006;13(5):489–492.

Ou YY, Chen SA, Gromiha MM. Prediction of membrane spanning segments and topology in beta-barrel membrane proteins at better accuracy. J Comp Chem 2009 (in press) DOI:10.1002/jcc.21281.

Ou YY, Gromiha MM, Chen SA, Suwa M. TMBETADISC-RBF: discrimination of beta-barrel membrane proteins using RBF networks and PSSM profiles. Comput Biol Chem. 2008;32(3):227–231.

Park KJ, Gromiha MM, Horton P, Suwa M. Discrimination of outer membrane proteins using support vector machines. Bioinformatics. 2005;21(23):4223–4229.

Pascarella S, De Persio R, Bossa F, Argos P. Easy method to predict solvent accessibility from multiple protein sequence alignments. Proteins. 1998;32(2):190–199.

Pasquier C, Promponas VJ, Palaios GA, Hamodrakas JS, Hamodrakas SJ. A novel method for predicting transmembrane segments in proteins based on a statistical analysis of the SwissProt database: the PRED-TMR algorithm. Protein Engin. 1999;12:381–385.

Paul C, Rosenbusch JP. Folding patterns of porin and bacteriorhodopsin. EMBO J. 1985;4(6):1593–1597.

Peng K, Radivojac P, Vucetic S, Dunker AK, Obradovic Z. Length-dependent prediction of protein intrinsic disorder. BMC Bioinformatics. 2006;7:208.

Persson B, Argos P. Topology prediction of membrane proteins. Protein Sci. 1996;5:363–371.

Petersen TN, Lundegaard C, Nielsen M, Bohr H, Bohr J, Brunak S, Gippert GP, Lund O. Prediction of protein secondary structure at 80% accuracy. Proteins. 2000;41(1):17–20.

Petrey D, Xiang Z, Tang CL, Xie L, Gimpelev M, Mitros T, Soto CS, Goldsmith-Fischman S, Kernytsky A, Schlessinger A, Koh IY, Alexov E, Honig B. Using multiple structure alignments, fast model building, and energetic analysis in fold recognition and homology modeling. Proteins. 2003;53(Suppl 6):430–435.

Pollastri G, Baldi P. Prediction of contact maps by GIOHMMs and recurrent neural networks using lateral propagation from all four cardinal corners. Bioinformatics. 2002;18:S62–S70.

Pollastri G, Przybylski D, Rost B, Baldi P. Improving the prediction of protein secondary structure in three and eight classes using recurrent neural networks and profiles. Proteins. 2002a;47(2):228–235.

Pollastri G, Baldi P, Fariselli P, Casadio R. Prediction of coordination number and relative solvent accessibility. Proteins. 2002b;47:142–153.

Ponnuswamy PK, Gromiha MM. Prediction of transmembrane helices from hydrophobic characteristics of proteins. Int J Pept Protein Res. 1993;42(4):326–341.

Ponnuswamy PK, Prabhakaran M, Manavalan P. Hydrophobic packing and spatial arrangement of amino acid residues in globular proteins. Biochim Biophys Acta. 1980;623(2):301–316.

Prilusky J, Felder CE, Zeev-Ben-Mordehai T, Rydberg EH, Man O, Beckmann JS, Silman I, Sussman JL. FoldIndex: a simple tool to predict whether a given protein sequence is intrinsically unfolded. Bioinformatics. 2005;21(16):3435–3438.

Promponas VJ, Palaios GA, Pasquier CM, Hamodrakas JS, Hamodrakas SJ. CoPreTHi: a Web tool which combines transmembrane protein segment prediction methods. In Silico Biol. 1999;1:159–162.

Przybylski D, Rost B. Alignments grow, secondary structure prediction improves. Proteins. 2002;46(2):197–205.

Punta M, Rost B. PROFcon: novel prediction of long-range contacts. Bioinformatics. 2005;21(13):2960–2968.

Qian N, Sejnowski TJ. Predicting the secondary structure of globular proteins using neural network models. J Mol Biol. 1988;202:865–884.

Randall A, Cheng J, Sweredoski M, Baldi P. TMBpro: secondary structure, beta-contact and tertiary structure prediction of transmembrane beta-barrel proteins. Bioinformatics. 2008;24(4):513–520.

Richardson CJ, Barlow DJ. The bottom line for prediction of residue solvent accessibility. Protein Eng. 1999;12(12):1051–1054.

Rose GD. Prediction of chain turns in globular proteins on a hydrophobic basis. Nature. 1978;272(5654):586–590.

Rost B. PHD: predicting one-dimensional protein structure by profile-based neural networks. Methods Enzymol. 1996;266:525–539.

Rost B, Sander C. Secondary structure prediction of all-helical proteins in two states. Protein Eng. 1993a;6(8):831–836.

Rost B, Sander C. Prediction of protein secondary structure at better than 70% accuracy. J Mol Biol. 1993b;232(2):584–599.

Rost B, Sander C. Combining evolutionary information and neural networks to predict protein secondary structure. Proteins. 1994a;19(1):55–72.

Rost B, Sander C. Conservation and prediction of solvent accessibility in protein families. Proteins. 1994b;20:216–226.

Rost B, Casadio R, Fariselli P, Sander C. Prediction of helical transmembrane segments at 95% accuracy. Protein Sci. 1995;4:521–533.

Rost B, Casadio R, Fariselli P. Topology prediction for helical transmembrane proteins at 86% accuracy. Protein Sci. 1996;5:1704–1718.

Ruan J, Wang K, Yang J, Kurgan LA, Cios K. Highly accurate and consistent method for prediction of helix and strand content from primary protein sequences. Artif Intell Med. 2005;35(1–2):19–35.

Russell RB, Barton GJ. The limits of protein secondary structure prediction accuracy from multiple sequence alignment. J Mol Biol. 1993;234(4):951–957.

Salamov AA, Solovyev VV. Protein secondary structure prediction using local alignments. J Mol Biol. 1997;268(1):31–36.

Sali A, Blundell TL. Comparative protein modelling by satisfaction of spatial restraints. J Mol Biol. 1993;234(3):779–815.

Salzberg S, Cost S. Predicting protein secondary structure with a nearest-neighbor algorithm. J Mol Biol. 1992;227(2):371–374.

Sasson I, Fischer D. Modeling three-dimensional protein structures for CASP5 using the 3D-SHOTGUN meta-predictors. Proteins. 2003;53(Suppl 6):389–394.

Schirmer T, Cowan SW. Prediction of membrane-spanning beta-strands and its application to maltoporin. Protein Sci. 1993;2(8):1361–1363.

Schirmer T, Keller TA, Wang YF, Rosenbusch JP. Structural basis for sugar translocation through maltoporin channels at 3.1 A resolution. Science. 1995;267:512–514.

Schlessinger A, Punta M, Yachdav G, Kajan L, Rost B. Improved disorder prediction by combination of orthogonal approaches. PLoS ONE. 2009;4(2):e4433.

Schulz GE. A critical evaluation of methods for prediction of protein secondary structures. Annu Rev Biophys Biophys Chem. 1988;17:1–21.

Schulz GE. β Barrel membrane proteins. Curr Opin Struct Biol. 2000;10:443–447.

Schulz GE. The structure of bacterial outer membrane proteins. Biochim Biophys Acta. 2002;1565:308–317.

Schwede T, Kopp J, Guex N, Peitsch MC. SWISS-MODEL: an automated protein homology-modeling server. Nucleic Acids Res. 2003;31(13):3381–3385.

Sellers PH. On the theory and computation of evolutionary distances. SIAM J Appl Math. 1974;26:787–793.

Sethi D, Garg A, Raghava GP. DPROT: prediction of disordered proteins using evolutionary information. Amino Acids. 2008;35(3):599–605.

Shao Y, Bystroff C. Predicting interresidue contacts using templates and pathways. Proteins. 2003;53(Suppl 6):497–502.

Shen HB, Yang J, Liu XJ, Chou KC. Using supervised fuzzy clustering to predict protein structural classes. Biochem Biophys Res Commun. 2005;334(2):577–581.

Shepherd AJ, Gorse D, Thornton JM. Prediction of the location and type of beta-turns in proteins using neural networks. Protein Sci. 1999;8(5):1045–1055.

Shimizu K, Hirose S, Noguchi T. POODLE-S: web application for predicting protein disorder by using physicochemical features and reduced amino acid set of a position-specific scoring matrix. Bioinformatics. 2007a;23(17):2337–2338.

Shimizu K, Muraoka Y, Hirose S, Tomii K, Noguchi T. Predicting mostly disordered proteins by using structure-unknown protein data. BMC Bioinformatics. 2007b;8:78.

Skolnick J, Kihara D. Defrosting the frozen approximation: PROSPECTOR–a new approach to threading. Proteins. 2001;42(3):319–331.

Skolnick J, Kolinski A, Kihara D, Betancourt M, Rotkiewicz P, Boniecki M. Ab initio protein structure prediction via a combination of threading, lattice folding, clustering, and structure refinement. Proteins. 2001;Suppl 5:149–156.

Skolnick J, Zhang Y, Arakaki AK, Kolinski A, Boniecki M, Szilagyi A, Kihara D. TOUCHSTONE: a unified approach to protein structure prediction. Proteins. 2003;53(Suppl 6):469–479.

Sonego P, Kocsor A, Pongor S. ROC analysis: applications to the classification of biological sequences and 3D structures. Brief Bioinform. 2008;9(3):198–209.

Taguchi YH, Gromiha MM. Application of amino acid occurence for discriminating different folding types of globular proteins. BMC Bioinformatics. 2007;8:404.

The NMITLI-BioSuite Team. BioSuite: a comprehensive bioinformatics software package (a unique industry-academia collaboration). Curr Sci. 2007;92:29–38.

Thompson MJ, Goldstein RA. Predicting solvent accessibility: higher accuracy using Bayesian statistics and optimized residue substitution classes. Proteins. 1996;25(1): 38–47.

Tomii K, Hirokawa T, Motono C. Protein structure prediction using a variety of profile libraries and 3D verification. Proteins. 2005;61(Suppl 7):114–121.

Tusnady GE, Simon I. Principles governing amino acid composition of integral membrane proteins: application to topology prediction. J Mol Biol. 1998;283:489–506.

Vogel H, Jahnig F. Models for the structure of outer-membrane proteins of Escherichia coli derived from Raman spectroscopy and prediction methods. J Mol Biol. 1986;190(2):191–199.

von Heijne G. The distribution of positively charged residues in bacterial inner membrane proteins correlates with the trans-membrane topology. EMBO J. 1986;5:3021–3027.

von Heijne G. Membrane protein structure prediction. J Mol Biol. 1992;225:487–494.

von Heijne G. Control of topology and mode of assembly of a polytopic membrane protein by positively charged residues. Nature. 1989;341(6241):456–458.

Wallin E, von Heijne G. Genome-wide analysis of integral membrane proteins from eubacterial, archaean, and eukaryotic organisms. Protein Sci. 1998;7(4):1029–1038.

Wang JY, Ahmad S, Gromiha MM, Sarai A. Look-up tables for protein solvent accessibility prediction and nearest neighbor effect analysis. Biopolymers. 2004;75(3):209.

Wang ZX, Yuan Z. How good is prediction of protein structural class by the component-coupled method? Proteins. 2000;38(2):165–175.

Ward JJ, McGuffin LJ, Buxton BF, Jones DT. Secondary structure prediction with support vector machines. Bioinformatics. 2003;19(13):1650–1655.

Welte W, Weiss MS, Nestel U, Weckesser J, Schiltz E, Schulz GE. Prediction of the general structure of OmpF and PhoE from the sequence and structure of porin from Rhodobacter capsulatus. Orientation of porin in the membrane. Biochim Biophys Acta. 1991;1080(3):271–274.

Wimley WC. Toward genomic identification of beta-barrel membrane proteins: composition and architecture of known structures. Protein Sci. 2002;11:301–312.

Wimley WC. The versatile β barrel membrane protein. Curr Opin Struct Biol. 2003;13: 404–411.

Yarov-Yarovoy V, Baker D, Catterall WA. Voltage sensor conformations in the open and closed states in ROSETTA structural models of K(+) channels. Proc Natl Acad Sci U S A. 2006;103(19):7292–7297.

Yi TM, Lander ES. Protein secondary structure prediction using nearest-neighbor methods. J Mol Biol. 1993;232(4):1117–1129.

Yuan Z, Burrage K, Mattick JS. Prediction of protein solvent accessibility using support vector machines. Proteins. 2002;48(3):566–570.

Yuan Z, Huang B. Prediction of protein accessible surface areas by support vector regression. Proteins. 2004; 57:558–564.

Zemla A, Venclovas C, Fidelis K, Rost B. A modified definition of Sov, a segment-based measure for protein secondary structure prediction assessment. Proteins. 1999;34(2):220–223.

Zhang CT, Chou KC. An optimization approach to predicting protein structural class from amino acid composition. Protein Sci. 1992;1(3):401–408.

Zhang Z, Sun ZR, Zhang CT. A new approach to predict the helix/strand content of globular proteins. J Theor Biol. 2001;208(1):65–78.

Zhou G, Xu X, Zhang CT. A weighting method for predicting protein structural class from amino acid composition. Eur J Biochem. 1992;210(3):747–749.

Zhou GP, Assa-Munt N. Some insights into protein structural class prediction. Proteins. 2001;44(1):57–59.

CHAPTER 6

Protein Stability

Protein stability is the net balance of forces, which determine whether a protein will be its native folded conformation or a denatured (unfolded or extended) state. The net stability of proteins is quite small and is the difference between two large opposing forces. The folded native state of protein structures is stabilized by various atomic/group interactions, such as hydrophobic, electrostatic, hydrogen bonding, van der Waals, and disulphide, and the unfolded state is dominated by entropic and nonentropic free energies. It is something of a paradox, given the apparently very complicated three-dimensional structure of proteins and their rapid spontaneous folding, that their net stability is so small, typically 5–10 kcal/mol (Pace, 1990).

6.1   Determination of protein stability

Protein stability is measured with several experiments such as circular dichroism (CD), differential scanning calorimetry (DSC), absorbance (Abs), fluorescence (Fl), nuclear magnetic resonance (NMR), gel filtration, isothermal calorimetry, and light scattering. Among these methods, CD, DSC, Fl, and Abs are the widely used ones to measure stability.

The Gibbs free energy change (ΔG) for the equilibrium reaction is referred to as the conformational stability of a protein and is defined by

$$\text{Folded (N)} \Leftrightarrow \text{unfolded (D).} \tag{6.1}$$

Furthermore, related thermodynamic quantities such as enthalpy change (ΔH), entropy change (ΔS), and heat capacity change (ΔC_p) at unfolding can be determined that allow to gain deeper insight into the forces that stabilize the unique three-dimensional structure.

6.1.1   Differential scanning calorimetry

DSC is one of the most widely used tools for studying the thermodynamics of protein unfolding, and it does not need any models of *a priori* assumptions. DSC enables precise determination of the thermodynamic values T_m, ΔH_{unf}, and ΔC_{punf}, and these values can be determined with sufficient accuracy, within the deviations of ± 0.1 K, $\pm 2\%$ in ΔH_{unf} and $\pm 5\%$ in ΔC_{punf}.

In DSC, the heat capacity of the protein in solution is directly measured versus temperature. An example obtained with DSC is shown in **Figure 6.1**. From this figure, the enthalpy change can be obtained from the area under the peak. This is the calorimetric enthalpy change ΔH_{unf}^{cal}, which is free of any assumptions. In the mean

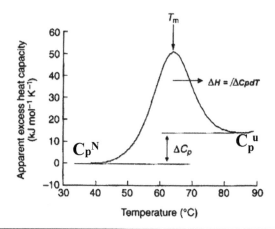

FIGURE 6.1   Heat capacity versus temperature obtained by differential scanning calorimetry; ΔC_p is the difference of heat capacity between the native (C_p^N) and unfolded (C_p^U) states. Figure was taken as a screenshot from http://www.microcal.com/functional-application-areas/stability/default.asp

time, from peak height or peak half width the van't Hoff enthalpy change ΔH_{unf}^{vH} can be obtained from the same calorimetric reading. This makes DSC a valuable tool for testing the validity of the two-state assumption.

The heat capacity change ΔC_{punf} can be determined as follows: (i) by extrapolation of pre- and postdenaturational heat capacity (C_p^N, C_p^U) into the transition region as shown in **Figure 6.1**; (ii) plotting ΔH_{unf}^{cal} versus transition temperature, T_{trs} or T_m; (iii) by direct heat capacity measurements of native and denatured protein, since $\Delta C_{punf} = C_p^U - C_p^N$.

6.1.2   Calculation of thermodynamic functions

Thermal unfolding usually takes place at elevated temperature, whereas the conformational stability ΔG_{unf} is generally of interest at the standard temperature $T = 25°C$ (298.16 K) or at the physiological temperature. Hence, thermodynamic functions such as $\Delta G_{unf}(T)$ have to be determined. At the transition temperature, T_{trs} (transition midpoint, T_m),

$$\Delta G_{unf}(T_{trs}) = \Delta H_{unf}(T_{trs}) - T_{trs}\, \Delta S_{unf}(T_{trs}) = 0 \tag{6.2}$$

ΔG_{unf}, ΔH_{unf}, and ΔS_{unf} are temperature dependent.

$$\Delta H_{unf}(T) = \Delta H_{unf}(T_{trs}) - \int_T^{T_{trs}} \Delta C_{punf}(T)\mathrm{d}T \tag{6.3}$$

$$\Delta S_{unf}(T) = [\Delta H_{unf}(T_{trs})/T_{trs}] - \int_T^{T_{trs}} \Delta C_{punf}(T)\, \mathrm{d}\ln T \tag{6.4}$$

$$\Delta G_{unf}(T) = \Delta H_{unf}(T) - T\Delta S_{unf}(T)$$

$$= \Delta H_{unf}(T_{trs})[(T_{trs} - T)/T_{trs}] - \int_T^{T_{trs}} \Delta C_{punf}(T)\mathrm{d}T + T\int_T^{T_{trs}} (\Delta C_{punf}(T)/T)\mathrm{d}T.$$

$$\tag{6.5}$$

In case the heat capacity change $\Delta C_{\text{punf}}(T)$ at protein unfolding does not depend significantly on temperature, the above equations can be simplified as follows:

$$\Delta H_{\text{unf}}(T) = \Delta H_{\text{unf}} - \Delta C_{\text{punf}}(T_{\text{trs}} - T) \tag{6.6}$$

$$\Delta S_{\text{unf}}(T) = (\Delta H_{\text{unf}}/T) - \Delta C_{\text{punf}} \ln(T_{\text{trs}}/T) \tag{6.7}$$

$$\Delta G_{\text{unf}}(T) = \Delta H_{\text{unf}}[(T_{\text{trs}} - T)/T_{\text{trs}}] - \Delta C_{\text{punf}}(T_{\text{trs}} - T) + \Delta C_{\text{punf}}T \ln(T_{\text{trs}}/T). \tag{6.8}$$

Equation 6.8 is important for transforming ΔG_{unf} from one temperature to another. At the same time, this equation is the main equation for determining $\Delta G_{\text{unf}}(T)$ from calorimetrically measured transition temperature, enthalpy change, and heat capacity change (Pfeil, 1998).

6.1.3   Denaturant-induced unfolding

This approach includes the assumption that the protein unfolds via two-state transition and the assumptions on the mechanism of protein denaturant interaction. The procedure for the determination of conformational stability, however, is rather simple and sensitive, in particular, in different ΔG_{unf} determination. The approach is based on the determination of equilibrium data for protein unfolding in the denaturant-containing solution (GdnHCl, urea, etc.). The extrapolation to zero denaturant concentrations is of critical importance for the evaluation of ΔG_{unf} values obtained in this way.

Free energy change can be computed with three different models: denaturant binding model, linear extrapolation model, and transfer model. However, the linear extrapolation model is the most widely used technique for obtaining ΔG_{unf} from denaturant-induced unfolding transitions. It is given by

$$\Delta G_{\text{unf}} = \Delta G_{\text{unf}}(H_2O) - m \, (\text{denaturant}), \tag{6.9}$$

where m is the slope of the ΔG_{unf} versus denaturant. An example is shown in **Figure 6.2**. The ΔG_{unf} values obtained at various concentrations of urea have been extrapolated to zero concentration. This resultant value is the free energy of unfolding $[\Delta G_{\text{unf}}(H_2O)]$ in the absence of urea. In **Figure 6.2**, the value of $\Delta G_{\text{unf}}(H_2O)$ is estimated to be about 4 kcal/mol, and m-value is 2.6 kcal/mol/M.

6.2   Thermodynamic database for proteins and mutants

Thermodynamic data for proteins are important for understanding the mechanism of protein stability. Several experiments have been carried out during the past few decades, and the thermodynamic data on protein stability have been published in several journals, which are rather scattered in the literature. The collection of these data and make the data available to research community would help to develop novel methods for understanding and predicting protein stability. Pfeil (1998, 2001) collected a set of thermodynamic data for protein folding and stability. Gromiha et al. (1999a, 2002a) developed an electronically accessible thermodynamic database for proteins and mutants, ProTherm, which includes several thermodynamic data, structural information, measuring methods, experimental conditions, reversibility of folding, and literature information. At present, ProTherm contains more than 23,000 entries. This database has been

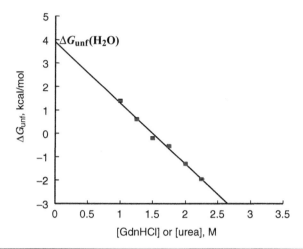

Figure 6.2   The extrapolation to zero denaturant (GdnHCl/urea) concentration of denaturant (GdnHCl/urea) induced protein unfolding.

widely used to understand the mechanism of proteins stability as well as for predicting the stability of protein mutants. A WWW interface has been developed to facilitate searching the database and sorting outputs. It is available at http://gibk26.bse.kyutech.ac.jp/jouhou/protherm/protherm.html.

6.2.1   Organization of ProTherm

ProTherm is implemented into 3DinSight, a relational database system for structure, function, and property of biomolecules (An et al. 1998). This facilitates more efficient search and retrieval of data by flexible queries and enables users to gain insight into the relationship among structure, thermodynamics, and function of proteins. **Figure 6.3** illustrates the organization of ProTherm. The information about the name and source of the protein, details about thermodynamic data, and other supplementary information are put together in ProTherm. The data in ProTherm can be retrieved through a form-based interface with several search, display, and sorting options.

6.2.2   Contents of ProTherm

Each entry in the database is identified by a serial number and includes the following information. Furthermore, each entry is linked with its related entries.

Sequence and structural information:

(1) Name of the protein and its source
(2) PIR and SWISS-PROT (UniProt) codes for identifying the amino acid sequence
(3) Length and molecular weight of the protein calculated from the sequence information given in PIR
(4) PDB codes for the wild and mutant structures. Each PDB code is linked with its homologous proteins, which have the sequence identity of more than 95%

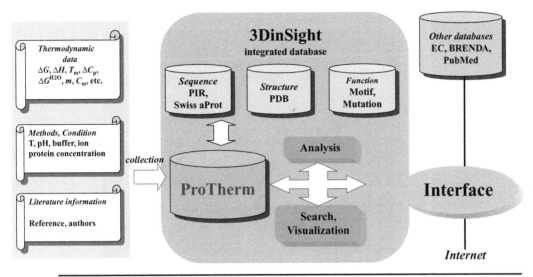

FIGURE 6.3   Organization of ProTherm. Contents of data and links to other databases are shown. Figure was adapted from Gromiha et al. (2002).

(5) Enzyme commission number (Schomburg et al. 2002)

(6) Number of molecules in the protein (biological units)

(7) Number of transition states for folding/unfolding, which indicates the presence of intermediate states

(8) For mutants, the information about the residue(s) in wild type protein, mutant residue, and the residue number are given. Furthermore, the secondary structure and solvent accessibility of the wild type residue at the mutant position have also been provided. The information about secondary structure has been obtained from DSSP (Kabsch and Sander, 1983), and the solvent accessibility has been computed using the program, ASC (Eisenhaber and Argos, 1993)

(9) The residues that are surrounded by each mutant within the distance of 4 Å and 8 Å are given along with the display of central and surrounding residues. **Figure 6.4** shows the mutant residue Lys91 in 2RN2 (Ribonuclease H) and its surrounding residues obtained with Rasmol (Sayle and Milner-White, 1995). The central residue, surrounding residues within the distance of 4 Å, and the distance 4 to 8 Å are shown in red, yellow, and green, respectively.

Thermodynamic data obtained from denaturant denaturation experiments:

(1) Unfolding Gibbs free energy change in the absence and presence of denaturant (ΔG^{H_2O} and ΔG),

(2) Difference of unfolding Gibbs free energy due to mutations in the absence and presence of denaturant ($\Delta\Delta G^{H_2O}$ and $\Delta\Delta G$),

(3) Midpoint of denaturant concentration (C_m),

(4) Slope of denaturation curve (m), and

(5) Reversibility of folding.

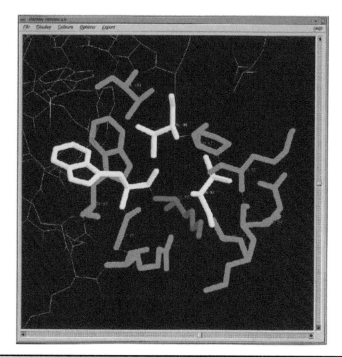

Figure 6.4 Display of protein mutants and surrounding residues by RasMol. As an example, the surrounding residues of Lys 91 in Ribonuclease H (2RN2) are shown. The central residue, surrounding residues within 4 Å, and surrounding residues between 4 and 8 Å are shown in red, yellow, and green, respectively.

Thermodynamic data obtained from thermal denaturation experiments:

(1) Transition temperature (T_m),
(2) Change in transition temperature for the mutants (ΔT_m),
(3) Unfolding Gibbs free energy change (ΔG),
(4) Difference in Gibbs free energy change for mutants ($\Delta\Delta G$),
(5) van't Hoff enthalpy change (ΔH_{vH}),
(6) Calorimetric enthalpy change (ΔH_{cal}),
(7) Heat capacity change (ΔC_p), and
(8) Reversibility of folding

Experimental methods and conditions:

(1) Temperature (T)
(2) pH
(3) Measurement
(4) Method
(5) Buffers and their concentrations
(6) Ions and their concentrations
(7) Additives
(8) Protein concentration

NO.	2		***** Thermodynamic data *****	
***** Sequence and structural information*****			dG_H2O	6.50 kcal/mol
PROTEIN NAME	Phospholipase A2		ddG_H2O	-3.00 kcal/mol
SOURCE	Bovine		dG	
LENGTH	130		ddG	
MOLECULAR WEIGHT	14513.08		Tm	
PIR_ID	PSBOA		dTm	
SWISSPROT_ID	PA21_BOVIN (P00593)		dHvH	
E C NUMBER			dHcal	
PMD NO	A930651		m	1.20 kcal/mol/M
PDB_wild	1BP2 (GO to JPD 1BP2) Homologous PDB Entries		Cm	5.40 M
PDB_mutant			dCp	
MUTATION	H 48 N (Go to STING REPORT wild type H48N)		STATE	2
NO. OF MOLECULE	1		REVERSIBILITY	Unknown
SECONDARY STRUCTURE	Helix (Go to PDBcartoon wild type)		ACTIVITY	
ACCESSIBLE SURFACE AREA	17.1 A**2		ACTIVITY_Km	2.6
***** Experimental condition *****			ACTIVITY_Kcat	0.04
TEMPERATURE	30.0 C		ACTIVITY_Kd	
pH	8.00		***** Literature *****	
BUFFER_NAME	borate		KEY_WORDS	catalytic triad; PLA2; conformational stability; structural role
BUFFER_CONC	10 mM		REFERENCE	J AM CHEM SOC 115, 8523-8526 (1993) PMID:
ION_NAME_1			AUTHOR	Li Y. & Tsai M.-D.
ION_CONC_1			REMARKS	additive : EDTA(0.1 mM),
PROTEIN_CONC	5 mM		RELATED_ENTRIES	1 , 3 , 4 ,
MEASURE	CD			
METHOD	GdnHCl			

FIGURE 6.5 Contents of ProTherm. The details about sequence and structural information, experimental condition, thermodynamic data and literature are shown.

Functional information: Enzyme activity and binding constants
Literature information

(1) Keywords
(2) Reference
(3) Authors
(4) Remarks

Figure 6.5 shows the sample data sheet for entry 2, which is for the protein Bovine phospholipase A2. It includes all the available information about the sequence and structural information (PIR: PSBOA; SWISS-PROT: PA21_BOVIN; PDB: 1BP2; etc.), experimental conditions (pH $= 8$, $T = 30°C$, etc.) and the thermodynamic data ($\Delta G^{H_2O} = 6.50$ kcal/mol; $\Delta\Delta G^{H_2O} = -3.0$ kcal/mol; $m = 1.20$ kcal/mol/M; $C_m = 5.40$M; etc.) along with activity and literature information.

At present ProTherm contains more than 23,000 entries from 955 proteins. **Table 6.1** shows the statistics on May 29, 2009, based on mutations, secondary structures, solvent accessibilities, experimental measurements, and methods.

6.2.3   Search option and access to ProTherm

ProTherm can be directly accessed online at http://gibk26.bse.kyutech.ac.jp/jouhou/protherm/protherm.html. Both quick search and advanced search options have been implemented in this database. **Figure 6.6a** shows the various search options available in ProTherm.

(1) Retrieving data for a particular protein and/or source or PDB code.
(2) Specifying the type of the mutant as single, double, multiple, or wild type. Further searching with specific mutation (e.g., Ala to Val) is also possible.
(3) Mutations at different secondary structures: helix, strand, turn, and coil.

TABLE 6.1 ProTherm statistics*

Features	Number of data
Total entries	23,576
Total proteins	955
Unique proteins	679
Proteins with mutants	279
Wild type	9318
Single mutants	11,688
Double mutants	1605
Multiple mutants	965
Mutations in helix	5286
Mutations in strand	3877
Mutations in turn	2012
Mutations in coil	2973
Buried mutations	6015
Partially buried mutations	3868
Exposed mutations	3793
Absorbance	891
CD	10,318
DSC	6131
Fluorescence	5135
NMR	977
Thermal	14,508
GdnHCl	5422
Urea	3487
References	1779

*as on May 29, 2009.

(4) Searching the data based on solvent accessibility, buried, partially buried, or exposed. Furthermore, one can search the data with any ranges of solvent accessibility (e.g., 10 to 20 Å^2 or 0%–5%).
(5) Extracting the data obtained with any particular measurement (CD, DSC, Fl, etc.) or any specific method (Thermal, GdnHCl, urea, etc.).
(6) Limiting the data for a particular range of T, T_m, ΔT_m, ΔG, $\Delta\Delta G$, ΔG^{H_2O}, $\Delta\Delta G^{H_2O}$, ΔH, ΔC_p, and pH. The options of energy units, kJ and kcal, are available.
(7) Searching with reversibility.
(8) Number of transitions.
(9) Keywords, authors, and the year of publication.

6.2.4 Display and sorting options in ProTherm

Several options are available for displaying the results obtained from the search conditions. The users have the flexibility to select the options to display the results. **Figure 6.6b** shows the availability of display and sorting options. In the sorting option, one can sort the data based on the year of publication, wild and mutant

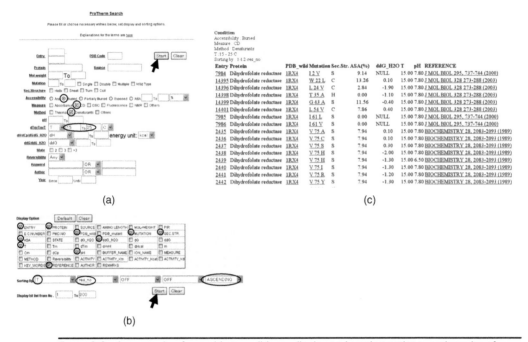

(a)

(b)

(c)

Figure 6.6 An example of searching conditions, display and sorting options, and results of ProTherm. (a) Main menu for the search options of ProTherm. In this example, items, buried (accessibility), CD (measure), denaturant (method) are selected from the menu, and T ($\Delta T_m/T_m/T$) is specified by filling the boxes for the values from 15°C to 25°C. (b) Display and sorting options of ProTherm. In this example, entry, protein, PDB wild, mutation, secondary structure, ASA, $\Delta\Delta G^{H_2O}$, T, pH, and reference are selected for the output. T and residue number are chosen for sorting the results in the order of priority. (c) Part of the results obtained from ProTherm.

residues, residue number, solvent accessibility, secondary structures, and thermodynamic parameters.

6.2.5 Tutorials and the usage of ProTherm

Detailed tutorials describing the usage of ProTherm are available at the home page. As an example, the necessary items to be filled or selected to search data for mutations in buried regions by denaturant denaturation and CD measurement at temperatures between 15°C and 25°C are shown in **Figure 6.6a**. In **Figure 6.6b**, the items to be selected for the output and sorting options are shown. In the sorting procedure, the first item has the topmost priority. In this figure, entry, protein, PDB wild, mutation, secondary structure, ASA, $\Delta\Delta G^{H_2O}$, T, pH, and reference are selected for the output. The selected outputs are sorted with temperature as the first priority and residue number as the second priority. The final results obtained from the search conditions (**Figure 6.6a**) and sorting options of necessary items (**Figure 6.6b**) are shown in **Figure 6.6c**. Selecting the data in buried, partially buried, and exposed regions has been used to understand the stability of protein mutants at different locations (Gromiha et al. 1999b,c; 2000; 2002b).

	To	Gly	Ala	Val	Leu	Ile	Cys	Met	Phe	Tyr	Trp	Pro	Ser	Thr	Asn	Gln	Asp	Glu	Lys	Arg	His
	Gly	---	230	57	5	0	12	0	10	2	6	14	55	2	5	26	28	24	6	19	17
	Ala	113	---	131	61	20	29	24	21	5	8	85	74	44	10	18	14	13	41	5	11
	Val	73	474	---	258	288	47	64	99	51	24	6	46	89	28	0	26	14	13	23	12
	Leu	34	350	114	---	88	55	40	86	2	4	18	16	14	4	4	13	32	12	23	5
	Ile	41	244	430	236	---	24	62	93	12	8	7	22	58	3	1	9	20	10	9	12
	Cys	4	137	17	26	1	---	2	2	2	1	1	105	86	0	0	0	4	0	0	0
	Met	18	65	50	124	45	1	---	17	2	0	0	0	6	1	0	6	6	18	21	0
	Phe	5	149	24	117	18	3	10	---	95	63	0	24	5	21	2	4	4	7	0	7
	Tyr	31	99	4	25	2	20	1	185	---	72	4	11	4	6	8	17	1	4	2	9
From	Trp	0	31	1	18	0	3	1	133	69	---	0	2	0	1	5	3	3	0	3	13
	Pro	66	187	7	17	2	9	0	4	2	4	---	68	10	4	4	6	5	5	8	3
	Ser	28	212	19	18	10	29	2	13	6	2	8	---	28	20	4	66	6	20	37	25
	Thr	37	204	126	32	103	38	11	24	28	2	3	118	---	20	27	18	92	7	40	21
	Asn	34	155	6	6	51	6	12	4	0	1	1	33	10	---	5	97	22	13	11	41
	Gln	42	74	3	23	5	12	3	3	4	0	9	5	1	14	---	10	34	33	14	6
	Asp	58	199	8	9	9	33	4	14	7	8	12	20	10	158	11	---	64	67	16	54
	Glu	50	299	67	48	6	10	16	25	29	17	8	29	12	13	103	32	---	134	30	13
	Lys	85	227	23	12	35	18	46	46	18	27	23	13	18	22	41	13	99	---	79	37
	Arg	40	161	14	13	0	19	18	3	0	1	1	8	2	0	20	1	71	26	---	67
	His	38	112	10	47	0	7	0	10	69	2	24	6	21	47	51	21	17	6	12	---

0-50:Blue ; 51-100:Magenta ; >100:Red

FIGURE 6.7   Frequency of occurrence of mutants in ProTherm.

6.2.6   Frequency of occurrence of mutants

The frequency of occurrence of mutants in protein structures is presented in **Figure 6.7**. The hydrophobic substitutions Val → Ala, Leu → Ala, Val → Ile, Tyr → Phe, etc. have been carried out frequently along with other substitutions, Glu → Ala, Asp → Asn, Thr → Ser, Cys → Ser, etc. These substitutions have been done to understand the influence of specific interactions, such as hydrophobic, electrostatic, and disulfide bonds, for the stability of protein mutants.

6.3   Relative contribution of noncovalent interactions to protein stability

The partitioning of free energy components contributing toward protein stability provides deep insights to understand protein stability. Ponnuswamy and Gromiha (1994) computed various free energy terms, contributing toward the folded state and unfolded state. The conformational stability has been computed using the difference between the folded and unfolded state free energies.

6.3.1   Computation of free energy terms

The folded state free energy is given by the sum of various free energies, such as hydrophobic, hydrogen bonding, electrostatic, van der Waals, and disulfide bonding. The unfolded state free energy is represented by entropic and nonentropic components.

6.3.2   Folded state free energies

The folded state free energy is given by

$$G_F = G_{hy} + G_{el} + G_{hb} + G_{vw} + G_{ss}, \tag{6.10}$$

where G_{hy}, G_{el}, G_{hb}, G_{vw}, and G_{ss} are, respectively, hydrophobic, electrostatic, hydrogen bonding, van der Waals, and disulfide bonding free energies. The computations of G_{hy}, G_{el}, G_{hb}, and G_{vw} have been illustrated in **Section 3.9**.

The site-directed mutagenesis experiments showed that the contribution of a disulfide bond to protein stability is in the range of 1.5 to 3.5 kcal/mol (Gromiha et al. 1999a). Furthermore, from the analysis of disulfide bonds in protein structures, Thornton (1981) reported that the contribution of a disulfide bond is approximately 2.3 kcal/mol. Accordingly, the value of 2.3 kcal/mol was used for each disulfide bond, and the free energy due to disulfide bond (G_{ss}) is given by

$$G_{ss} = 2.3 N_{ss}, \tag{6.11}$$

where N_{ss} is the number of disulfide bonds in a protein.

The van der Waals free energy was computed using the linear relationship between the van der Waals free energy and the number of residues ($G_{vw} = 8.885 + 0.1413N$), as well as with the AMBER force field (see **sections 3.9.4 and 3.9.5**).

6.3.3   Unfolded state free energies

The unfolded state free energy is given by

$$G_U = G_{en} + G_{ne} \tag{6.12}$$

The free energy of the unfolded protein mainly depends on configurational entropy, and it could be assumed as $-T \Delta S$ (Tanford, 1962). It is approximately estimated as 1.2 kcal/mol/residue (Kauzmann, 1959). It has been suggested that the unfolded protein contains some of the hydrogen bonds that are presented in the folded state, and it is approximated as half of the hydrogen bonds to that of the folded protein. Hence, the unfolded state free energy is given by

$$G_u = (1.2N + (1/2) G_{hb}) \, \text{kcal/mol}. \tag{6.13}$$

6.3.4   Prediction of protein stability

The folded and unfolded state free energies have been combined using a multiple regression technique, and the coefficients for each term have been evaluated. Interestingly, the coefficients of all the terms are close to unity. The conformational stability is computed using the following equation:

$$\Delta G = 1.584 + 1.008 G_{hy} + 1.016 E_{el} + 1.022 G_{hb} + 0.842 G_{vw} + 0.981 G_{ss} - G_U. \tag{6.14}$$

TABLE 6.2   Relative contributions of free energies to protein stability

Free energy	Contribution (%)
Hydrophobic	50.8
Hydrogen bonding	27.1
van der Waals	27.1
Electrostatic	6.4
Disulfide	1.1

This equation has been tested with a set of the then available 16 proteins, and the stability has been predicted within the deviation of 0.276 kcal/mol.

The application of the method is illustrated with two examples. T4 lysozyme shows moderate stability, and the experimental free energy difference is 12.8 kcal/mol. The computed free energy terms are $G_{hy} = 102.5$ kcal/mol; $G_{el} = 26$ kcal/mol; $G_{hb} = 109.0$ kcal/mol; $G_{vw} = 32.1$ kcal/mol, $G_{ss} = 0$ kcal/mol, and $G_U = 256.8$ kcal/mol. The conformational stability has been computed using these free energy values and substituting them in Equation 6.14 and is estimated to be 12.91 kcal/mol. This result shows the very close agreement between the computed and experimental values, and the deviation is 0.11 kcal/mol.

Ooi and Oobatake (1988) reported a ΔG value of 40.8 kcal/mol for taka amylase, a higher value than that of other proteins. The contributions of various free energies have been computed, and the conformational stability is estimated to be 38.5 kcal/mol, close to the experimental value ($G_{hy} = 335.71$; $G_{el} = 34.4$ kcal/mol; $G_{hb} = 331.0$ kcal/mol; $G_{vw} = 76.43$ kcal/mol, $G_{ss} = 9.2$ kcal/mol, and $G_U = 748.1$ kcal/mol).

6.3.5   Relative contributions to protein stability

The computations of various free energy terms in 39 proteins showed that the hydrophobic and hydrogen bonding terms are two dominating members (**Table 6.2**) in imparting stability to the folded state of globular proteins (Ponnuswamy and Gromiha, 1994). As suggested by Pace (1990), the electrostatic contribution is much smaller than the contributions from hydrophobic and hydrogen bonding interactions. The analysis on different contributions shows that on an average the factors hydrophobic, hydrogen bonding, van der Waals, electrostatic, and disulfide interactions contribute, respectively, 50.8%, 27.1%, 14.6%, 6.4%, and 1.1% levels to protein stability. In this estimation, as the unfolded state is assumed to have half the number of hydrogen bonds seen in the folded state, the hydrogen bond energy is taken to be one-half that reported for the folded state. These relative figures present us a picture that when the unfolded state starts folding progressively, the hydrophobic contribution increases, steadily overcoming the entropic factor and thus driving the chain to the folded state, and the other contributions, especially hydrogen bonds and van der Waals forces, emerge to keep the folded state from falling apart.

6.4   Stability of thermophilic proteins

Proteins from thermophilic organisms exhibit high thermal stability, and they proliferate at temperatures around 80°C to 100°C. These proteins have structures that are very similar to their mesophilic homologues. Understanding the structural basis for the enhanced stability of thermophilic organisms relative to their mesophilic counterparts is a challenging problem. Several methods have been proposed for understanding the thermal stability of thermophilic proteins and reported that various factors such as increase in number of hydrogen bonds, balance between packing and solubility, helical propensity, salt bridges, ion-pairs, and van der Waals contacts enhance the stability. Detailed reviews have been available in literature for getting deep insights on these aspects (Jaenicke and Bohm, 1998; Ladenstein and Antranikian, 1998; Szilagyi and Zavodsky, 2000; Kumar and Nussinov, 2001; Yano and Poulos, 2003; Mozo-villiarias and Querol, 2006).

6.4.1 Prokaryotic growth temperature database

Huang et al. (2004) developed a "Prokaryotic Growth Temperature database (PGTdb)," which contains temperature data for prokaryotic organisms, bacteria, and archaea. PGTdb integrates microbial growth temperature data from literature with their nucleotide/protein sequence and protein structure data from related databases. In addition, the taxonomy and ribosomal RNA sequence(s) of an organism are linked through NCBI taxonomy and the Ribosomal RNA Operon Copy Number Database, respectively. PGTdb is available at http://pgtdb.csie.ncu.edu.tw. **Figure 6.8** shows the organization of PGTdb, which integrates the information on

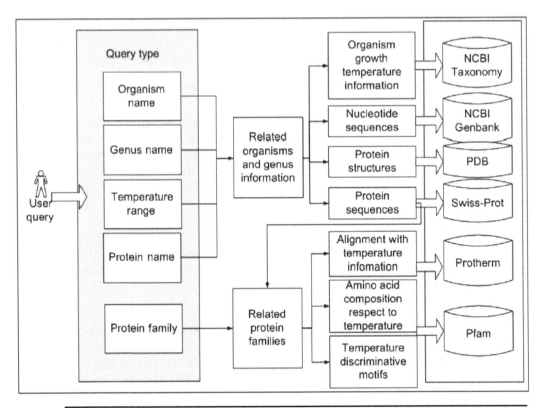

Figure 6.8 Organization of PGTdb. Figure was taken as a screenshot from http://pgtdb.csie.ncu.edu.tw/.

check the following check box(es) and press "search" to display detailed information:

☐ Nucleotide Sequence   ☐ Protein Sequence   ☐ Protein Structure   ☐ Temperature Class   ☐ Growth Temperature   ☑ Optimal Growth Temperature   [Search]

	Optimal Growth Temperature
Desulfurococcus	
Desulfurococcus saccharovorans	≥100°C
Hyperthermus	
Hyperthermus butylicus	95°C–106°C
Pyrobaculum	
Pyrobaculum aerophilum	100°C
Pyrobaculum islandicum	100°C
Pyrobaculum organotrophum	102°C
Pyrococcus	
Pyrococcus endeavori	100°C
Pyrococcus furiosus	100°C
Pyrococcus furiosus DSM 3638	100°C
Pyrococcus sp.	100°C
Pyrococcus sp. GB-3A	100°C
Pyrococcus sp. GB-D	100°C
Pyrococcus sp. GI-H	100°C
Pyrococcus sp. GI-J	100°C
Pyrococcus sp. JT1	100°C
Pyrococcus sp. MV1019	100°C
Pyrococcus sp. MZ14	100°C
Pyrococcus sp. MZ4	100°C
Pyrococcus sp. ST700	100°C
Pyrococcus woesei	100°C–103°C
Pyrodictium	
Pyrodictium brockii	105°C
Pyrodictium occultum	105°C
Pyrolobus	
Pyrolobus fumarii	106°C

Search information with keywords

○ Species name
◉ Genus name
○ NCBI Taxonomy ID
○ PDB ID
○ Pfam family
○ EC number
○ Swiss-Prot Entry Name

[Search]

Search species with temperature information

☐ Optimal growth temperature: ___ °C
☑ Optimal growth temperature range: 100 °C to 200 °C
☐ Growth temperature range: ___ °C to ___ °C
☑ Growth temperature class: Hyperthermophile(>80°C) ▾

[Search]

(a)   (b)

FIGURE 6.9 (a) Searching PGTdb for the hyperthermophilic organisms with optimal growth temperature between 100 and 120 degrees and (b) result obtained for the search.

names of the protein, organism, sequence, structure, thermodynamics, as well as environmental growth temperature. PGTdb can be searched with different options, including keywords and temperature. The search with optimal growth temperature between 100 and 120 degrees and classified as hyperthermophilic is shown in **Figure 6.9a**. The result obtained with the search is displayed in **Figure 6.9b**.

6.4.2 Relationship between melting temperature and average environmental temperature

Gromiha et al. (1999d) collected the melting temperature and the average environmental temperature for a set of mesophilic and thermophilic proteins and analyzed the relationship between them. The analysis showed a direct relationship between the average environmental temperature (T_{env}) and the melting temperature (T_m) of proteins in each family, and the correlation coefficient is 0.91; the corresponding regression equation is

$$T_m = 24.4 + 0.93T_{env}. \tag{6.15}$$

Furthermore, studies on elongation factors of the Tu family (EF-Tu) also showed a direct relationship between melting and environmental growth temperatures (Gaucher et al. 2003; 2008).

6.4.3 Sequence analysis

In early years, investigations have been carried out using the difference of amino acid sequences (and compositions) between mesophilic and thermophilic proteins for understanding the stability. Argos et al. (1979) observed that Gly, Ser, Lys, and Asp in mesophiles are generally substituted by Ala, Thr, Arg, and Glu, respectively, in thermophiles to enhance the stability. In addition, it has been reported that the substitution of Trp by Tyr increased the stability (Gromiha et al. 1999d). In thermophiles, the occurrence of Arg and Tyr is higher, and Cys and Ser are lower than that of mesophilic proteins (Kumar et al. 2000).

Ponnuswamy et al. (1982) analyzed the relationship between amino acid composition and thermal stability, and reported that the set of amino acid residues Asp, Cys, Glu, Lys, Leu, Arg, Trp, and Tyr enhances the stability and the set of residues Ala, Asp, Gly, Gln, Ser, Thr, Val, and Tyr decreases the stability. They found a correlation between the melting temperature and the amino acid composition of a stabilizing and destabilizing group of amino acids. An empirical relationship has been reported relating to the melting temperature and the amino acid composition. It is given by

$$T_m = 64.462 + 0.894 \, X1 - 0.591 \, X2, \tag{6.16}$$

where X1 and X2 are the amino acid compositions (in %) of the stabilizing and destabilizing groups. Later, Zeldovich et al. (2007) analyzed the relationship between amino acid compositions of 204 proteomes and the optimal growth temperature, and showed that there is a strong correlation ($r = 0.93$) between the composition of the combination of amino acids, Ile, Val, Tyr, Trp, Arg, Glu, and Leu, and the optimal growth temperature.

Fukuchi and Nishikawa (2001) performed systematic comparisons between proteins from thermophilic bacteria and mesophilic bacteria, in terms of the amino acid composition of the protein surface and the interior, as well as the entire amino acid chains, by using sequence information from the genome projects. They reported that there is a significant difference between the amino acid compositions of thermophilic and mesophilic proteins at the protein surface. On the other hand, the interior composition was not distinctive between the thermophilic and mesophilic proteins.

6.4.4   Amino acid properties and protein stability

Gromiha et al. (1999d) analyzed the thermostability of 16 different families by comparing mesophilic and thermophilic proteins with 48 various physicochemical, energetic, and conformational properties. They found that the increase in shape (s), which is defined as the location of the branch point in the side chain (van Gunsteren and Mark, 1992), increases the thermostability, whereas an opposite trend is observed for Gibbs free energy change of hydration for native proteins, G_{hN}. A good correlation is observed between these two properties, and the simultaneous increases of $-G_{hN}$ and s is necessary to enhance the thermostability from mesophiles to thermophiles. These results suggest that the stability of thermophilic proteins may be achieved by a balance between better packing and solubility. Facchiano et al. (1998) compared the structures of 13 thermophilic proteins with their homologues and reported that the intrinsic helical propensities of amino acids play a major role to the stability of thermophilic proteins.

Pakula and Sauer (1990) demonstrated that the increase in side-chain hydrophobicity has a reverse relationship with the protein stability, which was termed as "reverse hydrophobic effect." Recently, Gromiha (2009) addressed this problem with several examples of mutants that span at different locations in the protein structure based on the secondary structure and the solvent accessibility. The

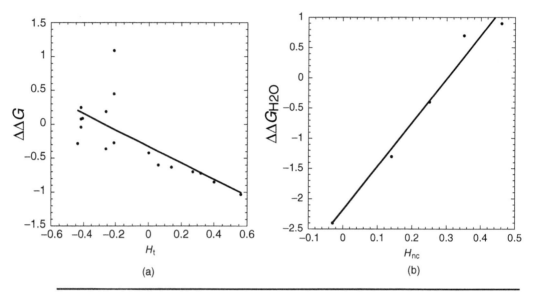

Figure 6.10   (a) Relationship between thermodynamic transfer hydrophobicity and free energy change due to the replacement of Lys15 in bovine trypsin inhibitor; (b) Correlation between normalized consensus hydrophobicity and protein mutant stability (E49 in 1WQ5). Figure was adapted from Gromiha (2009).

results showed that the stability change upon single coil mutation at exposed region is reversely correlated with hydrophobicity with a single exception. **Figure 6.10a** shows the inverse relationship between thermodynamic transfer hydrophobicity and free energy change due to the replacement of Lys15 in bovine trypsin inhibitor ($r = -0.73$). The physical basis for this effect has been interpreted as follows: (i) the amino acid residue is more exposed to water in the folded state than in the unfolded state (Pakula and Sauer, 1990), (ii) greater burial of nonpolar surface in the denatured state than in the native state (Shortle et al. 1992), and (iii) due to the existence of nonnative interactions in the folding transition state (Zarrine-Afsar et al. 2008). In addition, this relationship exists in partially buried coil mutants. The stability of exposed helical mutants is governed by conformational properties. In buried and partially buried helical and strand mutants, properties reflecting hydrophobicity have a direct relationship with stability (**Figure 6.10b**), whereas an opposite relationship was obtained with entropy and flexibility.

6.4.5   Dipeptide composition and protein stability

Ding et al. (2004) analyzed the influence of dipeptide composition on protein thermostability. They have compared the normalized dipeptide composition between mesophilic proteins and thermophilic proteins in archael and bacterial proteins. For archaeal proteins, the compositions of VK, KI, YK, IK, KV, KY, and EV increased significantly, and the compositions of DA, AD, TD, DD, DT, HD, DH, DR, and DG decreased significantly; and for bacterial proteins, the compositions of KE, EE, EK, YE, VK, KV, KK, LK, EI, EV, RK, EF, KY, VE, KI, KG, EY, FK, KF, FE, KR, VY, MK, WK, and WE increased significantly, and the compositions of WQ, AA,

QA, MQ, AW, QW, QQ, RQ, QH, HQ, AD, AQ, WL, QL, HA, and DA decreased significantly. They suggested that these characteristic dipeptides are correlative to protein thermostability.

6.4.6 Structural analysis

The availability of the three-dimensional structure of thermophilic proteins and their mesophilic counterparts paved a way to understand the structural principles governing the stability of thermophilic proteins. The comparative studies on mesophilic and thermophilic proteins suggest that ion pairs, cation-π interactions, hydrogen bonds, hydrophobic packing, etc. are important for the stability.

6.4.7 Electrostatic interactions

Szilagyi and Zavodszky (2000) compiled a database of mesophilic and thermophilic proteins, which comprises 64 mesophilic and 29 thermophilic subunits, representing 25 protein families. A systematic comparison has been made between mesophilic and thermophilic proteins using 13 different structural parameters. They reported that different protein families adapt to higher temperatures by different sets of structural devices. Furthermore, there is an increase in the number of ion pairs with the increasing growth temperature.

The importance of ion pairs for enhanced thermal stability has also been stressed by other investigators. Kumar et al. (2000) analyzed a set of 18 nonredundant families of thermophilic and mesophilic proteins and observed that the salt bridges and side chain to side chain hydrogen bonds increase in majority of the thermophilic proteins. Grimsley et al. (1999) suggested that protein stability can be increased by improving the columbic interactions among charged groups on the protein surface. Das and Gerstein (2000) compared a set of 12 genomes (from four thermophilic archaeons, one eukaryote, six mesophilic eubacteria, and one thermophilic eubacteria) and showed that thermophiles have a greater content of charged residues than mesophiles, both at the overall genomic level and in α-helices. Chakravarty and Varadarajan (2002) reported that salt bridges are significantly higher in thermophiles than mesophiles. The additional salt bridges in thermophiles are almost exclusively in solvent-exposed regions, and 35% are in the same element of secondary structure. Helices in thermophiles are stabilized by intrahelical salt bridges and by an increase in negative charge at the N-terminus.

Xiao and Honig (1999) calculated the electrostatic contributions to the folding free energy of several hyperthermophilic proteins and their mesophilic homologs and reported that the electrostatic interactions are more favorable in the hyperthermophilic proteins. The electrostatic free energy is found not to correlate directly with the number of ionizable amino acid residues, ion pairs, or ion pair networks in a protein but rather depends on the location of these groups within the protein structure. Furthermore, ion pairs located on the protein surface also provide stabilizing interactions in number of cases. Essentially they suggested that many hyperthermophilic proteins enhance electrostatic interactions through the optimum placement of charged amino acid residues within the protein structure.

Dominy et al. (2004) suggested that the electrostatic interactions of charged residues in the folded state and the dielectric response of the folded protein are the two key factors for enhancing the stability of thermophilic proteins. The dielectric response for proteins in a "thermophilic series" globally modulates the thermal stability of its members, with the calculated dielectric constant for the protein increasing from mesophiles to hyperthermophiles. Furthermore, the contribution of electrostatic interactions to the stability of the folded state is more favorable for thermophilic proteins than for their mesophilic homologues. This reveals that electrostatic interactions play an important role in determining the stability of proteins at high temperatures. Greaves and Warwicker (2009) showed that electrostatic properties distinguish hyperthermophilic, psychrophilic, and mesophilic proteins, and protein stability is determined by the location of charges.

6.4.8 Cation-π interactions

Gromiha et al. (2002c) analyzed the influence of cation-π interactions to the stability of thermophilic proteins and reported that thermophilic proteins have more number of cation-π interactions than their mesophilic counterparts. Furthermore, Tyr has a greater number of such interactions with Lys in thermophilic proteins. Chakravarty and Varadarajan (2002) also reported similar results that cation-π interactions play an important role to the stability of thermophilic proteins. Ibrahim and Pattabhi (2004) analyzed the influence of weak noncanonical interactions and found that a cumulative effect of weak interactions seems to be important in the thermal stability of proteins. Furthermore, the number of aromatic amino acids in the thermophiles is more than mesophiles, and hence a large number of aromatic clusters were observed in this class.

6.4.9 Hydrogen bonds

Vogt et al. (1997) examined 16 families of proteins with different thermal stability by comparing their respective fractional polar atom surface areas and the number and type of hydrogen bonds and salt links between explicit protein atoms. In over 80% of the families, correlations were found between the thermostability of the familial members and an increase in the number of hydrogen bonds as well as an increase in the fractional polar surface, which results in added hydrogen bonding density to water. They suggested that the increased hydrogen bonding may provide the most general explanation for thermal stability in proteins. Furthermore, it has been reported that there is an increase of main-chain hydrogen bonds in thermophilic proteins (Sadeghi et al. 2006).

Gromiha (2001) analyzed the medium- and long-range contacts in mesophilic and thermophilic proteins of 16 different families and found that the thermophiles prefer to have contacts between residues with hydrogen-bond-forming capability. Apart from hydrophobic contacts, more contacts are observed between polar and nonpolar residues in thermophiles than mesophiles. Residue-wise analysis showed that Tyr has good contacts with several other residues, and Cys has considerably higher long-range contacts in thermophiles compared with mesophiles.

6.4.10 Hydrophobic interactions

Imanaka et al. (1986) proposed that the single amino acid substitutions based on an increase in hydrophobicity and stabilization of helices enhance the thermal stability of enzymes. Menendez-Arias and Argos (1989) reported that the decrease in flexibility and increase in hydrophobicity in α-helical regions are the major factors to enhance the thermostability of proteins. Kannan and Vishveshwara (2000) analyzed the contribution of aromatic interactions to stability of thermophilic proteins and reported that the thermophilic protein families have additional aromatic clusters or enlarged aromatic networks than their mesophilic homologues. The additional aromatic clusters identified in the thermophiles are smaller in size and are largely found on the protein surface. The aromatic clusters are found to be relatively rigid regions of the surface, and often the additional aromatic cluster is located close to the active site of the thermophilic enzyme. Further analysis of the packing geometry of the pairwise aromatic interaction in the additional aromatic clusters shows a preference for a T-shaped orthogonal packing geometry in thermophilic proteins.

Saraboji et al. (2005b) analyzed the role of different free energy contributions to the stability of thermophilic proteins and observed that the main-chain hydrophobic free energy plays an important to the enhanced stability of thermophilic proteins. Recently, Baldasseroni and Pascarella (2009) analyzed a set of oligomeric thermophilic and hyperthermophilic enzymes and showed that enhanced compactness at subunit interfaces is an important factor for the increased stability of hyperthermophilic enzymes.

6.4.11 Discrimination of mesophilic and thermophilic proteins

Zhang and Feng (2006) analyzed the distribution of two neighboring residues in the sequences of thermophilic and mesophilic proteins and reported that the occurrence of EE, KK, RR, PP, KI, VV, VE, KE, and VK in thermophilic proteins is significantly higher than that of mesophilic proteins, and an opposite trend was observed for QQ, AA, EQ, LL, QA, QL, NN, KQ, QG, RQ, QT, and AQ. On the basis of the information of dipeptide composition, a statistical method has been developed for discriminating thermophilic and mesophilic proteins, which discriminated the mesophilic and thermophilic proteins with the accuracy of 86.3%.

Recently, a method based on neural networks using amino acid composition was developed for discriminating mesophilic and thermophilic proteins (Gromiha and Suresh, 2008). This method could discriminate the mesophilic and thermophilic proteins with the accuracy of 89%. Interestingly, the hyperthermophilic proteins are better discriminated than thermophilic proteins (**Figure 6.11**). A Web server has been developed to discriminate mesophilic and thermophilic proteins, which takes the amino acid sequence as input and displays the type of the protein in the output (**Figure 6.12**).

6.4.12 Identifying the stabilizing residues in protein structures

Protein structures are stabilized by various interactions, including hydrophobic, hydrogen bonding, electrostatic, and van der Waals. On the basis of the importance of hydrophobic interactions, long-range contacts, and conservation of residues among proteins, Gromiha et al. (2004) proposed an algorithm for locating the stabilizing residues in proteins structures. The method was developed mainly for TIM barrel proteins, and the same was extended to be applicable to all globular protein structures.

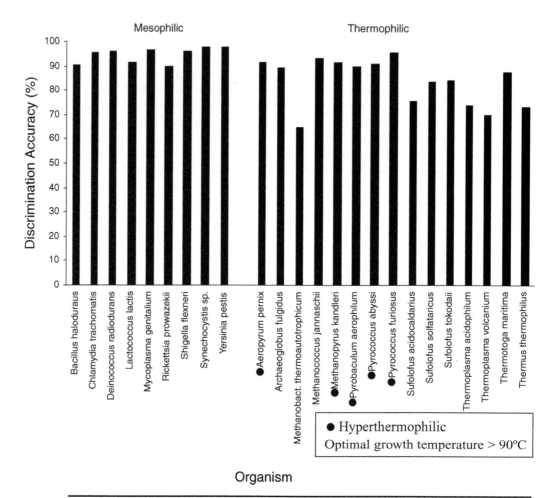

Figure 6.11 Discrimination accuracy in different mesophilic and thermophilic organisms. Figure was adapted from Gromiha and Suresh (2008).

Gromiha et al. (2004) computed the parameters, surrounding hydrophobicity (H_p, **Section 3.10.3**), long-range order (LRO, **Section 3.11.2**), stabilization center (SC, **Section 3.11.3**), and conservation score (**Section 2.3.6**) for all the residues in protein structures and imposed cutoff values for each term ($H_p \geq 20$ kcal/mol; LRO ≥ 0.02; SC ≥ 1; and conservation score ≥ 6) to identify the stabilizing residues. This method picked up 4.6% of residues in a considered set of 71 TIM barrel proteins. An example is shown in **Figure 6.13**. The identified stabilizing residues Cys126 and Gly229 in 1BTM (Triose Phosphate Isomerase from *Bacillus stearothermophilus*) are confirmed as stabilizing ones through thermodynamic and kinetic experiments (Kursula et al. 2002; Gonzalez-Mondragon et al. 2004). However, one can relax the conditions (give weight to each term and/or reduce the cutoff value) and get a larger number of stabilizing residues, which will be useful for selecting potential candidates for protein engineering experiments and protein design. A Web server

DISC-THERMAL: Discrimination of Mesophilic and Thermophilic Proteins.

Discriminating thermophilic proteins from their mesophilic counterparts is a challenging task, and it would help to design stable proteins.

We have systematically analyzed the amino acid compositions of 3075 mesophilic and 1609 thermophilic proteins belonging to 9 and 15 families, respectively. We found that the charged residues Lys, Arg, and Glu as well as the hydrophobic residues, Val and Ile have higher occurrence in thermophiles than mesophiles. We have developed a neural network-based method, which could discriminate the thermophiles from mesophiles at the 5-fold cross-validation accuracy of 89% in a dataset of 4684 proteins using amino acid composition. Moreover, this method is tested with 325 mesophiles in *Xylella fastidosa* and 382 thermophiles in *Aquifex aeolicus*, and it could successfully discriminate them with the accuracy of 91%.

DISC-THERMAL discriminates the mesophilic and thermophilic proteins using neural networks with amino acid composition as attributes. For discrimination, enter your sequence in a single letter code in the following box.

```
MDVIPTEAIWWRLFLLFTAVGVLAAGTVTAFFIYSLFKYRSSGQALGEDQGTAGRIYRI
MVESFVSGKSKYLLFVTGIIVMGLIVATIDETLYLEKSPPVEDALVVMVIGFQFGWQFE
YSVGGETVTTLNYLVVPSDTLIEFRVTSRDVFHAFGIPEFKNKIDAIPGILNSMWIKTP
DEPGKVYNAYCYELCGIGHSLMVGKVIVVDKEEFYNAYNSGPDVFSEYVNNVISKYK
```

Predict Reset

Residue	Occurrence	Composition(%)
A	13	5.56
D	10	4.27
C	2	0.85
E	15	6.41
F	15	6.41
G	20	8.55
H	2	0.85
I	19	8.12
K	12	5.13
L	18	7.69
M	6	2.56
N	8	3.42
P	10	4.27
Q	4	1.71
R	6	2.56
S	15	6.41
T	14	5.98
V	27	11.54
W	4	1.71
Y	14	5.98

Amino acid sequence seems to be a THERMOPHILIC Protein

FIGURE 6.12 Discrimination of mesophilic and thermophilic proteins using DISC-THERMAL.

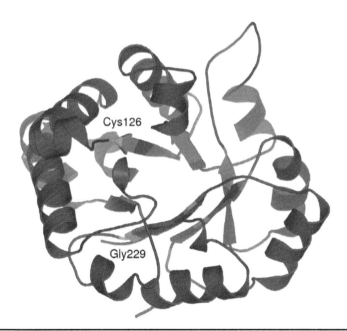

FIGURE 6.13 Stabilizing residues in a typical TIM barrel protein, 1btm. The α-helices are shown as spiral ribbons and the β-strands are drawn as arrows from the amino end to the carboxyl end of the β-strand. The stabilizing segments are highlighted with red color. The two stabilizing residues Gly229 and Cys126 observed by thermodynamic, and kinetic experiments are also indicated. The picture was generated using PyMOL program (DeLano, 2002). Figure was adapted from Gromiha et al. (2004).

has been developed for locating the stabilizing residues in protein structures, and the results are available at http://sride.enzim.hu (Magyar et al. 2005).

6.5   Analysis and prediction of protein mutant stability

At present, plenty of thermodynamic data are collectively available for the stability of proteins and mutants due to the development of ProTherm database. Several investigations have been carried out to understand and predict the stability of proteins upon amino acid substitutions. They are based on amino acid properties, database-derived potentials, multiple regression techniques, energy calculations, neural networks, support vector machines, etc.

The performance of these methods has been generally tested with the measures, accuracy, correlation coefficients, and mean-absolute error. The accuracy of distinguishing the stability of mutants (stabilizing/destabilizing) has been determined by using the following expression:

$$\text{Accuracy } (\%) = p^*100.0/N, \qquad (6.17)$$

where p is the total number of correctly discriminated residues, and N is the total number of data used for discrimination (also see Eqn. 5.13).

The correlation between the experimental and assigned stability (ΔT_m) has been calculated using Equation 5.14.

The mean-absolute error (MAE) is defined as the absolute difference between predicted and experimental stability values:

$$\text{MAE} = \frac{1}{N} \sum_i |X_i - Y_i|, \qquad (6.18)$$

where X_i and Y_i are the experimental and predicted stability values, respectively, and i varies from 1 to N, N being the total number of mutants.

6.5.1   Amino acid properties

Gromiha et al. (1999b,c; 2000; 2002b) analyzed the correlation between stability changes caused by buried, partially buried, and exposed mutations and changes in 49 physicochemical, energetic, and conformational properties. They found that properties reflecting hydrophobicity strongly correlated with stability of buried mutations, and there was a direct relation between the property values and the number of carbon atoms. Classification of mutations based on their location within helix, strand, turn, or coil segments improved the correlation of mutations with stability. Buried mutations within β-strand segments correlated better than those in α-helical segments, suggesting stronger hydrophobicity of the β-strands (Gromiha et al. 1999b,c).

The stability changes caused by partially buried mutations in ordered structures (helix, strand, and turn) correlated most strongly and were mainly governed by hydrophobicity. Because of the disordered nature of coils, the mechanism underlying their stability differed from that of the other secondary structures: The stability changes due to mutations within the coil were mainly influenced by the effects of entropy. Further classification of mutations within coils, based on their hydrogen-bond forming capability, led to much stronger correlations (Gromiha et al. 1999b, 2000).

In exposed mutations, the hydration entropy was the major contributor to the stability in helical segments; other properties responsible for size and volume of molecule also correlated significantly with stability. Classification of coil mutations based on their locations in the Ramachandran ($\Phi - \Psi$) map improved the correlation significantly, demonstrating the existence of a relationship between stability and strain energy, which indicates that the role of strain energy is very important for the stability of surface mutations (Gromiha et al. 2002b). **Table 6.3** shows the properties that have significantly high correlation with protein mutant stability.

The influence of neighboring residues in the amino acid sequence and the residues surrounded by the mutated residue in protein structures have been analyzed for understanding the protein mutant stability. It has been observed that the inclusion of sequence and structural information raised the correlation between amino acid properties and the protein mutant stability of partially buried and exposed mutations, indicating the influence of surrounding residues on the stability of partially buried and surface mutations.

TABLE 6.3 Correlation between amino acid properties and protein mutant stability

Mutation	Highest absolute single property correlation (r)	Properties
Buried mutations		
Hydrophobic	0.72	ΔC_{ph}, R_f, $-T\Delta S$
Helix	0.73	ΔC_{ph}, R_f
Strand	0.81	E_l, M_w, ΔC_{ph}
Partially buried mutations		
Hydrophobic	0.66	H_{gm}, R_f, V^0
Helix	0.63	ΔASA, $-T\Delta S_h$, ΔC_{Ph}, F
Strand	0.71	P_β, pHi, H_p, ΔASA
Turn	0.63	R_a, H_p, B_r
Exposed mutations		
Helix	0.93	ΔASA, V^0
Strand	0.62	R_a, $-T\Delta S_h$
Turn	0.61	P_t, $P_{\Phi-\Psi}$, B_l, M_w
Coil	0.71	ΔH_c, P_β, μ, P_c, N_m
Coil (left-handed)	0.81	
H-bond	0.86	H_t, v

H_t, thermodynamic transfer hydrophobicity; H_p, surrounding hydrophobicity; pHi, isoelectric point; M_w, molecular weight; B_1, bulkiness; R_f, chromatographic index; μ, refractive index; E_1, long-range nonbonded energy; P_β, P_t, and Pc are, respectively, β-structure, turn, and coil tendencies; F, mean r.m.s. fluctuational displacement; B_r, buriedness; R_a, solvent-accessible reduction ratio; V°, partial specific volume; N_m, average medium-range contacts; H_{gm}, combined surrounding hydrophobicity (globular and membrane); ΔASA, solvent accessible surface area for unfolding; $-T\Delta S_h$, unfolding entropy change of hydration; ΔC_{ph}, unfolding hydration heat capacity change; ΔH_c, unfolding enthalpy changes of chain; $-T\Delta S$, unfolding entropy change; v, volume (number of nonhydrogen side-chain atoms); $P_{\Phi-\Psi}$, backbone dihedral probability.

H_t, H_p, H_{gm}, $-T\Delta S_h$, ΔH_c, and $-T\Delta S$ in kcal/mol; pHi in pH units; E_1 in kcal/mol/atom; B_1 and ΔASA in Å^2; F in Å; V° in m³/mol ($\times 10^{-6}$); ΔC_{ph} in cal/mol/K; and the rest are dimensionless quantities.

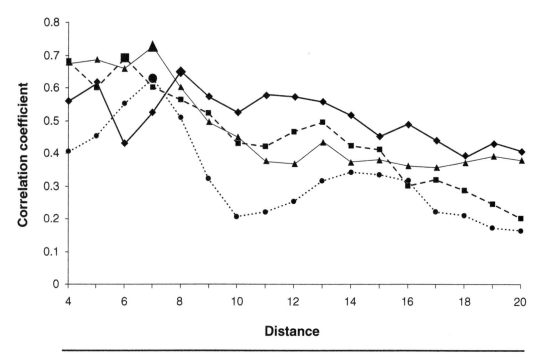

FIGURE 6.14 The single correlation coefficients obtained for different distances from the mutant residue (radii of the spheres around the mutant residue to accommodate the surrounding residues); continuous thick line: helix; dashed thick line: strand; continuous thin line: coil; dashed thin line: hydrophobic mutations. The highest correlation coefficient for each set of mutations is highlighted with big symbols. Figure was adapted from Gromiha et al. (2000).

The correlation coefficients obtained using structural information for different distances are shown in **Figure 6.14**. The observed highest correlation coefficient for each set of mutations showed that all the mutations prefer the distance of 6 to 8 Å and specifically, helical mutations prefer 8 Å; strand mutations, 6 Å; and coil mutations, 7 Å. The distance of 6 to 8 Å is sufficient to accommodate the nearest neighboring residues and several residues that are far in sequence level. On the other hand, the influence of sequence and structure is not significant in buried mutations. These results revealed that hydrophobicity was the major factor in determining the stability of buried mutations, whereas hydrogen bonds, other polar interactions, and hydrophobic interactions were all important determinants of the stability of partially buried mutations.

Further analysis on the stability of specific proteins showed that the mechanism behind protein stability varies for different proteins (Gromiha, 2003). In lysozyme T4, the properties reflecting hydrophobicity, flexibility, turn and coil tendency, and long-range interactions show a strong correlation with stability. Entropy plays an important role, and the contribution of hydrophobicity is minimal in barnase. The stability of human lysozyme is attributed with both hydrophobicity and secondary structure. The stability of buried mutants in staphylococcal nuclease is influenced by hydrophobicity and physical properties. Furthermore, the classification of mutants based on secondary structure and solvent accessibility improved the correlation between amino acid properties and protein mutant stability (Saraboji et al. 2005a).

6.5.2 Effective potentials

The information about inter-residue interactions in protein structures has been successfully used to predict the stability of protein mutants. Gilis and Rooman (1996) analyzed a set of 106 surface mutations and reported that the correlation coefficient between experimental and computed free energy change using backbone torsion potentials (see **Section 3.6.2**) is 0.67 for all the mutations, and it rose up to 0.87 for a subset of 96 mutations. This torsion potential is mainly based on the neighboring residues in a sequence, and hence the local interactions along the chain are more dominant than hydrophobic interactions at the protein surface (Gilis and Rooman, 1996). Furthermore, the buried and partially buried mutations have been systematically analyzed with distance potentials (see **Section 3.6.3**; Gilis and Rooman, 1997). They reported that for a set of completely buried mutations, the combination of distance potential and torsion potential weighted by a factor of 0.4 yielded the correlation coefficient of 0.80 between the computed and measured changes in folding free energy. For mutations of partially buried residues, the best potential is a combination of torsion potential and a distance potential weighted by a factor of 0.7 and the correlation coefficient is 0.82. These results indicate that the distance potentials, dominated by hydrophobic interactions, represent best the main interactions stabilizing the protein core.

Topham et al. (1997) proposed an approach based on the knowledge of amino acid replacements that are tolerated within the families of homologous proteins of the known 3D structure. Amino acid variations in families of homologous proteins are converted into propensity and substitution tables. These tables provide quantitative information about the existence of an amino acid in a structural environment and the probability of replacement by any amino acids, and this approach is similar to the development of knowledge-based potentials. Hence, the information about local and long-range contacts is included in these environment-dependent propensity and substitution tables. It has been reported that the stability score obtained with these tables to a set of 159 mutations in T4 lysozyme showed a correlation of 0.83 with differences in melting temperatures. Furthermore, 86% of the mutants were correctly classified as stabilizing or destabilizing. Khatun et al. (2004) derived contact potentials and applied them to predict the stability of 1356 mutants in 11 different proteins. They reported that the correlation between experimental and predicted free energy changes is 0.47, and it increased when the potential has been applied to specific proteins.

Hoppe and Schomburg (2005) developed a knowledge-based discrimination function for predicting protein mutant stability, which consists of two parts: a pairwise energy function based on a distance- and direction-dependent atomic description of the amino acid environment and a torsion angle energy function. This method showed a correlation of 0.81 in a training set of 561 mutants and 0.74 in test set of 747 mutants.

Parthiban et al. (2007) analyzed protein stability upon point mutations using distance-dependant pair potential, representing mainly through-space interactions

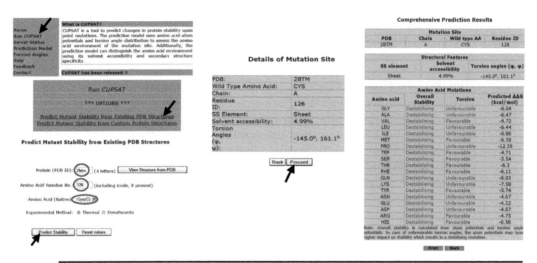

FIGURE 6.15   Prediction of the protein mutant stability using CUPSAT. The detailed procedure is illustrated in the figure with arrows.

and torsion angle potential, representing mainly neighboring effects. They have developed the potentials at various ranges of solvent accessibility and at different secondary structures. The method was trained and tested with two datasets of point mutations obtained from thermal ($\Delta\Delta G$) and denaturant denaturation ($\Delta\Delta G^{H_2O}$) experiments. This method showed the maximum correlation of 0.87 with a standard error of 0.71 kcal/mol between predicted and measured $\Delta\Delta G$ values and a prediction accuracy of 85.3% for discriminating the stabilizing and destabilizing protein mutants. For $\Delta\Delta G^{H_2O}$, they reported a correlation of 0.78 (standard error 0.96 kcal/mol) with a prediction efficiency of 84.65%. Furthermore, a Web server, CUPSAT, was developed for predicting the stability of protein mutants, and it is available at http://cupsat.uni-koeln.de/ (Parthiban et al. 2006). An example is shown in **Figure 6.15**. It takes the coordinates from protein three-dimensional structures and calculates the secondary structure, solvent accessibility, and other required parameters. In the output, it provides the stability of the mutant for all possible 19 substitutions.

6.5.3   Energy functions

Guerois et al. (2002) developed a computer algorithm, FOLDEF (for FOLD-X energy function), to provide a fast and quantitative estimation of the importance of the interactions contributing to the stability of proteins. It includes terms that have been found to be important for protein stability. The free energy of unfolding (ΔG) of a target protein is calculated using the equation:

$$\Delta G = W_{vdw}\Delta G_{vdw} + W_{solvH}\Delta G_{solvH} + W_{solvP}\Delta G_{solvP} + \Delta G_{wb} + \Delta G_{hbond}$$
$$+\Delta G_{el} + W_{mc}T\Delta S_{mc} + W_{sc}T\Delta S_{sc}, \qquad (6.19)$$

where ΔG_{vdw} is the sum of the van der Waals contributions of all atoms. ΔG_{solvH} and ΔG_{solvP} is the difference in solvation energy for apolar and polar groups, respectively, when going from the unfolded to the folded state. ΔG_{hbond} is the free energy difference between the formation of an intramolecular hydrogen-bond compared with intermolecular hydrogen-bond formation (with solvent). ΔG_{wb} is the extra stabilizing free energy provided by a water molecule making more than one hydrogen bond to the protein (water bridges) that cannot be taken into account with nonexplicit solvent approximations. ΔG_{el} is the electrostatic contribution of charged groups interactions. ΔS_{mc} is the entropy cost for fixing the backbone in the folded state. This term is dependent on the intrinsic tendency of a particular amino acid to adopt certain dihedral angles. ΔS_{sc} is the entropic cost of fixing a side-chain in a particular conformation. The energy values of ΔG_{vdw}, ΔG_{solvH}, ΔG_{solvP}, and ΔG_{hbond} attributed to each atom type have been derived from a set of experimental data, and ΔS_{mc} and ΔS_{sc} have been taken from theoretical estimates. The terms W_{vdw}, W_{solvH}, W_{solvP}, W_{mc}, and W_{sc} correspond to the weighting factors applied to the raw energy terms.

First, a training database of 339 mutants in nine different proteins was considered and optimized the set of parameters and weighting factors that best accounted for the changes in the stability of the mutants. The predictive power of the method was then tested using a blind test mutant database of 667 mutants, which showed a correlation of 0.83 with a standard deviation of 0.81 kcal/mol. This energy function uses a minimum of computational resources and can therefore easily be used in protein design algorithms, and in the field of protein structure and folding pathways prediction where one requires a fast and accurate energy function. FOLDEF is available via a Web-interface at http://fold-x.embl-heidelberg.de/.

Furthermore, Bordner and Abagyan (2004) developed empirical energy functions for predicting the stability of protein mutants and reported an accuracy of 0.79 between experimental and predicted stabilities. Masso and Vaisman (2008) combined computational energy-based approaches based on a four-body, knowledge-based, and statistical contact potentials, and machine learning techniques for predicting protein stability upon point mutation. They developed a Web server, AUTOMUTE, for prediction, and it is available at http://proteins.gmu.edu/automute. **Figure 6.16** shows the utility of AUTOMUTE, which can be used to predict the stability change based on ΔT_m, $\Delta\Delta G$, and $\Delta\Delta G^{H_2O}$. The stability change ($\Delta\Delta G^{H_2O}$) due to the mutation of D25E in 3PHV is illustrated in **Figure 6.16a**. The two-state prediction (stabilizing or destabilizing) has been selected with Random Forest method. The server predicted that the mutation destabilize the protein. The stability change upon mutation has been carried out with SVM regression (**Figure 6.16b**), and it predicted the stability change of -0.75 kcal/mol.

6.5.4 Machine learning techniques

Capriotti et al. (2004) proposed a neural-network-based method to predict if a given mutation increases or decreases the protein thermodynamic stability with respect to the native structure. Using a dataset consisting of 1615 mutations, this method correctly classifies >80% of the mutations in the database. Furthermore, when this method is coupled with energy-based methods, the joint prediction accuracy increases up to 90%, suggesting that it can also be used to increase the performance

AUTO–MUTE

AUTOmated server for predicting...
...functional consequences of amino acid MUTations in protEins

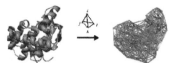

Stability Changes ($\Delta\Delta G^{H2O}$)

AUTO–MUTE Home

- Stability Changes ($\Delta\Delta G$)
- Stability Changes ($\Delta\Delta G^{H2O}$)
- Stability Changes (ΔT_m)

Activity Changes

Disease Potential of Human nsSNPs

Drug Susceptibility Changes

Structural Bioinformatics at
George Mason University

Questions or Comments?
mmasso@gmu.edu

	PDB ID (e.g., 3PHV)	Chain (use @ if null)	Mutation (e.g., D25E)	Temperature (°C, 0-100)	pH (-log[H+], 0-14)
Mutant #1	3PHV	@	D25E	25	7
Mutant #2				25	7
Mutant #3				25	7
Mutant #4				25	7
Mutant #5				25	7

Note: Use D25_ to obtain predictions for all 19 substitutions at the requested position.

Select a model for making predictions (details):

Classification (sign of $\Delta\Delta G\_H2O$): ⊙ Random Forest   ○ Support Vector Machine (SVM)
Regression (value of $\Delta\Delta G\_H2O$): ○ Tree Regression (REPTree)   ○ SVM Regression

[Submit Request]

PDB_ID	Chain	Mutation	Stability	Confid	Temp	pH	Vol	sT	Loc	Num	SS
3PHV	@	D25E	Decreased	0.51	25	7	14.6	0.16	B	0	C

When citing AUTO–MUTE please reference: Masso, M. & Vaisman, I. (2008) Accurate prediction of stability changes in protein mutants by combining machine learning with structure based computational mutagenesis, *Bioinformatics* 24, 2002-2009.

Glossary

Stability:	Increased (ddG_H2O >= 0) or Decreased (ddG_H2O < 0)
Confid:	Confidence (probability) in stability classification
Vol:	Mean volume of tessellation simplices for which mutated position is a vertex
sT:	Mean tetrahedrality measure of the simplices
Loc:	Location of the mutated position based on tessellation (S = surface, U = undersurface, B = buried)
Num:	Number of edge contacts with surface positions
SS:	Secondary structure of mutated position (H = helix, S = strand, T = turn, C = coil)

(a)

FIGURE 6.16 Applications of AUTOMUTE to predict protein mutant stability: (a) discrimination of stabilizing and destabilizing mutants and (b) prediction of protein stability change upon point mutation (D25E in 3PHV)

of preexisting methods, and generally to improve protein design strategies. They have also developed a method based on support vector machines for predicting the stability of protein mutants. In this method, the stability of protein mutants has been correctly assigned to an accuracy of 80%, and the correlation between experimental and computed stabilities is 0.71 (Capriotti et al. 2005). A Web server, I-Mutant2.0, has been developed for predicting the protein mutant stability and is available at http://gpcr.biocomp.unibo.it/cgi/predictors/I-Mutant2.0/I-Mutant2.0.cgi/.

Cheng et al. (2006) used support vector machines to predict protein stability changes for single amino acid mutations from both sequence and structural information. This method could discriminate the stabilizing and destabilizing protein mutants with an accuracy of 84%. They developed a Web server, Mupro, for predicting protein stability changes upon mutations, and it is available at http://www.igb.uci.edu/servers/servers.html.

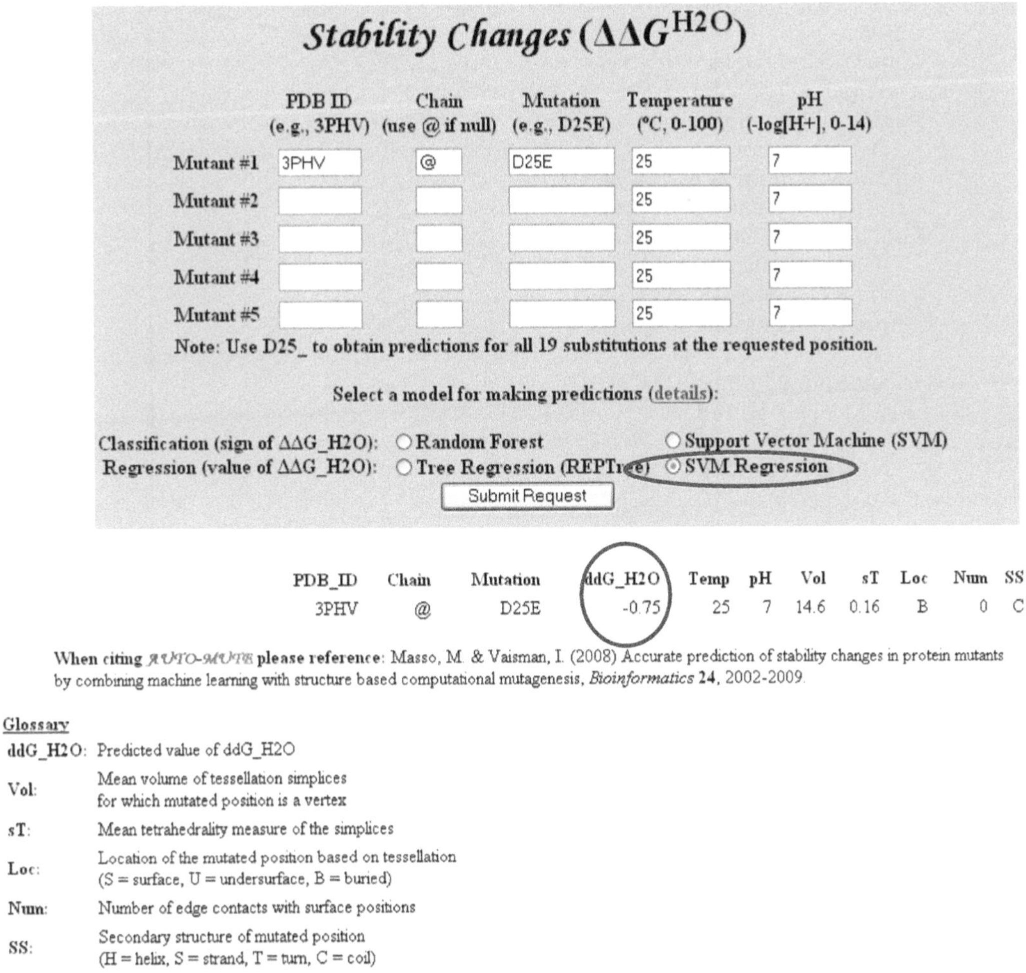

Stability Changes ($\Delta\Delta G^{H2O}$)

	PDB ID (e.g., 3PHV)	Chain (use @ if null)	Mutation (e.g., D25E)	Temperature (°C, 0-100)	pH (-log[H+], 0-14)
Mutant #1	3PHV	@	D25E	25	7
Mutant #2				25	7
Mutant #3				25	7
Mutant #4				25	7
Mutant #5				25	7

Note: Use D25_ to obtain predictions for all 19 substitutions at the requested position.

Select a model for making predictions (details):

Classification (sign of ΔΔG_H2O): ○ Random Forest   ○ Support Vector Machine (SVM)
Regression (value of ΔΔG_H2O): ○ Tree Regression (REPTree) ⊙ SVM Regression

Submit Request

PDB_ID	Chain	Mutation	ddG_H2O	Temp	pH	Vol	sT	Loc	Num	SS
3PHV	@	D25E	-0.75	25	7	14.6	0.16	B	0	C

When citing ꓫUTO-MUTE please reference: Masso, M. & Vaisman, I. (2008) Accurate prediction of stability changes in protein mutants by combining machine learning with structure based computational mutagenesis, *Bioinformatics* 24, 2002-2009.

Glossary

ddG_H2O:	Predicted value of ddG_H2O
Vol:	Mean volume of tessellation simplices for which mutated position is a vertex
sT:	Mean tetrahedrality measure of the simplices
Loc:	Location of the mutated position based on tessellation (S = surface, U = undersurface, B = buried)
Num:	Number of edge contacts with surface positions
SS:	Secondary structure of mutated position (H = helix, S = strand, T = turn, C = coil)

(b)

FIGURE 6.16 (Continued)

Huang et al. (2006, 2007a, 2007b) developed a method based on interpretable decision tree coupled with adaptive boosting algorithm and classification and regression tool for predicting protein stability upon amino acid substitutions. This method could correctly discriminate the stabilizing and destabilizing protein mutants at an accuracy of 82% in a dataset of 1859 single mutants. Furthermore, a correlation of 0.70 is obtained between the predicted and experimental stabilities. The performance of the method was also assessed with ROC (**Figure 6.17**), which showed an AUC of 0.83 (Gromiha et al. 2008). The main features of the method are as follows (Huang et al. 2007b): (i) It is based on the neighboring residues of short window length, (ii) it can predict the stability from amino acid sequence alone, (iii) developed different servers for both discrimination and prediction, and integrated them together, (iv) utilized the information about experimental conditions,

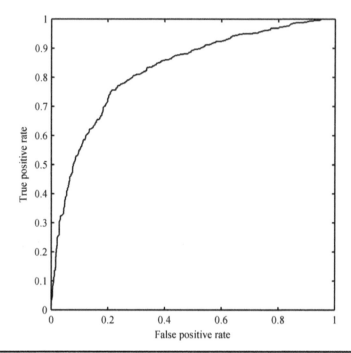

Figure 6.17   ROC curve for discriminating the stabilizing and destabilizing mutants.

pH and T, and (v) implemented several rules for discrimination and prediction from the knowledge of experimental stability and input conditions.

A Web server (iPTREE-STAB) has been set up for discrimination and prediction and is available at http://bioinformatics.myweb.hinet.net/iptree.htm. The utility of the method is illustrated in **Figure 6.18**. The program takes the information about the mutant and mutated residues, three neighboring residues on both sides of the mutant residue along with pH and T (**Figure 6.18a**). In the output, the predicted protein stability change upon mutation was displayed along with input conditions (**Figure 6.18b**). In the case of discrimination, the effect of the mutation to protein stability is shown as stabilizing or destabilizing. Both discrimination and prediction services offer an option for additional sequence composition information of neighboring residues (**Figure 6.18b**). The bar chart shows the number of amino acids of each type. The two pie charts represent the percentage of residues according to the polarity and the metabolic role of amino acids.

6.5.5   Average assignment method

Saraboji et al. (2006) developed a method based on average assignment for predicting the stability of protein mutants. In this method, the mutants have been classified into 380 possible substitutions, and the stability of each mutant has been assigned using the information obtained with similar type of mutations. This assignment could distinguish the stabilizing and destabilizing mutants to an accuracy of 70%

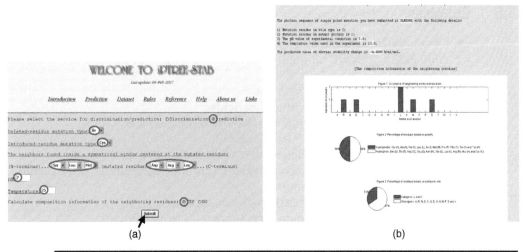

Figure 6.18 Discrimination and prediction of protein stability upon mutations using iPTREE-STAB: (a) snapshot showing the necessary items to be given as input for discrimination and prediction and (b) results obtained for predicting the stability change along with the related information of neighboring residues.

to 80% at different measures of stability (ΔT_m, $\Delta\Delta G$, and $\Delta\Delta G^{H_2O}$). Furthermore, the classification of mutants based on secondary structure and solvent accessibility (ASA) improved the accuracy of assigning stabilizing/destabilizing mutants to an accuracy of 84% to 89% for different measures of stability. This method is able to predict the free energy change ($\Delta\Delta G$) upon mutations within a deviation of 0.64 kcal/mol.

6.5.6 Human readable rules for understanding protein mutant stability

Most of the bioinformatics tools developed for predicting mutant protein stability appear as a black box, and the relationship between amino acid sequence/structure and stability is hidden to the users. Huang et al. (2009) addressed this problem and developed a human-readable rule generator for integrating the knowledge of amino acid sequence and experimental stability change upon single mutation. Using information about the original residue, substituted residue, and three neighboring residues, classification rules have been generated to discriminate the stabilizing and destabilizing mutants and to explore the basis for experimental data. These rules are human readable, and hence the method enhances the synergy between the expert knowledge and the computational system. Furthermore, the performance of the rules has been assessed on a nonredundant dataset of 1859 mutants, which showed an accuracy of 80% using cross-validation. The results revealed that the method could be effectively used as a tool for both knowledge discovery and predicting mutant protein stability. A Web server, iROBOT, has been developed for the classification rule generator, and it is freely available at http://bioinformatics.myweb.hinet.net/irobot.htm.

6.6 Exercises

1. Obtain the buried mutants reported with thermal denaturation method.
   Hint: Select "buried" in accessibility and "thermal" in method in ProTherm database.

2. Get the mutations from Ala to Val.
   Hint: Input Ala and Val in mutation fields.

3. Find the mutants that change the stability more than 5 kcal/mol.
   Hint: Input $\Delta\Delta G / \Delta\Delta G^{H_2O}$ as 5-kcal.mol.

4. Get the data obtained by denaturant denaturation in helical regions.
   Hint: Select "helix" in the secondary structure and "denaturant" in the method.

5. Get the latest data in ProTherm.
   Hint: Input the year since "2008."

6. Find the thermodynamic data for lysozyme T4 single mutants.
   Hint: Input protein name and select "single" in mutants.

7. Display the results obtained in previous question for wild type residue, mutant residue, secondary structure, solvent accessibility, PDB codes, $\Delta\Delta G$, and $\Delta\Delta G^{H_2O}$.
   Hint: Select the relevant fields in display option.

8. Find the mutants that have the data for ΔT_m and $\Delta\Delta G$.
   Hint: Find the mutants with ΔT_m by giving a range of ΔT_m (say, -100 to 100) and the same with $\Delta\Delta G$. Compare the same mutants in same protein.

9. What are the most frequently occurring mutants?
   Hint: Check statistics in ProTherm.

10. Analyze the relationship between ΔH_p and $\Delta\Delta G$ for the mutants A52 in 2RN2.
    Hint: Obtain the $\Delta\Delta G$ data from ProTherm. For each mutant, compute ΔH_p as the difference between mutant and wild type. H_p data can be obtained from **Figure 2.15**.

11. Estimate the correlation between ΔT_m and $\Delta\Delta G$ in single mutants.
    Hint: Obtain the single mutant data for ΔT_m and $\Delta\Delta G$ separately using ProTherm. Select identical mutants. Compute the correlation.

12. Analyze the relationship between ΔH_t and ΔT_m for the mutants Thr157 in 2LZM.
    Hint: Obtain the ΔT_m data from ProTherm. For each mutant, compute ΔH_t as the difference between mutant and wild type. H_t data can be obtained from **Figure 2.15**.

13. Discuss the relationship between hydrophobicity and stability in buried, partially buried, and exposed mutations.
    Hint: Collect the data in buried, partially buried, and exposed mutations. Compute difference in hydrophobicity values with different parameters (H_p, H_t, H_{nc}, etc.). Relate with correlation coefficient.

14. Compare the optimum growth temperature of mycobacterium and thermos thermophilus.
    Hint: Search with these keywords in PGTdb.

15. Identify the stabilizing residues in 2BTM.
    Hint: Use the server SRIDE.

16. Predict the stability change ($\Delta\Delta G$) of R156L in 1TSR using CUPSAT.
Hint: Follow **Figure 6.15**.

17. Predict the stability change ($\Delta\Delta G$) of R156L in 1TSR using AUTOMUTE.
Hint: Follow **Figure 6.16**.

18. Predict the stability change ($\Delta\Delta G$) of D225Q in 1TIM using iPTREE-STAB.
Hint: Follow **Figure 6.17**.

19. What is the stability change ($\Delta\Delta G$) for the mutation from Ala to Val?
Hint: Get the data for all the Ala to Val mutants from ProTherm. Take the average.

References

An J, Nakama T, Kubota Y, Sarai A. 3DinSight: an integrated relational database and search tool for the structure, function and properties of biomolecules. Bioinformatics. 1998;14(2):188–195.

Argos P, Rossman MG, Grau UM, Zuber H, Frank G, Tratschin JD. Thermal stability and protein structure. Biochemistry. 1979;18(25):5698–5703.

Baldasseroni F, Pascarella S. Subunit interfaces of oligomeric hyperthermophilic enzymes display enhanced compactness. Int J Biol Macromol. 2009;44(4):353–360.

Bordner AJ, Abagyan RA. Large-scale prediction of protein geometry and stability changes for arbitrary single point mutations. Proteins. 2004;57(2):400–413.

Capriotti E, Fariselli P, Calabrese R, Casadio R. Predicting protein stability changes from sequences using support vector machines. Bioinformatics. 2005;21:ii54–ii58.

Capriotti E, Fariselli P, Casadio R. A neural-network-based method for predicting protein stability changes upon single point mutations. Bioinformatics. 2004;20:I63–I68.

Chakravarty S, Varadarajan R. Elucidation of factors responsible for enhanced thermal stability of proteins: a structural genomics based study. Biochemistry. 2002;41:8152–8161.

Cheng J, Randall A, Baldi P. Prediction of protein stability changes for single-site mutations using support vector machines. Proteins. 2006;62(4):1125–1132.

Das R, Gerstein M. The stability of thermophilic proteins: a study based on comprehensive genome comparison. Funct Integr Genomics. 2000;1(1):76–88.

DeLano WL. The PyMOL molecular graphics system. San Carlos, CA: DeLano Scientific; 2002. Available online at http://www.pymol.org.

Ding Y, Cai Y, Zhang G, Xu W. The influence of dipeptide composition on protein thermostability. FEBS Lett. 2004;569:284–288.

Dominy BN, Minoux H, Brooks III CL. An electrostatic basis for the stability of thermophilic proteins. Proteins Struct Funct Bioinfo. 2004;57:128–141.

Eisenhaber F, Argos P. Improved strategy in analytical surface calculation for molecular system: handling of singularities and computational efficiency. J Comp Chem. 1993;14:1272–1280.

Facchiano AM, Colonna G, Ragone R. Helix stabilizing factors and stabilization of thermophilic proteins: an X-ray-based study. Protein Eng. 1998;11:753–760.

Fukuchi S, Nishikawa K. Protein surface amino acid compositions distinctively differ between thermophilic and mesophilic bacteria. J Mol Biol. 2001;309(4):835–843.

Gaucher EA, Govindarajan S, Ganesh OK. Palaeotemperature trend for Precambrian life inferred from resurrected proteins. Nature. 2008;451(7179):704–707.

Gaucher EA, Thomson JM, Burgan MF, Benner SA. Inferring the palaeoenvironment of ancient bacteria on the basis of resurrected proteins. Nature. 2003;425(6955):285–288.

Gilis D, Rooman M. Predicting protein stability changes upon mutation using database-derived potentials: solvent accessibility determines the importance of local versus non-local interactions along the sequence. J Mol Biol. 1997;272(2):276–290.

Gilis D, Rooman M. Stability changes upon mutation of solvent-accessible residues in proteins evaluated by database-derived potentials. J Mol Biol. 1996;257(5):1112–1126.

González-Mondragón E, Zubillaga RA, Saavedra E, Chánez-Cárdenas ME, Pérez-Montfort R, Hernández-Arana A. Conserved cysteine 126 in triosephosphate isomerase is required not for enzymatic activity but for proper folding and stability. Biochemistry. 2004;43:3255–3263.

Greaves RB, Warwicker J. Stability and solubility of proteins from extremophiles. Biochem Biophys Res Commun. 2009;380(3):581–585.

Grimsley GR, Shaw KL, Fee LR, Alston RW, Huyghues-Despointes BM, Thurlkill RL, Scholtz JM, Pace CN. Increasing protein stability by altering long-range coulombic interactions. Protein Sci. 1999;8(9):1843–1849.

Gromiha MM. Important inter-residue contacts for enhancing the thermal stability of thermophilic proteins. Biophys Chem. 2001;91:71–77.

Gromiha MM. Factors influencing the thermal stability of buried protein mutants. Polymer. 2003;44:4061–4066.

Gromiha MM. Revisiting "reverse hydrophobic effect": applicable only to coil mutations at the surface. Biopolymers. 2009;91(7):591–599.

Gromiha MM, Suresh MX. Discrimination of mesophilic and thermophilic proteins using machine learning algorithms. Proteins. 2008;70(4):1274–1279.

Gromiha MM, An J, Kono H, Oobatake M, Uedaira H, Sarai A. ProTherm: thermodynamic database for proteins and mutants. Nucl Acids Res. 1999a;27:286–288.

Gromiha MM, Oobatake M, Kono H, Uedaira H, Sarai A. Role of structural and sequence information for predicting protein stability changes: comparison between buried and partially buried mutations. Protein Engineering. 1999b;12:549–555.

Gromiha MM, Oobatake M, Kono H, Uedaira H, Sarai A. Relationship between amino acid properties and protein stability: buried mutations. J Protein Chem. 1999c;18:565–578.

Gromiha MM. Oobatake M, Sarai A. Important amino acid properties for enhanced thermostability from mesophilic to thermophilic proteins. Biophysical Chemistry. 1999d;82:51–67.

Gromiha MM, Oobatake M, Kono H, Uedaira H, Sarai A. Importance of surrounding residues for predicting protein stability of partially buried mutations. J Biomol Str Dyn. 2000;18:281–295.

Gromiha MM, Uedaira H, An J, Selvaraj S, Prabakaran P, Sarai A. ProTherm, thermodynamic database for proteins and mutants: developments in version 3.0. Nucleic Acids Res. 2002a;30(1):301–302.

Gromiha MM, Oobatake M, Kono H, Uedaira H, Sarai A. Importance of mutant position in Ramachandran plot for predicting protein stability of surface mutations. Biopolymers. 2002b;64:210–220.

Gromiha MM, Thomas S, Santhosh C. Role of cation-pi interactions to the stability of thermophilic proteins. Prep Biochem Biotechnol. 2002c;32:355–362.

Gromiha MM, Pujadas G, Magyar C, Selvaraj S, Simon I. Locating the stabilizing residues in (alpha/beta) 8 barrel proteins based on hydrophobicity, long-range interactions, and sequence conservation. Proteins Struct Funct Bioinf. 2004;55:316–329.

Gromiha MM, Huang L-T, Lai L-F. Sequence based prediction of protein mutant stability and discrimination of thermophilic proteins. Lect Note Bioinf. 2008;5265:1–12.

Guerois R, Nielsen JE, Serrano L. Predicting changes in the stability of proteins and protein complexes: a study of more than 1000 mutations. J Mol Biol. 2002;320(2):369–387.

Hoppe C, Schomburg D. Prediction of protein thermostability with a direction- and distance-dependent knowledge-based potential. Protein Sci. 2005;14(10):2682–2692.

Huang LT, Gromiha MM, Hwang SF, Ho SY. Knowledge acquisition and development of accurate rules for predicting protein stability changes. Comput Biol Chem. 2006;30(6):408–415.

Huang LT, Saraboji K, Ho SY, Hwang SF, Ponnuswamy MN, Gromiha MM. Prediction of protein mutant stability using classification and regression tool. Biophys Chem. 2007a;125(2–3):462–470.

Huang LT, Gromiha MM, Ho SY. iPTREE-STAB: interpretable decision tree based method for predicting protein stability changes upon mutations. Bioinformatics. 2007b;23(10):1292–1293.

Huang L, Lai L, Gromiha MM. Human-readable rule generator for integrating amino acid sequence information and stability of mutant poteins. IEEE Trans Comp Biol Bioinformatics. 2009 (in press). DOI:10.1109/TCBB.2008.128.

Huang SL, Wu LC, Liang HK, Pan KT, Horng JT, Ko MT. PGTdb: a database providing growth temperatures of prokaryotes. Bioinformatics. 2004;20(2):276–278.

Ibrahim BS, Pattabhi V. Role of weak interactions in thermal stability of proteins. Biochem Biophys Res Commun. 2004;325(3):1082–1089.

Imanaka T, Shibazaki M, Takagi M. A new way of enhancing the thermostability of proteases. Nature. 1986;324(6098):695–697.

Jaenicke R, Bohm G. The stability of proteins in extreme environments. Curr Opin Struct Biol. 1998;8:738–748.

Kabsch W, Sander C. Dictionary of protein secondary structure: pattern recognition of hydrogen-bond and geometrical features. Biopolymers. 1983;22:2577–2637.

Kannan N, Vishveshwara S. Aromatic clusters: a determinant of thermal stability of thermophilic proteins. Protein Eng. 2000;13(11):753–761.

Kauzmann W. Sulfur in Proteins. New York: Academic Press; 1959.

Khatun J, Khare SD, Dokholyan NV. Can contact potentials reliably predict stability of proteins? J Mol Biol. 2004;336(5):1223–1238.

Kumar S, Nussinov R. How do thermophilic proteins deal with heat? Cell Mol Life Sci. 2001;58:1216–1233.

Kumar S, Tsai CJ, Nussinov R. Factors enhancing protein thermostability. Protein Eng. 2000;13:179–191.

Kursula I, Partanen S, Lambeir AM, Wierenga RK. The importance of the conserved Arg191-Asp227 salt bridge of triosephosphate isomerase for folding, stability, and catalysis. FEBS Lett. 2002;518:39–42.

Ladenstein R, Antranikian G. Proteins from hyperthermophiles: stability and enzymatic catalysis close to the boiling point of water. Adv Biochem Eng Biotechnol. 1998;61:37–85.

Magyar C, Gromiha MM, Pujadas G, Tusnady GE, Simon I. SRide: a server for identifying stabilizing residues in proteins. Nucleic Acids Res. 2005;33(Web Server issue):W303–W305.

Masso M, Vaisman II. Accurate prediction of stability changes in protein mutants by combining machine learning with structure based computational mutagenesis. Bioinformatics. 2008;24(18):2002–2009.

Menendez-Arias L, Argos P. Engineering protein thermal stability. Sequence statistics point to residue substitutions in alpha-helices. J Mol Biol. 1989;206(2):397–406.

Mozo-Villiarías A, Querol E. Theoretical analysis and computational predictions of protein thermostability. Current Bioinformatics. 2006;1:25–32.

Ooi T, Oobatake M. Intermolecular interactions between protein and other molecules including hydration effects. J Biochem (Tokyo). 1988;104(3):440–444.

Ponnuswamy PK, Gromiha MM. On the conformational stability of folded proteins. J Theor Biol. 1994;166:63–74.

Pace CN. Conformational stability of globular proteins. Trends Biochem Sci. 1990;15(1):14–17.

Pakula AA, Sauer RT. Reverse hydrophobic effects relieved by amino-acid substitutions at a protein surface. Nature. 1990;344(6264):363–364.

Parthiban V, Gromiha MM, Hoppe C, Schomburg D. Structural analysis and prediction of protein mutant stability using distance and torsion potentials: role of secondary structure and solvent accessibility. Proteins. 2007;66(1):41–52.

Parthiban V, Gromiha MM, Schomburg D. CUPSAT: prediction of protein stability upon point mutations. Nucleic Acids Res. 2006;34(Web Server issue):W239–W242.

Pfeil W. Protein Stability and Folding: A Collection of Thermodynamic Data. New York: Springer; 1998.

Pfeil W. Protein Stability and Folding, Supplement 1: A Collection of Thermodynamic Data. New York: Springer; 2001.

Ponnuswamy PK, Muthusamy R, Manavalan P. Amino acid composition and thermal stability of globular proteins. Int J Biol Macromol. 1982;4:186–190.

Sadeghi M, Naderi-Manesh H, Zarrabi M, Ranjbar B. Effective factors in thermostability of thermophilic proteins. Biophys Chem. 2006;119:256–270.

Saraboji K, Gromiha MM, Ponnuswamy MN. Relative importance of secondary structure and solvent accessibility to the stability of protein mutants: a case study with amino acid properties and energetics on T4 and human lysozymes. Comp Biol Chem. 2005a;29:25–35.

Saraboji K, Gromiha MM, Ponnuswamy MN. Importance of main-chain hydrophobic free energy to the stability of thermophilic proteins. Int J Biol Macromol. 2005b;35(3–4):211–220.

Saraboji K, Gromiha MM, Ponnuswamy MN. Average assignment method for predicting the stability of protein mutants. Biopolymers. 2006;82(1):80–92.

Sayle RA, Milner-White EJ. RASMOL: biomolecular graphics for all. Trends Biochem Sci. 1995;20:374–376.

Schomburg I, Chang A, Hofmann O, Ebeling C, Ehrentreich F, Schomburg D. BRENDA: a resource for enzyme data and metabolic information. Trends Biochem Sci. 2002;27(1):54–56.

Shortle D, Chan HS, Dill KA. Modeling the effects of mutations on the denatured states of proteins. Protein Sci. 1992;1:201–215.

Szilagyi A, Zavodszky P. Structural differences between mesophilic, moderately thermophilic and extremely thermophilic protein subunits: results of a comprehensive survey. Structure Fold Des. 2000;8:493–504.

Tanford C. Contribution of hydrophobic interactions to the stability of the globular conformation of proteins. J Amer Chem Soci. 1962;84:4240–4247.

Thornton JM. Disulphide bridges in globular proteins. J Mol Biol. 1981;151:261–287.

Topham CM, Srinivasan N, Blundell TL. Prediction of the stability of protein mutants based on structural environment-dependent amino acid substitution and propensity tables. Protein Eng. 1997;10(1):7–21.

Van Gunsteren WF, Mark AE. Prediction of the activity and stability effects of site-directed mutagenesis on a protein core. J Mol Biol. 1992;227(2):389–395.

Vogt G, Woell S, Argos P. Protein thermal stability, hydrogen bonds, and ion pairs. J Mol Biol. 1997;269:631–643.

Xiao L, Honig B. Electrostatic contributions to the stability of hyperthermophilic proteins. J Mol Biol. 1999;289:1435–1444.

Yano JK, Poulos TL. New understandings of thermostable and peizostable enzymes. Curr Opin Biotechnol. 2003;14:360–365.

Zarrine-Afsar A, Wallin S, Neculai AM, Neudecker P, Howell PL, Davidson AR, Chan HS. Theoretical and experimental demonstration of the importance of specific non-native interactions in protein folding. Proc Natl Acad Sci U S A. 2008;105:9999–10004.

Zeldovich KB, Berezovsky IN, Shakhnovich EI. Protein and DNA sequence determinants of thermophilic adaptation. PLoS Comput Biol. 2007;3(1):e5.

Zhang G, Feng B. Application of amino acid distribution along the sequence for discriminating mesophilic and thermophilic proteins. Process Biochem. 2006;41:1792–1798.

CHAPTER 7

Protein Interactions

Along with protein folding, understanding molecular recognition, especially protein interactions, is one of the major challenges in computational/molecular biology. Predicting the binding interfaces, binding strengths, kinetics, and sensitivity to environmental change or to mutations are all important for understanding the functions of living systems. In recent years, rapid progress has been shown for understanding and predicting the binding site residues and binding specificities in protein–protein, protein–nucleic acid, and protein–ligand complexes. These studies pave ways to understand several functions.

7.1 Protein–protein interactions

Protein–protein interactions play a key role in many biological processes, such as signal transduction, gene expression and control, and antibody-antigen complex. Several investigations have been carried out to understand the details of interactions between the residues at protein–protein interface and the identification of binding sites. The development of methods for understanding the concepts of protein–protein interactions in terms of experimental techniques, databases, and prediction of protein–protein and domain interactions have been reviewed in detail (Valencia and Pazos, 2002; Salwinski and Eisenberg, 2003; Wodak and Mendez, 2004; Aloy et al. 2005; Shoemaker and Panchenko, 2007a,b; Keskin et al. 2005, 2008; Tuncbag et al. 2009; Ezkurdia et al. 2009).

7.1.1 Databases for protein–protein interactions

The accumulation of data on protein–protein interactions would help to understand the role of different factors for the specificity of protein–protein interactions and predicting the binding sites. Salwinski et al. (2004) developed a database of interacting proteins, DIP. It aims to integrate the diverse body of experimental evidence on protein–protein interactions into a single, easily accessible online database. The DIP database is implemented as a relational database using an open source PostgreSQL database management system: (i) The key tables, PROTEIN, SOURCE, and EVIDENCE, provide data on individual proteins, sources, and individual experiments, respectively; (ii) the information on protein–protein interactions is stored in two tables, INTERACTION and INT_PRT. Such arrangement of tables enables description of binary interactions as well as of multiprotein complexes; (iii) the METHOD table provides a list of controlled vocabulary terms, together with references to the corresponding PSI ontology entries, which are used to annotate the experiments; (iv) available information on the details of the topology of a molecular complex that was inferred from each experiment is stored in the TOPOLOGY and LOCATION tables. The LOCATION table describes regions of proteins participating in interactions, whereas the TOPOLOGY table pairs them into records that describe observed binary interactions; and (v) it also specifies the type of interaction inferred from each experiment as one of aggregate (both partners shown to be present in the same complex but not necessarily in direct contact), contact, or covalent bonds. This DIP database can be used as a reference when evaluating the reliability of high-throughput, protein–protein interaction data sets, for the development of prediction methods, as well as in the studies of the properties of protein interaction networks. It is freely available at http://dip.doe-mbi.ucla.edu/.

Zanzoni et al. (2002) developed the molecular interaction database, MINT, a database designed to store data on functional interactions between proteins, including enzymatic modifications of one of the partners. MINT focuses at being exhaustive in the description of the interaction, information about kinetic and binding constants and about the domains participating in the interaction. MINT consists of entries extracted from the scientific literature by expert curators assisted by "MINT Assistant," a program that targets abstracts containing interaction information and presents them to the curator in a user-friendly format. The interaction data can be easily extracted and viewed graphically through "MINT Viewer." It is available at http://mint.bio.uniroma2.it/mint/ (Chatr-aryamontri et al. 2007). Alfarano et al. (2005) developed the Biomolecular Interaction Network Database (BIND), which archives biomolecular interaction, reaction, complex, and pathway information. BIND aims to curate the details about molecular interactions that arise from published experimental research and to provide this information to researchers worldwide. BIND data are curated into a comprehensive machine-readable archive of computable information and provide users with methods to discover interactions and molecular mechanisms. BIND has worked to develop new methods for visualization that amplify the underlying annotation of genes and proteins to facilitate the study of molecular interaction networks. The BIND Query and Submission interface, a Standard Object Access Protocol service, and the Small Molecule Interaction Database (http://smid.blueprint.org) allow users to determine probable small molecule binding sites of new sequences and examine conserved binding residues. BIND database is available at http://bind.ca/.

Kerrien et al. (2007) developed an open source database and software, IntAct, for modeling, storing, and analyzing molecular interaction data. The data were obtained from published literature and are manually annotated by expert biologists to a high level of detail, including experimental methods, conditions, and interacting domains. The database features more than 126,000 binary interactions, and it is freely available from http://www.ebi.ac.uk/intact. **Figure 7.1** shows the search options in IntAct database and the results obtained with the species, "human" and interaction type, "association." The output shows the details about the interacting partners, proteins, organisms, types of interactions, and so on.

Prasad et al. (2009) developed a Human Protein Reference Database (HPRD) to serve as a comprehensive collection of protein features, enzyme–substrate

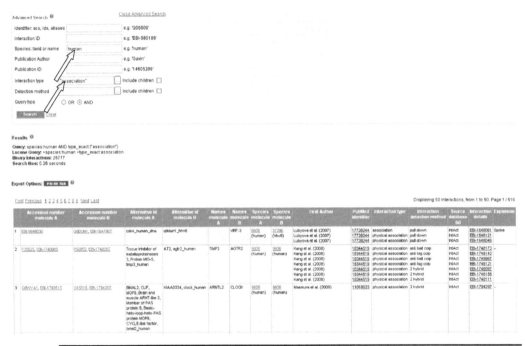

Figure 7.1 Searching data using IntAct. The search options and results are shown.

relationships, disease associations, post-translational modifications, and protein–protein interactions. This database is a resource of genomic and proteomic information and providing an integrated view of sequence, structure, function, and protein networks in health and disease. It is available at http://www.hprd.org. **Figure 7.2** shows the features of HPRD. It is searched with "prion protein" that is located in "mitochondrial membrane" (**Figure 7.2a**). The output is displayed in **Figure 7.2b**. The details about the protein are shown in **Figure 7.2c**. It is linked with diseases, sequences, interactions, etc.

Breitkreutz et al. (2008) developed a Biological General Repository for Interaction Datasets (BioGRID database), which is a collection of protein and genetic interactions from major model organism species. An internally hyperlinked Web interface allows for rapid search and retrieval of interaction data. Furthermore, full or user-defined datasets are freely downloadable as tab-delimited text files and PSI-MI XML. BioGRID is available at (http://www.thebiogrid.org).

The comparison of data available in terms of protein, protein–protein interactions, and publications in five different protein–protein interactions databases is presented in **Table 7.1**.

Kinetic database for protein–protein interactions

Ji et al. (2003) developed the database of Kinetic Data of Bio-molecular Interactions (KDBI), which provides experimentally determined kinetic data of protein–protein, protein–RNA, protein–DNA, protein–ligand, RNA–ligand, and DNA–ligand binding described in the literature. KDBI contains information about binding or reaction

Protein Name	prion
Accession Number	RefSeq ▾
HPRD Identifier	
Gene Symbol	
Chromosome Locus	
Molecular Class	See List
PTMs	See List
Cellular Component	Mitochondrial membrar See List
Domain Name	See List
Motif	See List
Expression	See List
Length of Protein Sequence	From ___ to ___ in amino acids
Molecular Weight	From ___ to ___ in kDa
Diseases	

Search   Clear

(a)

Your search for **prion** in **Protein** reports **42** matches.
Results are sorted by relevance, indicating the number of occurrences of items searched for.
Showing matches **1** to **42**

↓ Sort by   Relevance | Number of PTMs | Protein Length | Number of Interactions

1   Name : Succinate dehydrogenase cytochrome b   Molecule Function : Catalytic activity

Number of Interactions : 0

2   Name : Uncoupling protein 4   Molecule Function : T cell receptor activity

Number of Interactions : 0

3   Name : Translocase of inner mitochondrial membrane 9   Molecule Function : Transporter activity

Number of Interactions : 0

4   Name : TIM10   Molecule Function : Chaperone activity

(b)

FIGURE 7.2   Retrieving protein–protein interaction data using HPRD: (a) input options, (b) search results, and (c) details of the protein.

Isoform 1

(c)

FIGURE 7.2   *(Continued)*

event, participating molecules (name, synonyms, molecular formula, classification, SWISS-PROT, or CAS number), binding or reaction equation, kinetic data, and related references. The kinetic data include the following quantities: association/dissociation rate constant, first/second/third/order rate constant, equilibrium rate constant, catalytic rate constant, equilibrium association/dissociation constant,

TABLE 7.1   Statistics of protein–protein interaction data in different databases*

Database	Proteins	Protein–protein interactions	Publications
DIP	20,728	57,683	3915
MINT	29,222	110,958	2944
IntAct	57,909	129,658	10,376
HPRD	25,661	38,167	270,466
BioGRID	529,018	165,562	20,690

*Data were taken from their respective Web sites (accessed on June 1, 2009).

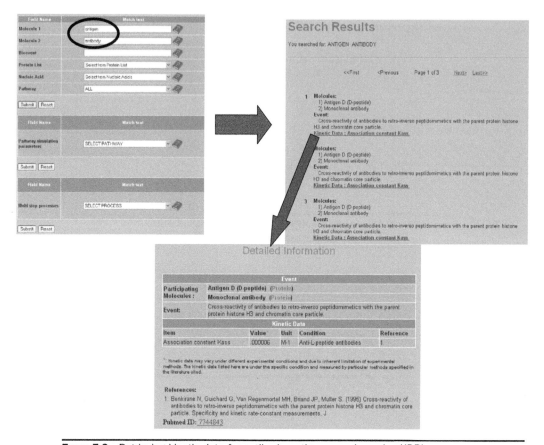

inhibition constant, and binding affinity constant. Each entry can be retrieved through protein or nucleic acid or ligand name, SWISS-PROT number, ligand CAS number and full-text search of a binding or reaction event. This database can be accessed at http://xin.cz3.nus.edu.sg/group/kdbi/kdbi.asp. **Figure 7.3** shows the applications of KDBI. The search with "antigen" and "antibody" is illustrated in this figure.

Thermodynamic database for protein–protein interactions

Kumar and Gromiha (2006) developed the Protein-protein Interactions Thermodynamic Database (PINT), which contains several thermodynamic parameters along with sequence and structural information, experimental conditions, and literature information. Each entry contains numerical data for the free energy change, dissociation constant, association constant, enthalpy change, and heat capacity change of the interacting proteins upon binding, which are important for understanding the mechanism of protein–protein interactions. PINT also includes the name and source of the proteins involved in binding, PIR, SWISS-PROT, and PDB codes, secondary structure and solvent accessibility of residues at mutant positions, measuring methods, experimental conditions, and literature information.

A WWW interface facilitates users to search data based on various conditions, feasibility to select the terms for output, and different sorting options. Furthermore, PINT is cross-linked with other related databases, PIR, SWISS-PROT, PDB, and NCBI PubMed literature database. The database is freely available at http://www.bioinfodatabase.com/pint/index.html/.

Detailed tutorials describing the usage of the present PINT are available at the homepage. As an example, the necessary items to be filled/selected to search the thermodynamic data, dissociation constant, and free energy change for protein–protein complexes obtained in the temperature range of 15 to 30 degrees C and pH more than 5 are shown in **Figure 7.4a**. In the same figure, the display items

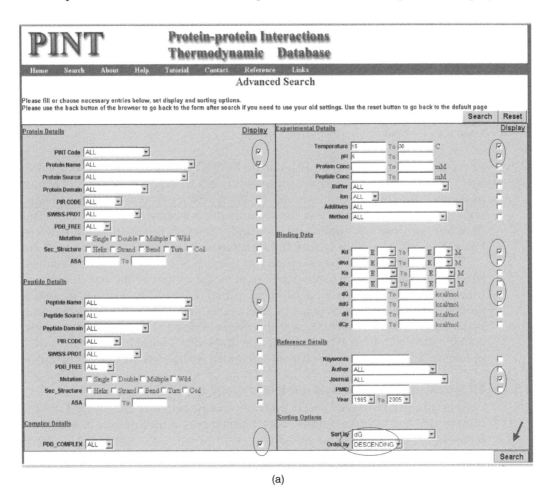

(a)

FIGURE 7.4 An example of searching conditions, display and sorting options, and the results of PINT search. (a) Search, display, and sorting options: The search is performed for obtaining K_d and ΔG for protein–protein complexes obtained in the temperature range of 15–30 degrees C and pH > 5. All these search items were selected to display in the output along with PINT code, protein name, peptide name, PDB complex, and journal name. The data are sorted with ΔG in descending order; (b) Partial results obtained from PINT under the conditions specified in **Figure 7.4a**. (*continued*)

PINT Database: Search Results

Your query resulted in 1174 data.

If you want to modify your query, please use the "back" button of your browser. If you want to try another new query, please go to the search page.

NUMBER	PINT_CODE	PROT_NAME	PEPT_NAME	PDB_COMPLEX	T (°C)	pH	Kd (M)	dG (kcal/mol)	REFERENCE
1021	PINT_CYCP_CYC1	Cytochrome c peroxidase	Cytochrome c iso-1	2PCC	25.0	6.0	4.05e-2	-1.90	Biochemistry. 2001 Jan 16;40(2):422-8
1042	PINT_CALM_RS20	Calmodulin	RS20	NULL	25.0	7.5	8.33e-4	-4.21	J Biol Chem. 1999 Jun 25;274(26) 18161-4
733	PINT_CYCP_CYCC	Cytochrome c peroxidase	Cytochrome C	2PCB	26.0	6.0	5.00e-4	-4.52	Biochemistry. 1995 Jul 4;34(26):8398-405
1182	PINT_GFR2_SOS	Growth factor receptor-bound protein 2	SOS-A	NULL	30.0	7.0	4.55e-4	-4.60	Biochemistry. 1994 Nov 22;33(46):13531-9
725	PINT_GFR2_PY317	Growth factor receptor protein 2	pY317	1QG1	20.0	7.5	3.59e-4	-4.62	Biochemistry. 1997 Aug 19;36(33):10006-14
1279	PP_PCAF_ACK	Histone acetyltransferase PCAF	ACK peptide	NULL	25.0	7.4	3.46e-4	-4.72	Nature. 1999 Jun 3;399(6735)491-6
1043	PINT_CALM_RS20	Calmodulin	RS20	NULL	25.0	7.5	2.78e-4	-4.85	J Biol Chem. 1999 Jun 25;274(26) 18161-4
1041	PINT_CALM_RS20	Calmodulin	RS20	NULL	25.0	7.5	2.38e-4	-4.95	J Biol Chem. 1999 Jun 25;274(26) 18161-4
1264	PP_CD2_CD48	T-cell surface antigen CD2	CD48 antigen	NULL	25.0	7.4	2.32e-4	-4.96	Proc Natl Acad Sci U S A. 1998 May 12;95 (10):5490-4
633	PINT_SRC_HMT	Tyrosine-protein kinase Src	hmT peptide	1SPS	25.0	7.5	2.27e-4	-5.00	J Mol Biol. 2002 Feb 15;316(2) 291-304
1269	PP_CD2_CD48	T-cell surface antigen CD2	CD48 antigen	NULL	25.0	7.4	1.92e-4	-5.07	Proc Natl Acad Sci U S A. 1998 May 12;95 (10):5490-4
1267	PP_CD2_CD48	T-cell surface antigen CD2	CD48 antigen	NULL	25.0	7.4	1.84e-4	-5.10	Proc Natl Acad Sci U S A. 1998 May 12;95 (10):5490-4
1265	PP_CD2_CD48	T-cell surface antigen CD2	CD48 antigen	NULL	25.0	7.4	1.76e-4	-5.12	Proc Natl Acad Sci U S A. 1998 May 12;95 (10):5490-4

(b)

FIGURE 7.4 (Continued)

in the output were shown by tick marks. In PINT, it is possible to sort the data by T, pH, K_d, ΔG, etc., and the sorting with ΔG in descending order is shown in **Figure 7.4a**. This search picked up 1174 data and a part of the results obtained with the search conditions, and the sorting option is shown in **Figure 7.4b**. This database has potential applications for understanding the relationship between binding specificity and the factors that are influencing protein–protein interactions.

Thorn and Bogan (2001) developed an Alanine Scanning Energetics database (ASEdb), which is a searchable database of single alanine mutations in protein–protein, protein–nucleic acid, and protein–small molecule interactions for which binding affinities have been experimentally determined. It also includes the information about ASA for the free type and complex proteins if structures are available. ASEdb can be accessible at http://nic.ucsf.edu/asedb/. **Figure 7.5** shows the search options in ASEdb and the search results obtained for the barnase–barstar complex.

7.1.2 Structural analysis of protein–protein interactions

Glaser et al. (2001) analyzed a set of protein–protein complexes and derived residue composition and residue–residue contact preferences at the interface based on the following criteria: a pair of amino acids i and j (i, $j = 1, 2, 3, \ldots 20$) was considered to be in contact if the distance between their C_β atoms (C_α for Gly) was smaller than a certain cutoff distance.

The normalized frequency of residue i (also referred to as the amino acid composition), W_i, is defined as

$$W_i = F_i / \sum_i F_i, \tag{7.1}$$

where F_i is the number of residues of type i having at least one contact with any residue across the interface. Similarly, if C_{ij} is the total number of contacts observed

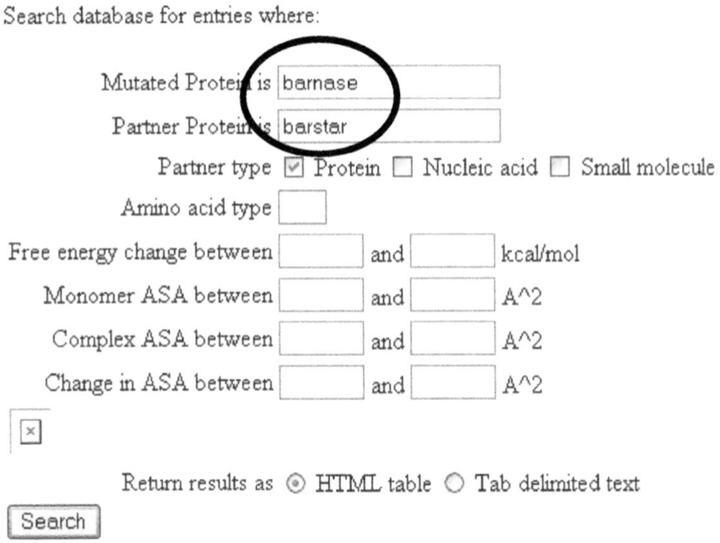

Search database for entries where:

Mutated Protein is [barnase]

Partner Protein is [barstar]

Partner type ☑ Protein ☐ Nucleic acid ☐ Small molecule

Amino acid type [ ]

Free energy change between [ ] and [ ] kcal/mol

Monomer ASA between [ ] and [ ] A^2

Complex ASA between [ ] and [ ] A^2

Change in ASA between [ ] and [ ] A^2

[×]

Return results as ⦿ HTML table ◯ Tab delimited text

[Search]

8 records match your query

Mutated Protein	Partner (Type)	PDB	AA	Pos.	ddG	Accesible Surface Area			RefID
						Monomer	Complex	Delta	
Barnase	Barstar (P)	1brs (D)	K	27	5.400	75.69	21.06	54.63	48
Barnase	Barstar (P)	1brs (D)	D	54	-0.800	30.15	30.15	0.00	47
Barnase	Barstar (P)	1brs (D)	N	58	3.100	18.10	15.48	2.62	47
Barnase	Barstar (P)	1brs (D)	R	59	5.200	177.78	33.55	144.23	48
Barnase	Barstar (P)	1brs (D)	E	60	-0.200	107.71	42.28	65.43	47
Barnase	Barstar (P)	1brs (D)	E	73	2.800	18.82	9.35	9.47	47
Barnase	Barstar (P)	1brs (D)	R	87	5.500	5.75	0.59	5.16	48
Barnase	Barstar (P)	1brs (D)	H	102	6.000	96.17	0.90	95.27	47

FIGURE 7.5   Free energy changes upon Alanine substitutions in barnase–barstar complex using ASEdb.

between residue type i and j, the normalized number of contacts between these residues is defined as

$$Q_{ij} = C_{ij} / \sum_{k \leq l} C_{kl}. \qquad (7.2)$$

The expected number of contacts between residue i and j is the value that would have been obtained if there were no preferences between residues of different types, i.e., $W_i \times W_j$. The likelihood of contacts between a pair of residues i and j, G_{ij}, has been estimated as the log of the ratio of actual to expected number of contacts:

$$G_{ij} = A \times \log(Q_{ij}/W_i \times W_j), \tag{7.3}$$

where A is a constant. Considering the volume of the amino acids, this equation has been written as

$$G_{ij} = A \times \log(Q_{ij}(v)/W_i \times W_j), \tag{7.4}$$

where

$$Q_{ij}(v) = (C_{ij} \times V_i \times V_j)/\sum_{k,l}(C_{kl} \times V_k \times V_l). \tag{7.5}$$

The residue–residue contact preferences for the 400 pairs of residues at protein–protein interface are presented in **Table 7.2**. The largest residue–residue preferences at the interface were recorded for interactions between pairs of large hydrophobic residues, such as Trp and Leu, and the smallest preferences for pairs of small residues, such as Gly and Ala. On average, contacts between pairs of hydrophobic and polar residues were unfavorable, and the charged residues tended to pair subject to charge complementarity.

7.1.3 Solvent accessibility studies at the interface of protein–protein complexes

Janin et al. (1988) analyzed the solvent accessibility of residues at protein–protein interface using a set of oligomeric proteins. It has been shown that the surfaces involved in protein–protein contacts differ from the rest of the subunit interface. They are enriched in hydrophobic side chains, yet they contain a number of charged groups, especially from Arg residues, which are the most abundant residues at interfaces except for Leu. Buried Arg residues are involved in hydrogen bonds between subunits. The smaller interfaces cover about 700 Å^2 of the subunit surface and the larger ones cover 3000 to 10,000 Å^2. The lower value corresponds to an estimate of the accessible surface area loss required for stabilizing subunit association through the hydrophobic effect alone.

Bahadur et al. (2004) analyzed the ASA at the protein–protein interface (ASA of free protein 1 + ASA of free protein 2 – ASA of the complex) using the program, Naccess. The interface area in antibody–antigen and protease–inhibitor complexes is in the range of 1200 to 2000 Å^2. On average, 58% of the buried protein interface area is contributed by nonpolar (carbon containing) and 42% by polar (N, O, S containing) groups. In the average protein–protein complex, 34% to 36% of the interface residues are fully buried (Chakrabarthi and Janin, 2002). The information about the ASA of the residues at the protein–protein interface may be helpful for discriminating monomeric and oligomeric proteins.

7.1.4 Hydrogen bonds and salt bridges across protein–protein interfaces

Xu et al. (1997) carried out an analysis of the hydrogen bonds and the salt bridges in a collection of nonredundant protein–protein complexes and showed that the geometry of the hydrogen bonds across protein interfaces is generally less optimal and has a wider distribution than typically observed within the chains. This difference reveals that hydrophilic side chains prefer to be buried in the binding interface than in the interior of the folded monomer. During folding, practically all

TABLE 7.2 Residue–residue contact preferences at protein–protein interface

	I	V	L	F	C	M	A	G	T	S	W	Y	P	H	E	Q	D	N	K	R
I	3.89	4.91	4.59	5.33	1.76	5.25	2.84	0.77	3.05	1.00	6.24	5.61	3.27	3.38	3.20	3.60	2.30	1.59	3.23	3.80
V	4.91	3.74	4.20	4.69	2.89	4.37	2.57	-0.41	2.83	1.42	2.92	3.95	2.90	3.21	3.22	3.22	1.93	1.36	4.45	4.18
L	4.59	4.20	4.03	4.86	2.93	5.32	2.77	-0.37	2.07	1.41	5.77	4.19	2.50	4.88	3.12	3.46	1.40	2.31	3.15	4.99
F	5.33	4.69	4.86	5.34	3.68	5.28	3.00	0.14	3.34	1.75	5.83	5.83	4.25	3.47	2.87	4.25	0.99	3.11	3.57	4.49
C	1.76	2.89	2.93	3.68	7.65	1.84	1.46	-0.25	1.03	2.48	2.14	2.47	2.74	4.12	2.51	1.33	0.24	-0.42	2.05	2.81
M	5.25	4.37	5.32	5.28	1.84	6.02	2.30	0.91	2.09	1.61	4.89	4.81	3.38	4.65	3.88	4.18	0.36	2.30	3.93	3.62
A	2.84	2.57	2.77	3.00	1.46	2.30	-0.52	-1.77	1.21	0.39	3.37	2.47	1.22	2.59	1.71	1.72	1.13	1.69	2.13	1.90
G	0.77	-0.41	-0.37	0.14	-0.25	0.91	-1.77	-4.40	0.21	-1.53	1.42	1.25	-0.51	1.08	-0.89	0.70	-0.08	-0.54	1.33	1.59
T	3.05	2.83	2.07	3.34	1.03	2.09	1.21	0.21	1.27	1.91	5.12	3.14	2.65	2.71	2.88	1.82	3.88	2.52	3.67	3.77
S	1.00	1.42	1.41	1.75	2.48	1.61	0.39	-1.53	1.91	-0.09	2.87	2.30	1.33	0.80	2.60	2.00	2.94	1.77	2.74	2.82
W	6.24	2.92	5.77	5.83	2.14	4.89	3.37	1.42	5.12	2.87	5.85	6.19	7.87	6.46	1.20	1.37	2.62	3.54	5.76	8.57
Y	5.61	3.95	4.19	5.83	2.47	4.81	2.47	1.25	3.14	2.30	6.19	5.93	4.22	6.05	4.54	2.05	1.76	3.66	5.26	5.28
P	3.27	2.90	2.50	4.25	2.74	3.38	1.22	-0.51	2.65	1.33	7.87	4.22	0.60	2.89	3.17	3.50	1.46	3.09	3.75	3.99
H	3.38	3.21	4.88	3.47	4.12	4.65	2.59	1.08	2.71	0.80	6.46	6.05	2.89	5.37	2.30	4.00	5.20	2.38	2.72	4.90
E	3.20	3.22	3.12	2.87	2.51	3.88	1.71	-0.89	2.88	2.60	1.20	4.54	3.17	2.30	1.65	1.95	0.08	2.68	5.32	5.75
Q	3.60	3.22	3.46	4.25	1.33	4.18	1.72	0.70	1.82	2.00	1.37	2.05	3.50	4.00	1.95	2.83	3.26	3.45	3.50	4.50
D	2.30	1.93	1.40	0.99	0.24	0.36	1.13	-0.08	3.88	2.94	2.62	1.76	1.46	5.20	0.08	3.26	0.13	3.85	3.90	4.94
N	1.59	1.36	2.31	3.11	-0.42	2.30	1.69	-0.54	2.52	1.77	3.54	3.66	3.09	2.38	2.68	3.45	3.85	2.92	3.17	3.85
K	3.23	4.45	3.15	3.57	2.05	3.93	2.13	1.33	3.67	2.74	5.76	5.26	3.75	2.72	5.32	3.50	3.90	3.17	3.24	2.29
R	3.80	4.18	4.99	4.49	2.81	3.62	1.90	1.59	3.77	2.82	8.57	5.28	3.99	4.90	5.75	4.50	4.94	3.85	2.29	2.87

Data were taken from Glaser et al. (2001).

degrees of freedom are available to the chain to attain its optimal configuration, whereas this is not the case for rigid binding that the protein molecules are already folded, with only six degrees of translational and rotational freedom available to the chains to achieve their most favorable configuration. These constraints enforce many polar/charged residues buried in the interface to form weak hydrogen bonds with protein atoms, rather than strong hydrogen bonding to the solvent. Since interfacial hydrogen bonds are weaker than the intrachain ones to compete with the binding of water, more water molecules are involved in bridging hydrogen bond networks across the protein interface than in the protein interior. Interfacial water molecules both mediate noncomplementary donor–donor or acceptor–acceptor pairs and connect nonoptimally oriented donor–acceptor pairs. These differences between the interfacial hydrogen bonding patterns and the intrachain ones further substantiate the notion that protein complexes formed by rigid binding may be far away from the global minimum conformations.

Pazos et al. (1997) developed a method for detecting correlated changes in multiple sequence alignments to a set of interacting protein domains and showed that positions where changes occur in a correlated fashion in the two interacting molecules tend to be close to the protein–protein interfaces. On the other hand, Audie (2009) developed an empirical free energy function using native geometry and physically realistic free energy minima for predicting the binding free energy change in protein–protein complexes.

7.1.5   Prediction of binding sites in protein–protein complexes

Binding sites are defined as the residues that are in contact with each other between the interacting partner proteins in each protein–protein complex. The distances between heavy atoms or C_α atoms and different distance cutoffs have been used to locate the binding sites from three-dimensional structures of protein–protein complexes (Glaser et al. 2001; Li et al. 2004; Keskin et al. 2004). On the other hand, several computational methods have been developed to predict the binding sites from amino acid sequence and/or parameters derived from protein structures. Prediction methods have also been proposed from the structure of a free protein. Jones and Thornton (1997) used surface patches for predicting protein–protein interaction sites. Potapov et al. (2008) utilized molecular architecture and naturally occurring template fragments in the Protein Data Bank (PDB) for identifying protein–protein interface. The combination of sequence and structural features as well as the information on nine consecutive residues, secondary structure of the central residue and average properties based on solvent accessibility, protrusion and depth has also been employed for detecting the binding sites from amino acid sequence (Sikic et al. 2009). Shulman-Pelag et al. (2008) constructed a method based on multiple alignment for detecting binding sites in protein–protein complexes. It recognizes the spatially conserved physico-chemical interactions, which often involve energetically important hot-spot residues that are crucial for protein–protein associations. Ertekin et al. (2006) proposed a method based on the fluctuation behavior of residues to predict the putative protein binding sites. Furthermore, machine learning techniques have been widely used to identify the binding sites in protein–protein complexes (Fariselli et al. 2002; Koike and Takagi, 2004; Res et al. 2005; Ofran and Rost, 2007). Recently, Ezkurdia et al. (2009) discussed the progress and challenges in predicting protein–protein interaction sites.

On the basis of different prediction methods, several Web servers have been developed to predict the binding sites from amino acid sequence (Tuncbag et al. 2009). The features used in these methods are the statistical parameters derived from amino acid sequences/structures, solvent accessibility, conservation of residues and amino acid properties. The Web servers for binding site prediction include cons-PPISP (http://pipe.scs.fsu.edu/ppisp.html; Zhou and Shan, 2001), PINEUP (http://sparks.informatics.iupui.edu/PINUP/; Liang et al. 2006), ProMate (http://bioportal.weizmann.ac.il/promate/; Neuvirth et al. 2004), PPI-Pred (http://bioinformatics.leeds.ac.uk/ppi-pred; Bradford and Westhead, 2005), SHARP2 (http://www.bioinformatics.sussex.ac.uk/SHARP2/sharp2.html Patch; Murakami and Jones, 2006), SPPIDER (http://sppider.cchmc.org/; Porollo and Meller, 2007), Firestar (http://firedb.bioinfo.cnio.es/Php/FireStar.php; Lopez et al. 2007), and InterProSurf (http://curie.utmb.edu/; Negi et al. 2007).

The prediction methods are illustrated with two examples. **Figure 7.6** shows the prediction of protein–protein interaction sites from amino acid sequence using the server, Firestar. It takes the amino acid sequence in FASTA format and

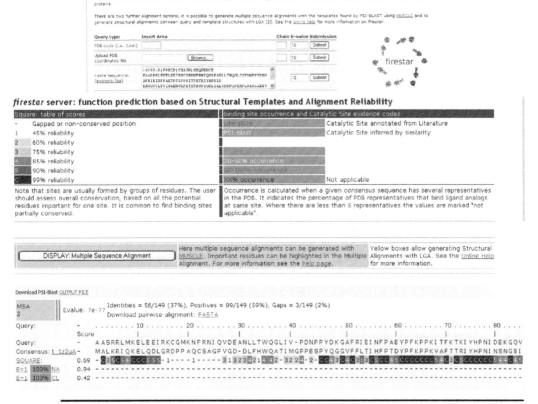

FIGURE 7.6 Prediction of protein–protein interaction sites from amino acid sequence using Firestar server.

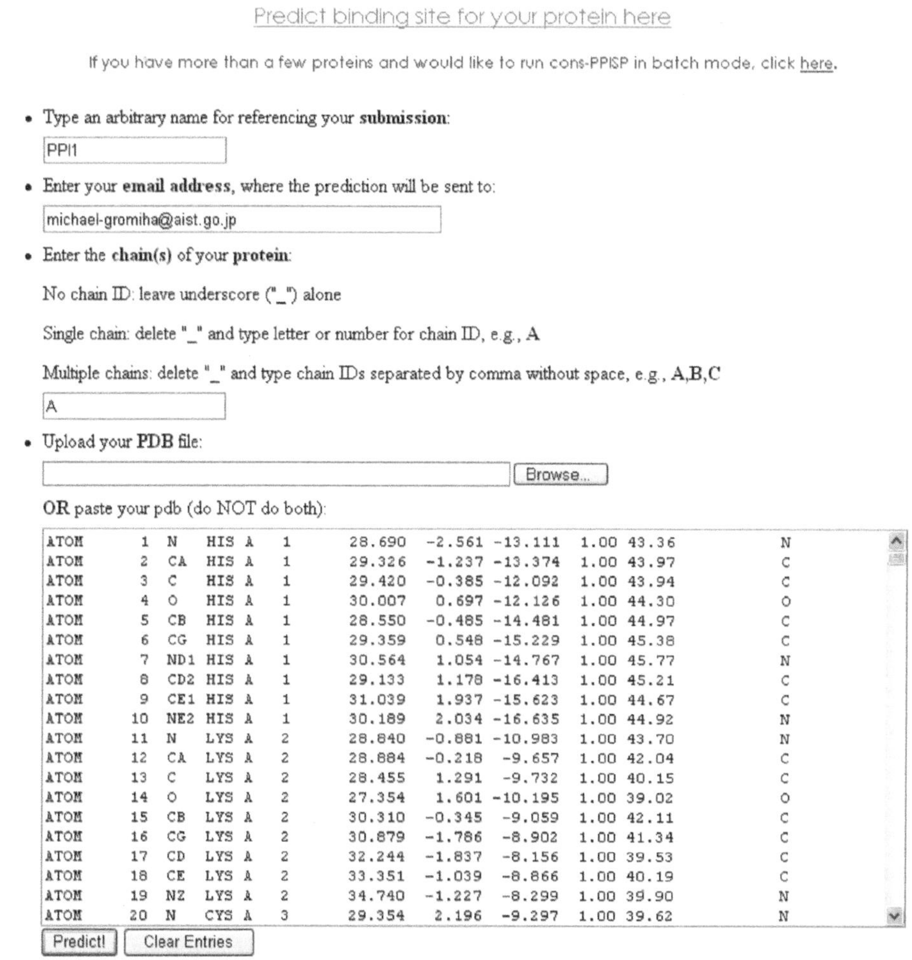

Predict binding site for your protein here

If you have more than a few proteins and would like to run cons-PPISP in batch mode, click here.

- Type an arbitrary name for referencing your **submission**:

  PPI1

- Enter your **email address**, where the prediction will be sent to:

  michael-gromiha@aist.go.jp

- Enter the **chain(s)** of your **protein**:

  No chain ID: leave underscore ("_") alone

  Single chain: delete "_" and type letter or number for chain ID, e.g., A

  Multiple chains: delete "_" and type chain IDs separated by comma without space, e.g., A,B,C

  A

- Upload your **PDB** file:

  ⬚ Browse...

  OR paste your pdb (do NOT do both):

```
ATOM      1  N   HIS A   1      28.690  -2.561 -13.111  1.00 43.36           N
ATOM      2  CA  HIS A   1      29.326  -1.237 -13.374  1.00 43.97           C
ATOM      3  C   HIS A   1      29.420  -0.385 -12.092  1.00 43.94           C
ATOM      4  O   HIS A   1      30.007   0.697 -12.126  1.00 44.30           O
ATOM      5  CB  HIS A   1      28.550  -0.485 -14.481  1.00 44.97           C
ATOM      6  CG  HIS A   1      29.359   0.548 -15.229  1.00 45.38           C
ATOM      7  ND1 HIS A   1      30.564   1.054 -14.767  1.00 45.77           N
ATOM      8  CD2 HIS A   1      29.133   1.178 -16.413  1.00 45.21           C
ATOM      9  CE1 HIS A   1      31.039   1.937 -15.623  1.00 44.67           C
ATOM     10  NE2 HIS A   1      30.189   2.034 -16.635  1.00 44.92           N
ATOM     11  N   LYS A   2      28.840  -0.881 -10.983  1.00 43.70           N
ATOM     12  CA  LYS A   2      28.884  -0.218  -9.657  1.00 42.04           C
ATOM     13  C   LYS A   2      28.455   1.291  -9.732  1.00 40.15           C
ATOM     14  O   LYS A   2      27.354   1.601 -10.195  1.00 39.02           O
ATOM     15  CB  LYS A   2      30.310  -0.345  -9.059  1.00 42.11           C
ATOM     16  CG  LYS A   2      30.879  -1.786  -8.902  1.00 41.34           C
ATOM     17  CD  LYS A   2      32.244  -1.837  -8.156  1.00 39.53           C
ATOM     18  CE  LYS A   2      33.351  -1.039  -8.866  1.00 40.19           C
ATOM     19  NZ  LYS A   2      34.740  -1.227  -8.299  1.00 39.90           N
ATOM     20  N   CYS A   3      29.354   2.196  -9.297  1.00 39.62           N
```

  Predict!   Clear Entries

(a)

FIGURE 7.7 Protein–protein interaction sites prediction using 3D structure with cons-PPISP: (a) input parameters, atomic coordinates, and chain information of the protein and (b) predicted binding sites; P: positive and N: negative.

displays the output with confidence levels. The prediction report obtained with cons-PPISP using protein three-dimensional structure is illustrated in **Figure 7.7**. It takes the PDB coordinates and chain information in the input (**Figure 7.7a**), and the score for each residue is displayed in the output along with the information on binding sites as positive/negative/not predicted (**Figure 7.7b**).

7.1.6 Energy-based approach for understanding the recognition mechanism in protein–protein complexes

Recently, Gromiha et al. (2009) developed an energy-based approach for identifying the binding sites and important residues for binding in protein–protein complexes. The new approach is different from the traditional distance-based contacts in which

```
Submission name: PPI1
Job submitted  : Mon Jun  1 02:26:25 EDT 2009
Job done       : Mon Jun  1 02:34:55 EDT 2009

Prediction by cons-PPISP : consensus Protein-Protein Interaction Site Predictor
Column 1: AA (Amino Acid code)
Column 2: Ch (Chain ID)
Column 3: AA# (Amino Acid number)
Column 4: Score (neural network score)
Column 5: Prediction of whether the residue contacts
          (P = Positive; N = Negative; - = Buried and not predicted)
*************************************************************************
AA Ch  AA#  Score    Prediction
H  A   1    0.940        P
K  A   2    0.543        P
C  A   3    0.936        P
D  A   4    0.719        P
I  A   5    0.020        N
T  A   6    0.062        N
L  A   7    0.000        -
Q  A   8    0.919        P
E  A   9    0.045        N
I  A   10   0.000        -
I  A   11   0.036        N
K  A   12   0.007        N
T  A   13   0.000        -
L  A   14   0.000        -
N  A   15   0.004        N
S  A   16   0.004        N
L  A   17   0.000        -
T  A   18   0.003        N
E  A   19   0.002        N
Q  A   20   0.003        N
```

(b)

FIGURE **7.7**   *(Continued)*

the repulsive interactions are treated as binding sites as well as the contacts within a specific cutoff have been treated in the same way. In a protein–protein complex, the interaction energy is computed for each residue in receptor with all residues in ligand using the nonbonded interactions in AMBER force field (**Section 3.9.5**). The calculations have been repeated for a set of 153 complexes and analyzed the interaction energies of all the residues in intervals of 0.1 from -15 to 5 kcal/mol. **Figure 7.8** shows the frequency of occurrence of residues in receptors at different intervals of interaction free energies (from -2 to 1 kcal/mol). In this figure, the results are presented for both the fraction of residues and total percentage of residues at each interval. The result showed that 7.7% of the residues have strong interactions with ligands, and the interaction free energy is less than -2 kcal/mol. On the other hand, 6.2% of residues have repulsive energies, and 77% of the residues have the interaction energy in the range of -0.3 to 0 kcal/mol, which might be due to the presence of residues that are far away in 3D structures. Among 48,657 residues, 5255 of them have the interaction free energy less than -1 kcal/mol. Interestingly,

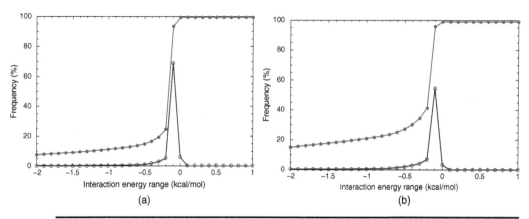

FIGURE 7.8 Occurrence of amino acid residues in different ranges of interaction energies: (a) receptors and (b) ligands. The open and closed circles show the fraction and total percentage of residues. Figure was adapted from Gromiha et al. (2009).

a similar number of residues (4957) are observed in ligands that are interacting with receptors.

The analysis of binding site residues revealed that the residues and residue-pairs with charged and aromatic side chains are important for binding (**Table 7.3**). These residues influence to form cation-π, electrostatic, and aromatic interactions.

TABLE 7.3 Binding propensity of amino acid residues in receptors

Residue	N_bind	N_tot	Propensity
ALA	264	3673	7.19
ASP	261	2831	9.22
CYS	82	800	10.25
GLU	323	3233	9.99
PHE	292	2057	14.20
GLY	270	3441	7.85
HIS	140	1066	13.13
ILE	304	2735	11.12
LYS	271	2840	9.54
LEU	482	4510	10.69
MET	132	1084	12.18
ASN	240	2161	11.11
PRO	262	2317	11.31
GLN	244	1917	12.73
ARG	366	2414	15.16
SER	272	2968	9.16
THR	287	2653	10.82
VAL	354	3505	10.10
TRP	132	764	17.28
TYR	277	1807	15.33

Data were taken from Gromiha et al. (2009)

The observation has been verified with the experimental binding specificity of protein–protein complexes and found good agreement with experiments. On the basis of these results, a novel mechanism has been proposed for the recognition of protein–protein complexes: The charged and aromatic residues in receptor and ligand initiate recognition by making suitable interactions between them; the neighboring hydrophobic residues assist the stability of complex along with other hydrogen bonding partners by the polar residues (Gromiha et al. 2009).

7.1.7   Prediction of protein–protein interaction pairs

The information about the conservation of protein sequences and structurally similar protein domains has been used to identify the protein–protein interaction sites along with the studies on other computational methods. Espadaler et al. (2005) combined both structural similarities among domains of known interacting proteins found in the Database of Interacting Proteins (DIP; see **Section 7.1.1**) and conservation of pairs of sequence patches involved in protein–protein interfaces to predict putative protein interaction pairs. The combination of docking algorithms with predictive approaches provides information on the interaction sites, infers intermolecular contacts, or exploits structural data on complexes of homologs. Docking procedures use computational methods to build atomic model of the complex between two proteins, starting from the 3D structures of the individual molecules.

A remarkable development in this field has been the establishment of CAPRI (Critical Assessment of PRedicted Interactions), a community-wide experiment analogous to CASP, but aimed at assessing the performance of protein–protein docking procedures.

Terashi et al. (2005) developed a method based on molecular dynamics, grid scoring, and the pairwise interaction potential of amino acid residues for identifying protein–protein interaction sites and docking. They have followed the below mentioned steps and are illustrated in **Figure 7.9**.

1. *Benzene cluster (BC) fitting*: The protein–protein interaction site (PPI site) is searched using "benzene cluster fitting," as reported in Komatsu et al. (2003). This method depends on the BC being easily formed around the hydrophobic surface of a globular protein in an aqueous solution. Molecular dynamics (MD) calculations were executed for a system consisting of a protein, such as a receptor or ligand, benzene molecules initially placed near hydrophobic amino acid residues, and water molecules, in order to form clusters of benzene molecules. Approximate fittings between the BCs formed on the surface of each receptor and ligand were executed to dock the ligand with the receptor.

2. *Docking based upon grid-scoring sum*: In the small areas around the interface regions identified by BCs, a grid-scoring method was used to search for energetically stable configurations, considering the 3D shape complementarity between the receptor and the ligand. Grids were generated on surfaces of protein structures based on the surface pattern data on protein–protein interactions created with protein quaternary structure file server, PQS (http://pqs.ebi.ac.uk/; Henrick and Thornton, 1998). In this step, the sum of the number of grids with characteristic surface patterns has been

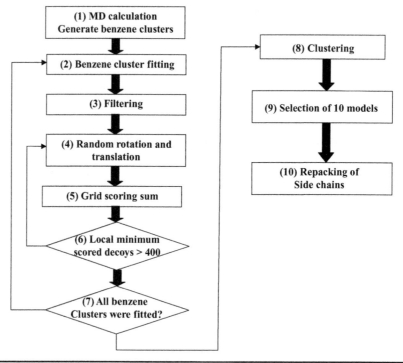

Figure 7.9 Flow chart of a protein–protein docking method. Figure was adapted from Terashi et al. (2005).

counted in the protein–protein contact area (Con_i) and on the whole protein surface (C_i). The grid score (Sco_i) is calculated using the equation:

$$Sco_i = -\log \left(\frac{\left(\dfrac{Con_i}{\sum\limits_{k=0}^{n} Con_k} \right)}{\left(\dfrac{C_i - Con_i}{\sum\limits_{k=0}^{n} (C_k - Con_k)} \right)} \right) \tag{7.6}$$

3. *Clustering and ranking by total score*: The most energetically stable protein–protein interaction structure was determined by giving scaled coefficients to the grid-scoring sum, the number of hydrogen bonds, and the pairwise interaction potential between 20×20 amino acid residue pairs. The pairwise interaction potential was represented by the following equation:

$$M(i,j) = -\log \frac{P(i,j)}{P(i)P(j)N} \tag{7.7}$$

where $P(i,j)$ is the probability of amino acid residues i and j being in contact in the interaction site. $P(i)$ and $P(j)$ represent the probabilities of residues i and j having at least 1 contact with any residue across the interface.

4. *Repacking of side chains*: The side chains were added, and the clashes between them were removed with the software FAMS Complex, a fully automated homology modeling system for protein complex structure.

5. *Normal mode analysis (NMA)*: Hinge regions that connect to domains are very important for considering domain movements of protein structures. Terashi et al. (2005) detected hinge regions that were defined by Φ and ψ around C_α atoms having torsionally large fluctuations using energy optimization and NMA algorithms (Weiner et al. 1984; Adachi et al. 2003; Kamiya et al. 2003).

Schueler-Furman et al. (2005) developed a program, RosettaDock, which uses real-space Monte Carlo minimization (MCM) on both rigid-body and side-chain degrees of freedom to identify the lowest free energy docked arrangement of two protein structures. An improved version of the method that uses gradient-based minimization for off-rotamer, side-chain optimization and includes information from unbound structures was used to create predictions for CAPRI. First, large numbers of independent MCM trajectories were carried out, and the lowest free energy docked configurations were identified. Second, new trajectories were started from these lowest energy structures to thoroughly sample the surrounding conformation space, and the lowest energy configurations were submitted as predictions. For all cases in which there were no significant backbone conformational changes, a small number of very low-energy configurations were identified in the global search and subsequently found to be close to the free energy landscape in the local search. Following the release of the experimental coordinates, it was found that the centers of these free energy minima were close to the native structures in not only the rigid-body orientation but also the detailed conformations of the side chains. The close correspondence of the lowest free energy structures found in their searches to the experimental structures suggests that the free energy function is a reasonable representation of the physical chemistry, and that the real space search with full side-chain flexibility to some extent solves the protein–protein docking problem in the absence of significant backbone conformational changes. Rosetta-Dock is available at http://rosettadock.graylab.jhu.edu/.

Wiehe et al. (2005) used ZDOCK and RDOCK programs for predicting protein–protein interactions. ZDOCK is a rigid-body protein-docking algorithm that explicitly searches rotational space and uses a Fast Fourier Transform (FFT) algorithm to significantly speed up searching in translational space. It includes pair-wise shape complementarity, electrostatics, and desolvation energy. RDOCK is a three-stage energy minimization algorithm designed as a refinement and reranking tool for ZDOCK's top predictions. RDOCK is implemented as a protocol in CHARMM involving the following three steps: (i) remove clashes that occur from the soft-shape complementarity parameter in ZDOCK that allows for small conformational change; (ii) optimize polar interactions; and (iii) optimize charge interactions. For steps (i) and (ii), the charges are turned off for ionic residues. This treatment of charges is particularly successful in antibody–antigen docking cases. After refinement, the predictions are rescored using both electrostatics and desolvation terms. The new scores are used to rerank the top ZDOCK predictions. ZDOCK is available at http://zlab.bu.edu/zdock/index.shtml.

Camacho (2005) showed that modeling key side-chains in explicit solvent improved the recognition of the binding region of a free energy-based computational docking method. Specifically, it has been shown that molecular dynamics (MD) is able to predict with relatively high accuracy the rotamer conformation of the anchor side-chains important for molecular recognition. The conformations are some of the most common rotamers for the given residue, while latch side-chains that undergo induced fit upon binding are forced into less common conformations. Furthermore, a free energy-based scoring function, consisting of the sum of van der Waals, Coulombic electrostatic with a distance-dependent dielectric, and desolvation free energy, successfully discriminates the nativelike conformation. The free energy approach emphasizes the critical role that thermodynamics plays on the algorithm to predict protein interactions.

Lee et al. (2005) applied conformational space annealing (CSA), an efficient global optimization method, to the study of protein–protein interaction. They have used an energy function for the protein–protein interaction that consists of electrostatic interaction, van der Waals interaction, and solvation energy terms represented by the occupancy desolvation method. Each energy term is calculated by precalculated grid potentials and B-spline method approximation. The ligand protein is placed inside a sphere of 50 Å radius centered at an appropriate location, and the CSA rigid docking studies are carried out to find stable complexes. Up to 10 complexes are selected using the K-mean clustering method and biological information when available. These complexes are energy-minimized for further refinement by considering the flexibility of interacting proteins. The results showed that the CSA method has potential applications for the study of protein–protein interactions. Recently, Yoshikawa et al. (2009) proposed an affinity evaluation and prediction method to identify the partners in protein–protein interactions based on the shape complementarity characteristics between proteins as well as the selection of protein complexes by a clustering procedure.

Furthermore, several methods have been developed for protein–protein docking, which includes 3D-Dock (http://www.sbg.bio.ic.ac.uk/docking/; Gabb et al. 1997), HEX (http://www.csd.abdn.ac.uk/hex/; Ritchie and Kemp, 2000), DOT (http://www.sdsc.edu/CCMS/DOT/; Mandell et al. 2001), Molfit (http://www.weizmann.ac.il/Chemical_Research_Support//molfit/home.html; Heifetz et al. 2002), HADDOCK (http://www.nmr.chem.uu.nl/haddock/; Dominguez et al. 2003), ClusPro (http://nrc.bu.edu/cluster/; Comeau et al. 2004), PatchDock (http://bioinfo3d.cs.tau.ac.il/PatchDock/; Schneidman-Duhovny et al. 2005), GRAMM-X (http://vakser.bioinformatics.ku.edu/resources/gramm/grammx; Tovchigrechko et al. 2006), FireDock (http://bioinfo3d.cs.tau.ac.il/FireDock/; Andrusier et al. 2007), PyDock (http://mmb.pcb.ub.es/PyDock/; Cheng et al. 2007), and BDOCK (http://www.biotec.tudresden.de/~bhuang/bdock/bdock.html; Huang and Schroeder, 2008).

7.2 Protein–DNA interactions

Protein–DNA interactions play a key role in many vital processes, including regulation of gene expression, DNA replication and repair, and packaging. The remarkable specificity with which proteins recognize target DNA sequences is of considerable theoretical and practical importance, and its basis has been demonstrated through the structural analysis of large numbers of protein–DNA complexes.

7.2.1   Structural analysis on protein–DNA interactions

Mandel-Gutfreund et al. (1998) analyzed the CH...O interactions in the protein–DNA interface and showed that the number of close intermolecular CH...O contacts involving the thymine methyl group and position C5 of cytosine is comparable to the number of protein–DNA hydrogen bonds involving nitrogen and oxygen atoms as donors and acceptors. A comprehensive analysis of the geometries of these close contacts revealed that they are similar to other CH...O interactions found in proteins and small molecules, as well as to classical NH...O hydrogen bonds. On the basis of these observations, it has been suggested that C5 of cytosine and C5-methyl of thymine form relatively weak CH...O hydrogen bonds with Asp, Asn, Glu, Gln, Ser, and Thr, contributing to the specificity of recognition.

Jones et al. (1999) analyzed the chemical and physical properties of the protein–DNA interface, such as polarity, size, shape, and packing. The DNA-binding sites shared common features, comprising many discontinuous sequence segments forming hydrophilic surfaces capable of direct and water-mediated hydrogen bonds. The comparison of binding sites between protein–DNA and protein–protein complexes showed that the binding sites in protein–DNA complexes are more polar and have more intermolecular hydrogen bonds and buried water molecules than the protein–protein interface sites.

Nadassy et al. (1999) analyzed the atomic structures of protein-nucleic acid complexes and reported that the interface area between the protein and nucleic acid vary between 1120 and 5800 Å^2. Despite this wide variation, the interfaces in complexes of transcription factors with double-stranded DNA could be broken up into recognition modules where 12 ± 3 nucleotides on the DNA side and 24 ± 6 amino acids on the protein side, with interface areas in the range 1600 ± 400Å^2. For enzymes acting on DNA, the recognition module is on average 600 Å^2 larger, due to the requirement of making an active site. As judged by its chemical and amino acid composition, the average protein surface in contact with the DNA is more polar than the solvent accessible surface or the typical protein–protein interface. The protein side is rich in positively charged groups from lysine and arginine side chains; on the DNA side, the negative charges from phosphate groups dominate. Calculations of Voronoi atomic volumes, performed in the presence and absence of water molecules, showed that protein atoms buried at the interface with DNA are on average as closely packed as in the protein interior. Water molecules contribute to the close packing, thereby mediating shape complementarity. On the DNA side, the extent of deformation showed some correlation with the size of the interface area; and on the protein side, the type and size of the structural changes spanned a wide spectrum.

Gromiha et al. (2004a) analyzed the cation-π interactions in protein–DNA complexes and showed that 73% of the studied complexes are involved in such interactions. The cation-π interactions are mainly formed by long-range contacts, and the preference of Arg is higher than Lys to form cation-π interactions. The pair-wise cation-π interaction energy between aromatic and positively charged residues shows that Arg-Tyr energy is the strongest among the possible six pairs. The structural analysis of cation-π interaction forming residues shows that Lys, Trp, and Tyr prefer to be in the binding site of protein–DNA complexes. Furthermore, the accessible surface areas of cation-π interaction forming cationic residues are significantly less than that of other residues. The preference of cation-π interaction forming residues in different secondary structures shows that Lys prefers to be in strand and Phe prefers to be in turn regions.

Ahmad and Sarai (2004) computed the net charge, electric dipole moment and quadrupole moment tensors for a set of DNA-binding proteins and showed that the magnitudes of the moments of electric charge distribution in DNA-binding proteins differ significantly from those of nonbinding proteins. Prabakaran et al. (2006) classified protein–DNA complexes by using a set of 11 descriptors, mainly characterizing protein–DNA interactions, including the number of atomic contacts at major and minor grooves, conformational deviations from standard B- and A-DNA forms, widths of DNA grooves, GC content, specificity measures of direct and indirect readouts, and buried surface area at the complex interface. The cluster analyses were carried out for a unique set of 62 complexes, including a variety of protein motifs, and 7 distinct clusters were revealed from the analyses. They found that some proteins with the same motif are classified into different clusters, whereas different proteins with distinct motifs are classified into the same cluster. These results suggest that the conventional motif-based classification of DNA-binding proteins may not necessarily correspond to structural and functional properties of protein–DNA complexes, and that the present classification will help to identify common properties and rules that govern protein–DNA recognition.

Ahmad et al. (2008) analyzed the thermodynamic and structural data of protein–DNA interactions and explored a relationship between free energy, sequence conservation, and structural cooperativity. They observed that the most of the stabilizing residues or putative hotspots are those which occur as clusters of conserved residues. These clusters have high packing density, which suggested cooperation between conserved residues in the clusters.

7.2.2 DNA stiffness and protein–DNA binding specificity

Hogan and Austin (1987) analyzed the factors influencing the binding specificity of 434 repressors and reported that the elastic properties of DNA may be an important factor for protein–DNA binding specificity. The analysis on the role of DNA elasticity to the binding affinity of Cro protein–DNA complexes showed that the free energy of complex formation increases with stiffness for the nonspecific interactions, and an opposite trend was observed for specific ones (Takeda et al. 1989; Gromiha et al. 1996). A decomposition of the energy terms suggests that binding energy in the nonspecific case is used mainly to compensate the free energy changes due to entropy lost by DNA, while the energy of specific interactions provide enough energy both to bend the DNA molecule and to change the conformation of the Cro protein upon binding (Gromiha et al. 1997).

Gromiha (2005) analyzed the influence of DNA stiffness to protein–DNA binding specificity with several examples. The average stiffness of DNA has been computed using the structure-based sequence dependent stiffness scale (Gromiha,

TABLE 7.4 Structure-based DNA stiffness (Young's modulus) scale for trinucleotides*

Trinucleotide	E (10^8 N/m^2)	Trinucleotide	E (10^8 N/m^2)
AAA/TTT	4.80	CAG/CTG	2.40
AAC/GTT	3.90	CCA/TGG	3.25
AAG/CTT	1.91	CCC/GGG	6.07
AAT/ATT	2.96	CCG/CGG	2.40
ACA/TGT	4.70	CGA/TCG	2.82
ACC/GGT	1.57	CGC/GCG	3.33
ACG/CGT	7.09	CTA/TAG	4.75
ACT/AGT	3.63	CTC/GAG	4.03
AGA/TCT	4.03	GAA/TTC	2.70
AGC/GCT	4.58	GAC/GTC	7.83
AGG/CCT	4.34	GCA/TGC	3.75
ATA/TAT	2.36	GCC/GGC	3.16
ATC/GAT	1.83	GGA/TCC	3.69
ATG/CAT	3.19	GTA/TAC	2.19
CAA/TTG	2.53	TAA/TTA	2.72
CAC/CTG	3.36	TCA/TGA	2.97

*Data were taken from Gromiha (2000)

2000). The numerical values for all the 32 trinucleotide units are given in **Table 7.4**. For calculating the average stiffness value of DNA, the DNA sequence has been represented by overlapping segments of trinucleotide units. The stiffness values for each trinucleotide were assigned appropriately from the structure-based stiffness scale (**Table 7.4**), and the average Young's modulus was computed using the equation

$$E = \sum E_i / n, \tag{7.8}$$

where E_i is the Young's modulus for the ith trinucleotide and n is the total number of trinucleotide units.

These average stiffness values obtained with the target sequences of protein–DNA complexes have been related with experimental protein–DNA binding specificity using correlation coefficient. The results showed that the correlations in the range of 0.65 to 0.97 between DNA stiffness and binding free energy change in several protein–DNA complexes (Gromiha, 2005). Furthermore, the DNA stiffness change due to systematic mutations in target sequence could correctly identify most of the bases in the target sequences of protein–DNA complexes. These results reveal the influence of DNA stiffness to protein–DNA binding specificity. In addition, the direct contacts between protein and DNA through hydrogen bonds and electrostatic and other interactions are also important for understanding the mechanism of protein–DNA recognition.

7.2.3 Inter- and intramolecular interactions in protein–DNA recognition

In protein–DNA complex structures, recognition involves, in part, direct contacts between amino acids and base pairs (direct readout mechanism). These contacts are both redundant and flexible, suggesting that there is no simple code for the specificity of DNA–protein interactions (Matthews, 1988; Pabo and Nekludova, 2000). In addition, the fact that mutation of bases not in direct contact with amino acids often affects the binding affinity implies that water molecules bridging between amino acids and bases (Schwabe, 1997), conformational changes in the DNA (e.g., bending) (Harrington and Winicov, 1994), and/or flexibility (Hogan and Austin, 1987; Sarai et al. 1989; Olson et al. 1998) also affects protein–DNA binding specificity (indirect readout mechanism). In terms of the energy contributed to the binding affinity, the direct readout and water-mediated contacts are intermolecular energies, whereas DNA deformation is associated with intramolecular energies.

Calculation of intermolecular interaction energies

Kono and Sarai (1999) developed a method for quantifying the specificity of direct readout based on the statistical analysis of the structures of protein–DNA complexes. In this method, a coordinate system has been defined by taking an origin N9 atom for A and G, and N1 atom for T and C. They considered the amino acids within a given box, and the box was divided into grids. Each amino acid residue has been represented by its C_α atom, and the distributions of C_α atoms of amino acid residues have been transformed into statistical potentials using the following equations (Sippl, 1990):

$$\Delta E^{ab}(s) = -RT \ln \frac{f^{ab}(s)}{f(s)}$$

$$f^{ab}(s) = \frac{1}{1 + m_{ab}w} f(s) + \frac{m_{ab}w}{1 + m_{ab}w} g^{ab}(s) \qquad (7.9)$$

where m_{ab} is the number of observed pairs (amino acid "a" and base "b"), w is the weight given to each observation, $f(s)$ is the relative frequency of occurrence of any amino acids at grid point s, and $g^{ab}(s)$ is the equivalent relative frequency of occurrence of amino acid "a" against base "b." R and T are gas constant and absolute temperature, respectively. A box of $|x| = |y| = 13.5$ Å and $|z| = 6$ Å, and a grid interval of 3 Å was used in the computation.

By threading a set of random DNA sequences onto the template structure, Kono and Sarai (1999) calculated the Z-score of the specific sequences against the random sequences, which represents the specificity of the complex. Assuming the additivity of potential energies, the sum of the potential energies $[E_{PD} = \Sigma_{ab,s} \Delta E^{ab}(s)]$ for a given DNA sequence in a complexed form was defined as the energy for the sequences. The energy for a particular sequence, in a crystal structure for example, was normalized to measure specificity by the Z-score against random sequences. The Z-score was defined as

$$\text{Z-score} = (X - m)/\sigma, \qquad (7.10)$$

where X is the energy of a particular sequence, m is the mean energy of 50,000 random DNA sequences, and σ is the standard deviation.

Calculation of intramolecular interaction energies

The statistical potential functions for intramolecular interactions have been derived from the conformational energy of DNA using protein–DNA complex structural data (Gromiha et al. 2004b). The sequence-dependent DNA conformational energy has been estimated based on the approach described in Olson et al. (1998). The conformation energies were approximated using a harmonic function, $E_{DNA} = 1/2$ $\sum\sum f_{ij}\Delta\theta_i\Delta\theta_j$, in which θ_i represents the base-step parameters, and f_{ij} are the elastic force constants impeding deformation of the given base step and $\Delta\theta_i = \theta_i - \theta_i^0$, in which θ_i^0 is the average base-step parameter. The base-step parameters used were shift, slide, rise, tilt, roll, and twist. The unknown parameters f_{ij} and θ_i^0 were determined by statistical analysis of the same nonredundant protein–DNA complexes. Setting up a covariance matrix from observed distributions of θ_i thus refers to an effective inverse harmonic force-constant matrix. Inversion of this matrix transformed it to a force-constant matrix in the original coordinate basis. All parameters of a base step for which one parameter exceeded three standard deviations were removed in an iterative manner. Then the final force field was calculated. The conformational energy of DNA in a given complex structure was calculated as the sum of all the base steps. Then, these potentials were used to quantify the specificity of intramolecular interactions of protein–DNA recognition, as a Z-score (Eqn. 7.10), by using the same threading procedure as that of intermolecular interactions.

Role of inter- and intramolecular interactions in protein–DNA recognition

The systematic comparison of direct and indirect readout specificities in a large number of protein–DNA complexes revealed that both intermolecular and intramolecular interactions contribute to the specificity of protein–DNA recognition, and their relative contributions vary depending upon the protein–DNA complex. Examples can be found in Gromiha et al. (2004b) and some of them are listed below: (i) Enzymes prefer to follow direct readout showing their importance in intermolecular interactions between protein and DNA, (ii) there is no significant influence with motifs, (iii) the Z-scores for the zinc finger proteins show that intermolecular interactions make large contribution to specificity, (iv) restriction enzyme, endonuclease EcoRV is influenced by intermolecular interactions, and the recognition is very strict and specific, (v) integration host factors show a severe bend, and the recognition is influenced with intramolecular interactions, (vi) although TATA binding protein has a bent, it is influenced with intermolecular interactions, and further analysis shows that the specificity is due to the nucleotides in the minor groove of DNA, and (vii) ETS proteins and Trp repressor are influenced with intramolecular interactions.

Combination of inter- and intramolecular interactions

The intermolecular and intramolecular energies obtained from statistical potentials have been combined to calculate the total energy using the equation:

$$E_{tot} = cE_{PD} + (1 - c)E_{DNA}, \tag{7.11}$$

where E_{PD} and E_{DNA} are the energies of the intermolecular and intramolecular interactions, respectively, and c is a weighting coefficient ranging between 0 and 1. This coefficient is determined by maximizing the total Z-score; that is, the Z-score is calculated from random sequences, and a value of c is sought that gives the highest total Z-score.

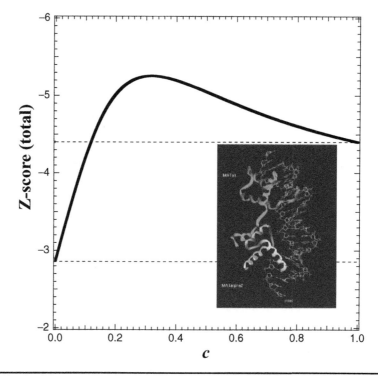

FIGURE 7.10 Total Z-score with respect to weight factor, c, for MATa1/α2-DNA complex. Total energy is given by $E_{tot} = cE_{PD} + (1 - c)E_{DNA}$. Shown in inset is the structure of MATa1/α2-DNA complex. Figure was adapted from Gromiha et al. (2005).

As an example, Gromiha et al. (2004b) considered 1YRN, a complex of DNA with MAT-α1 and α2, two proteins involved in determining mating type in yeast. **Figure 7.10** shows the total Z-score obtained from the combination of intermolecular and intramolecular interactions as a function of the weight factor c. Interestingly, the total Z-score (–5.3 at $c = 0.32$) was higher than either the Z-score for the intermolecular (–4.4) or intramolecular (–2.9) interactions. One interpretation of this result is that the energies of the intermolecular and intramolecular interactions each contain independent information that in combination enhances the specificity of the recognition.

Ahmad et al. (2006) developed a Web server for calculating the intermolecular (direct readout) and intramolecular interaction (indirect readout) Z-scores, and it is available at http://gibk26.bse.kyutech.ac.jp/jouhou/readout/. **Figure 7.11** shows an example with the complex, 1lat. The output displays the Z-scores for direct and indirect readouts along with other parameters used in the computation.

7.2.4 Discrimination of DNA-binding proteins/domains and prediction of DNA-binding sites

Ahmad et al. (2004) analyzed the relationship between DNA–binding and protein sequence composition, solvent accessibility, and secondary structure. Using nonredundant databases of transcription factors and protein–DNA complexes, neural

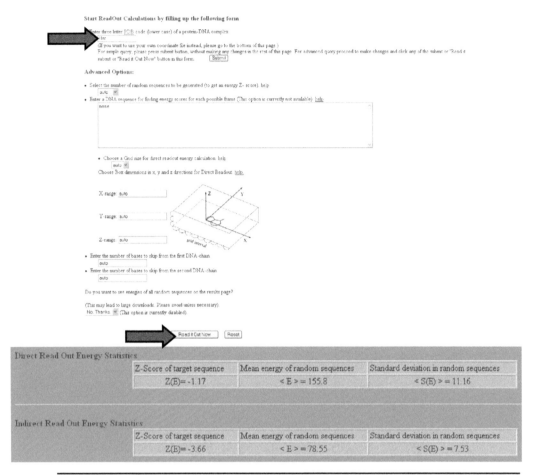

Figure 7.11   Illustrations to compute direct and indirect readout Z-scores using ReadOut.

network models were developed to utilize the information present in this relationship to predict DNA-binding proteins and their binding residues. It has been observed that sequence composition provides sufficient information to predict the probability of its binding to DNA; sequence neighborhood and solvent accessibility information aid to make binding site predictions. Detailed analysis of binding residues shows that some three- and five-residue segments frequently bind to DNA and that solvent accessibility plays a major role in binding. A Web server, DBS-PRED, has been developed for discriminating DNA-binding proteins and their binding residues and is available at http:/// www.netasa.org / dbs-pred/. **Figure 7.12** shows the input options and output obtained with DBS-PRED. Later, Ahmad and Sarai (2005) improved the accuracy of predicting DNA-binding sites using PSSM profiles (see **Section 2.2.4**).

Recently, several methods have been proposed for discriminating DNA-binding proteins and predicting their binding sites. Yu et al. (2006) developed a method based on support vector machines for discriminating DNA- and RNA-binding proteins from other globular proteins. Yan et al. (2006) used Naive Bayes classifier

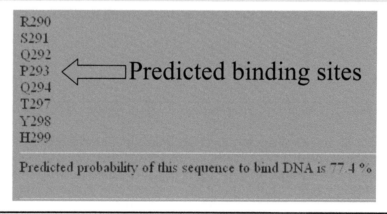

Welcome to DBS-PRED:
Prediction of DNA-binding in proteins using Neural Networks

This program predicts the DNA-binding in a protein from their sequence information. Two neighbours of each residue are taken as the input information and using information from known binding sites of proteins, probable binding sites are predicted. Amino acid composition is taken as the input information for the prediction of DNA-binding probability of a protein.

Reference: *Analysis and Prediction of DNA-binding proteins and their binding residues based on composition, sequence and structure information.*
by Shandar Ahmad, M. Michael Gromiha and Akinori Sarai *Bioinformatics* 20 (2004) 477-486

To predict probable binding sites for a protein sequence, choose a level of sensitivity. This allows for making predictions with low and high probabilities. Choosing strict will give few sites with high probability. Choosing sensitive will give too many over-predictions and choosing medium will give default predictions, which also has a bias towards some over-predictions. Composition based prediction of DNA-binding probability of your protein sequence will also be provided. **More Information**

○ Sensitive   ◉ Medium   ○ Strict

Please enter your sequence in a single letter code.

```
MRGSHHHHHTDPHASSVPLEWPLSSQSG
SYELRIEVQPKPHHRAHYETEGSRGAVKA
PTGGHPVVQLHGYMENKPLGLQ
```

[Predict] [Reset]

R290
S291
Q292
P293 ⇐ Predicted binding sites
Q294
T297
Y298
H299

Predicted probability of this sequence to bind DNA is 77.4 %

FIGURE 7.12 Discrimination of DNA-binding proteins and prediction of binding sites using DBS-PRED.

and the information about the binding site residue and four neighboring residues on both sides of the binding site residue to predict DNA-binding sites. Wang and Brown (2006) developed a method based on support vector machines for identifying DNA-binding sites using the attributes, side chain pK_a value, hydrophobicity index, and molecular mass of an amino acid. A Web server, BindN, has been developed for prediction purposes and is available at http://bioinformatics.ksu.edu/bindn/. Hwang et al. (2007) implemented several machine learning techniques for predicting DNA-binding sites from amino acid

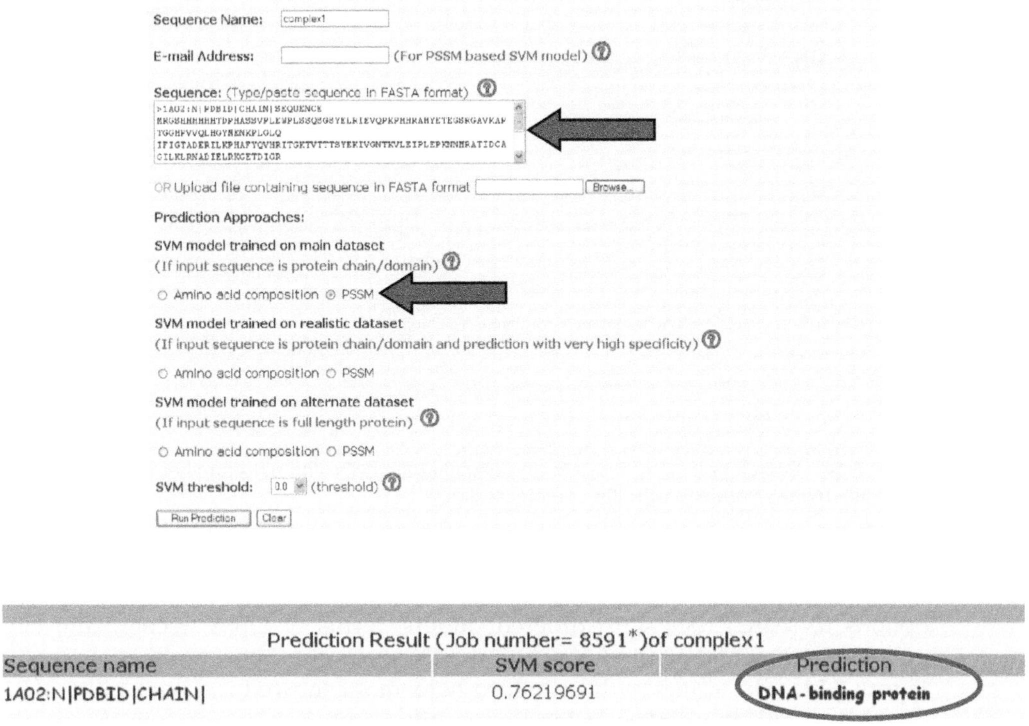

FIGURE 7.13 Discrimination of DNA-binding domains from amino acid sequence using PSSM with DNAbinder.

sequence with/without PSSM profiles and developed a Web server DP-Bind for prediction purposes (http://lcg.rit.albany.edu/dp-bind). Bhardwaj and Lu (2007) used support vector machines based approach to identify the DNA-binding residues with the features that include the residue's identity, charge, solvent accessibility, average potential, secondary structure, neighboring residues, and location in a cationic patch. In addition, mono and di-nucleotide-specific binding sites in proteins have been predicted with neural networks (Andrabi et al. 2009).

Kumar et al. (2007) developed various models using support vector machines and evolutionary profiles for discriminating DNA-binding domains and DNA-binding proteins. A Web server, DNAbinder, has been set up for discrimination, which has the options to select the discrimination method for DNA-binding domains or DNA-binding proteins. It is available at http://www.imtech.res.in/raghava/dnabinder/. **Figure 7.13** shows the prediction procedure using DNAbinder as well as the result obtained with discriminating DNA-binding domains.

7.2.5 Databases for protein–DNA interactions

A database of protein–DNA recognition has been maintained at http://gibk26.bse.kyutech.ac.jp/jouhou/3dinsight/recognition.html. It has two databases: protein–DNA complex database (ProNuc; http://gibk26.bse.kyutech.ac.jp/jouhou/

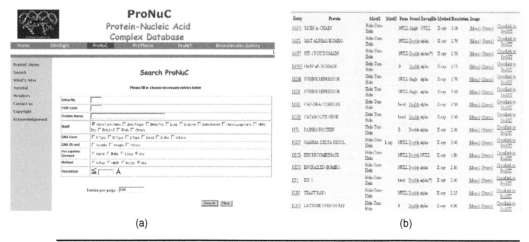

FIGURE 7.14   Search options in ProNuc: (a) searching the data with helix-turn-helix motif; and (b) partial results obtained with the search condition.

pronuc/pronuc.html) and database of base-amino acid interactions (BAInt; http://gibk26.bse.kyutech.ac.jp/jouhou/baint/baint.html). The ProNuc has the options to search data with different motifs, forms of DNA, recognizing elements, experimental method, resolution, PDB code, etc. An example is shown in **Figure 7.14a**. The search with helix-turn-helix motif showed the availability of 94 complexes with the details of binding motifs, resolution, images etc. A sample output is shown in **Figure 7.14b**.

The BAInt provides the information about the contacting DNA bases and amino acids for any specific distance. One can search the data for any protein–DNA complex using its respective PDB code and get the contacts. An example is shown in **Figure 7.15a** for the search options. The contacts for 1lmb within a distance of

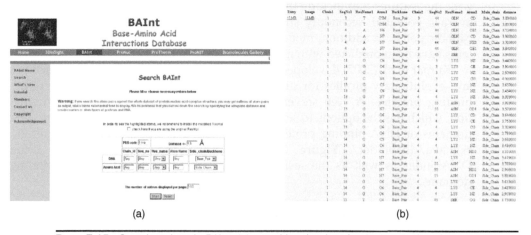

FIGURE 7.15   Search options in BAInt: (a) searching the data for the interactions between the amino acid residues and bases within the distance of 4.5 Å in 1lmb and (b) partial results obtained with the search condition.

4.5 Å are shown in **Figure 7.15b**. It shows the DNA and protein information (chain number, sequence number, name of the amino acid residue/nucleotide, and atom name) along with the distance between the atoms in protein and DNA.

Hoffman et al. (2004) created an Amino Acid–Nucleotide Interaction Database (AANT; http://aant.icmb.utexas.edu/) that categorizes all amino acid–nucleotide interactions from experimentally determined protein–nucleic acid structures and provides users with a graphic interface for visualizing these interactions in aggregate. AANT accomplishes this by extracting individual amino acid–nucleotide interactions from structures in the PDB, combining and superimposing these interactions into multiple structure files (e.g., 20 amino acids × 5 nucleotides) and grouping structurally similar interactions into more readily identifiable clusters. Using the Chime Web browser plug-in, users can view 3D representations of the superimpositions and clusters. The unique collection and representation of data on amino acid–nucleotide interactions facilitate understanding the specificity of protein–nucleic acid interactions at a more fundamental level and allow comparison of otherwise extremely disparate sets of structures. Moreover, by modularly representing the fundamental interactions that govern binding specificity, it may prove possible to better engineer nucleic acid binding proteins.

Prabakaran et al. (2001) developed a Thermodynamic Database for Protein-Nucleic Acid Interactions (ProNIT), which contains several important thermodynamic data for protein–nucleic acid binding, such as dissociation constant (K_d), association constant (K_a), Gibbs free energy change (ΔG), enthalpy change (ΔH), heat capacity change (ΔC_p), experimental conditions, structural information of proteins, nucleic acids and the complex, and literature information. These data are integrated into a relational database system together with structural and functional information to provide flexible searching facilities by using combinations of various terms and parameters. A www interface allows users to search for data based on various conditions, with different display and sorting options, and to visualize molecular structures and their interactions. ProNIT is freely accessible at http://gibk26.bse.kyutech.ac.jp/jouhou/pronit/pronit.html.

Several search and display options were implemented in ProNIT, which are feasible to the users. Detailed tutorials describing the usage of ProNIT are available at the homepage. As an example, the necessary items to be filled or selected to search data that have the values of dissociation constants (Kd_wild) in the range of 7×10^{-10} and 9×10^{-12} are shown in **Figure 7.16a**. In **Figure 7.16b**, the items to be selected for the output and sorting options are shown. In this figure, protein name, PDB_free, PDB_complex, T, pH, method, Kd_wild, and DG_wild are selected for the output. The selected outputs are sorted with Kd in ascending order. The final results obtained from the search conditions (**Figure 7.16a**), display and sorting options (**Figure 7.16b**), are shown in **Figure 7.16c**.

The search results are linked to their relevant sequence, structure, and literature databases. The structure of the protein–nucleic acid complex corresponding to the thermodynamic data can be visualized through the database of protein–nucleic acid complex structures (http://gibk26.bse.kyutech.ac.jp/jouhou/pronuc/pronuc.html) in the same database system. Here, users can also examine the conformational properties of DNA such as roll, tilt, slide, twist, rise, propeller twist of base pairs and dihedral angles of backbones (Dickerson, 1989), and

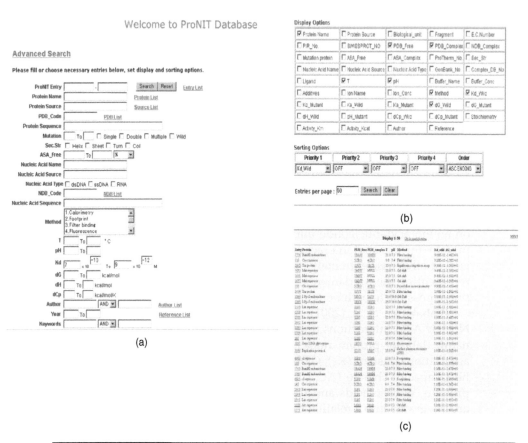

FIGURE 7.16   An example of searching conditions, display and sorting options, and results of ProNIT. (a) Main menu for the search options of ProNIT. In this example, Kd_wild is specified by filling the boxes for the values from 7×10^{-10} to 9×10^{-12}. (b) Display and sorting options of ProTherm. In this example, protein name, PDB_free, PDB_complex, T, pH, method, Kd_wild, and DG_wild are selected for the output and the selected outputs are sorted with Kd in ascending order. (c) Part of the results obtained from ProNIT.

sequence-dependent flexibility (Sarai et al. 1989) in the form of graphical plots, as shown in **Figures 7.17a** and **7.17b**, respectively. These parameters are helpful to understand sequence-dependent variations of local DNA geometry and conformational flexibility. If the users are interested in the specific base-amino acid interactions involved in the complex for comparison with the binding thermodynamic data, they can search for the pairs by specifying atom, residue and distance criteria. The specific base-amino acid pairs are automatically highlighted in the complex and visualized by 3D viewers, RasMol (Sayle and Milner-White, 1995) or VRML (see **Figure 7.18**).

7.3   Protein–RNA interactions

The binding specificity of protein–RNA complexes is somewhat different from that of protein–DNA complexes. The comparison of interactions at the interface

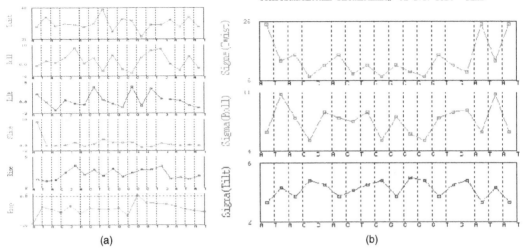

(a)   (b)

Figure 7.17 Conformational parameters and sequence dependent fluctuations for operator DNA in the λ repressor-operator complex (PDB code: 1LMB). (a) Graphical plots of major sequence dependent conformational parameters and (b) Graphical plots of conformational flexibility values for twist, roll, and tilt.

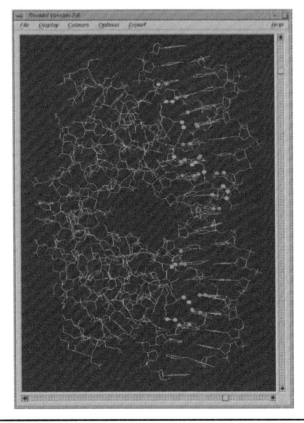

Figure 7.18 RasMol view of λ repressor-operator complex (PDB code: 1LMB) showing contacting atoms of base (red) and amino acids (green) within a distance of 3.5 Å.

would provide deep insights to understand the mechanisms of protein–DNA and protein–RNA interactions.

7.3.1   Structural analysis of protein–RNA complexes

Allers and Shamoo (2001) analyzed a set of protein–RNA complexes to explore the underlying chemical principles governing both sequence specific and nonsequence specific binding. To facilitate the analysis, they have constructed a database of interactions using ENTANGLE, a JAVA-based program that uses available structural models in their PDB format and searches for appropriate hydrogen bonding, stacking, electrostatic, hydrophobic, and van der Waals interactions. The resulting database of interactions reveals correlations that suggest the basis for the discrimination of RNA from DNA and for base-specific recognition. The data illustrate both major and minor interaction strategies employed by families of proteins such as tRNA synthetases, ribosomal proteins, or RNA recognition motifs with their RNA targets. The specific RNA recognition appears to be mediated largely by interactions of amide and carbonyl groups in the protein backbone with the edge of the RNA base. In cases where a base accepts a proton, the dominant amino acid donor is arginine, whereas in cases where the base donates a proton, the predominant acceptor is the backbone carbonyl group, not a side-chain group. This is in marked contrast to protein–DNA interactions, which are governed predominantly by amino acid side-chain interactions with functional groups that are presented in the accessible major groove. RNA recognition often proceeds through loops, bulges, kinks, and other irregular structures that permit use of all the RNA functional groups, and this is seen throughout the protein–RNA interaction database. ENTANGLE can be downloaded from http://www.bioc.rice.edu/~shamoo/resources.html.

Treger and Westhof (2001) analyzed three groups of protein–RNA complexes, tRNA synthetases, ribosomal complexes and other for the number of contacts between RNA components (phosphate, ribose and the four bases) and amino acid side chains, such as, ionic, neutral H-bond, C-H...O H-bond, or van der Waals interaction. They observed that (i) in all three groups of complexes, the most preferred amino acids (Arg, Asn, Ser, Lys) and the less preferred ones (Ala, Ile, Leu, Val) are the same; Trp and Cys are rarely observed (respectively 15 and 5 amino acids in the ensemble of interfaces); (ii) of the total number of amino acids located at the interfaces, 22% are hydrophobic, 40% charged (positive 32%, negative 8%), 30% polar, and 8% are Gly; (iii) in ribosomal complexes, phosphate is preferred over ribose, which is preferred over the bases, but there is no significant preference in the other two groups; (iv) there is no significant prevalence of a base type at protein–RNA interfaces, but specifically Arg and Lys display a preference for phosphate over ribose and bases; Pro and Asn prefer bases over ribose and phosphate; Met, Phe and Tyr prefer ribose over phosphate and bases. Furthermore, Ile, Pro, Ser prefer A over the others; Leu prefers C; Asp and Gly prefer G; and Asn prefers U.

The analysis on contact types revealed the following conclusions: (i) 23% of the contacts are via potential H-bonds (including CH...O H-bonds and ionic interactions), 72% belong to van der Waals interactions, and 5% are considered as short contacts; (ii) of all potential H-bonds, 54% are standard, 33% are of the C-H...O type, and 13% are ionic; (iii) the Watson-Crick sites of G, O6(G) and principally N2(G) and the hydroxyl group O2' is more often involved in H-bonds than expected;

the protein main chain is involved in 32% and the side chains in 68% of the H-bonds; considering the neutral and ionic H-bonds, the following couples are more frequent than expected pairs: base A-Ser, base G-Asp/Glu, base U-Asn. The RNA CH groups interact preferentially with oxygen atoms (62% on the main chain and 19% on the side chains); (iv) the bases are involved in 38% of all H-bonds, and more than 26% of the H-bonds have the H donor group on the RNA; (v) the atom O2′ is involved in 21% of all H-bonds, a number greater than expected; (vi) amino acids less frequently in direct contact with RNA components interact frequently via their main chain atoms through water molecules with RNA atoms; in contrast, those frequently observed in direct contact, except Ser, use instead their side chain atoms for water bridging interactions.

Jones et al. (2001) showed that the protein–RNA interface has diverse nature of binding sites. Van der Waals contacts played a more prevalent role than hydrogen bond contacts, and preferential binding to guanine (G) and uracil (U) was observed. The positively charged residue, Arg, and aromatic residues, Phe and Tyr, play key roles in the RNA-binding sites. A comparison between protein–RNA and protein–DNA complexes showed that while base and backbone contacts (both hydrogen bonding and van der Waals) were observed with equal frequency in the protein–RNA complexes, backbone contacts were more dominant in the protein–DNA complexes. Although similar modes of secondary structure interactions have been observed in RNA- and DNA-binding proteins, there exist differences between the two types of nucleic acid binding proteins at the atomic contact level.

Ellis et al. (2007) analyzed a set of protein–RNA complex structures and observed that van der Waals interactions are more numerous than hydrogen bonds, and the contacts made to the nucleic acid backbone occur more frequently than specific contacts to nucleotide bases. Baker and Grant (2007) analyzed the role of aromatic amino acids at protein–nucleic acid interface and reported that in protein–RNA complexes, Phe occurs less often and is instead replaced by Trp, which binds selectively to C and G, offering a possible mechanism for differentiation between the two nucleic acids. Phipps and Li (2007) analyzed the contacts at crystal packing interfaces of protein–RNA complexes and revealed the importance of electrostatic and hydrogen bonding interactions, facilitated by positively charged amino acids, in mediating both specific and nonspecific protein–RNA interactions.

7.3.2   Prediction of RNA-binding sites

Methods have been proposed for predicting the binding sites in RNA-binding proteins. Miyano's group (Jeong and Miyano, 2006; Jeong et al. 2004) developed a method based on neural networks and weighted profiles for identifying the RNA-interacting residues. Teribillini et al. (2006) developed a Web-based method, RNABindR for predicting RNA-binding sites from amino acid sequence using machine learning techniques. It is available at http://bindr.gdcb.iastate.edu. Kumar et al. (2008) combined evolutionary information and support vector machine and developed an improved method for predicting RNA-binding sites in a protein sequence. The prediction method is available at http://www.imtech.res.in/raghava/pprint/. **Figure 7.19** illustrates the utility of the method for identifying the binding site residues in a protein–RNA complex.

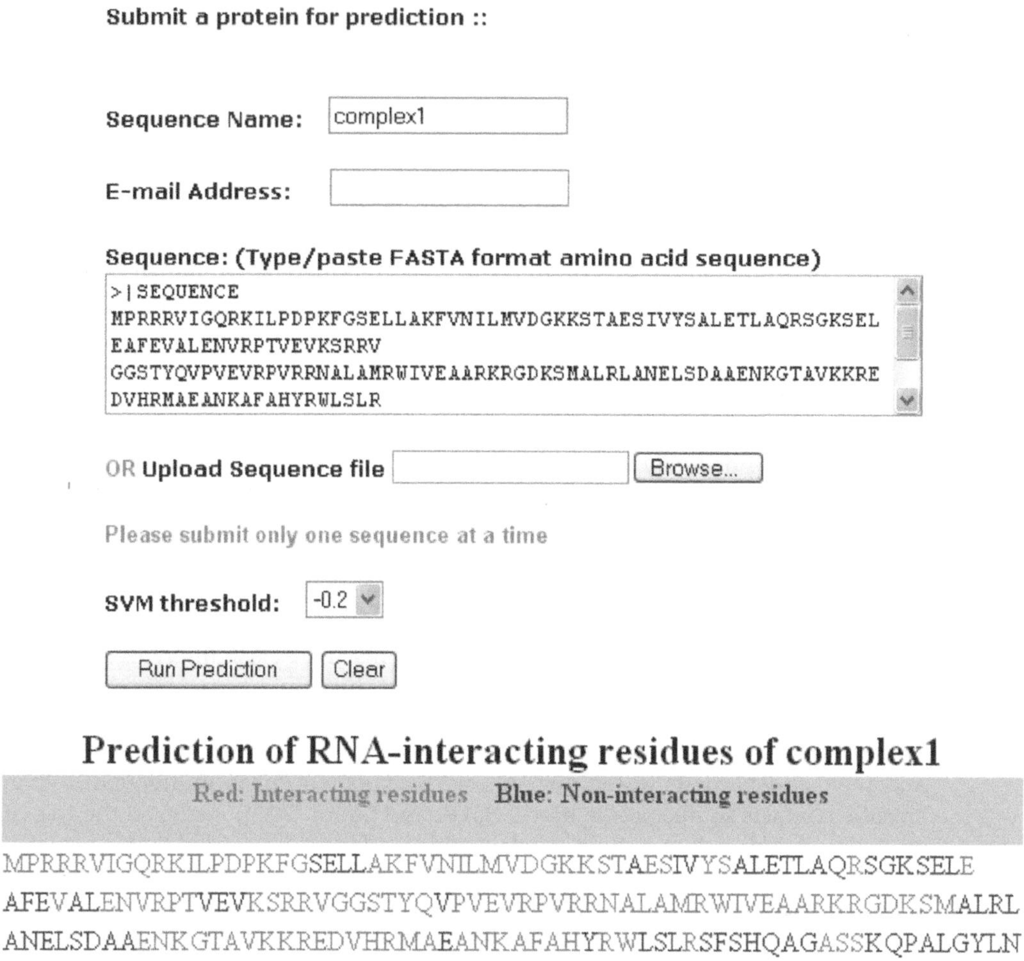

Submit a protein for prediction ::

Sequence Name: complex1

E-mail Address:

Sequence: (Type/paste FASTA format amino acid sequence)

```
>| SEQUENCE
MPRRRVIGQRKILPDPKFGSELLAKFVNILMVDGKKSTAESIVYSALETLAQRSGKSEL
EAFEVALENVRPTVEVKSRRV
GGSTYQVPVEVRPVRRNALAMRWIVEAARKRGDKSMALRLANELSDAAENKGTAVKKRE
DVHRMAEANKAFAHYRWLSLR
```

OR **Upload Sequence file** [   ] [Browse...]

Please submit only one sequence at a time

SVM threshold: -0.2 ▾

[Run Prediction] [Clear]

Prediction of RNA-interacting residues of complex1

Red: Interacting residues   Blue: Non-interacting residues

MPRRRVIGQRKILPDPKFGSELLAKFVNILMVDGKKSTAESIVYSALETLAQRSGKSELE
AFEVALENVRPTVEVKSRRVGGSTYQVPVEVRPVRRNALAMRWIVEAARKRGDKSMALRL
ANELSDAAENKGTAVKKREDVHRMAEANKAFAHYRWLSLRSFSHQAGASSKQPALGYLN

FIGURE 7.19   Prediction of RNA-binding site residues using Pprint.

7.4   Protein–ligand interactions

Protein–ligand interactions are involved in processes such as cell signalling, transport, metabolism, regulation, gene expression, and enzyme activity. Understanding the interactions between these molecules is of importance in searching for new pharmaceuticals, industrial compounds, and functional molecules for food products. The prediction of ligand binding sites is an essential part of the drug discovery process. Knowing the location of binding sites greatly facilitates the search for hits, the lead optimization process, the design of site-directed mutagenesis experiments, and the hunt for structural features that influence the selectivity of binding in order to minimize the drug's adverse effects.

7.4.1   Prediction of ligand-binding sites

Chang et al. (2005) designed a Web server, MEDock, which provides an efficient utility for predicting ligand-binding sites. The MEDock Web server incorporates a global search strategy that exploits the maximum entropy property of the Gaussian probability distribution in the context of information theory. As a result of the global search strategy, the optimization algorithm incorporated in MEDock performs well when dealing with very rugged energy landscapes, which usually have insurmountable barriers. MEDock is available at http://medock.csie.ntu.edu.tw/ and http://bioinfo.mc.ntu.edu.tw/medock/.

This server takes the input in the PDBQ format, which is an extension of the PDB format. The PDBQ format for ligands can be generated by many chemical software suites or Web servers. For example, Dundee's PRODRG server (Schuttelkopf and van Aalten, 2004; http://davapc1.bioch.dundee.ac.uk/programs/prodrg/) provides a convenient visual interface to generate this file format from the PDB file (or from other file formats) of a ligand. Although computationally more demanding, the quantum chemical calculation procedure used in the relaxed complex scheme (Lin et al. 2003) may be invoked for a more accurate assignment of partial charges on each atom of the ligand molecule. It should be emphasized that the accurate assignment of the ligand's partial charges is critically important when dealing with ligand–receptor interactions that are dominated by electrostatics. The ligand files for these four benchmarks were prepared using the relaxed complex procedures. The PDBQ file for proteins can be derived from the PDB2PQR server (Dolinsky et al. 2004; http://agave.wustl.edu/pdb2pqr/) and a simple awk or perl script. The PDB2PQR server also provides a prediction of the protonation states of the ionizable residues in a protein, which is an important issue for the correct description of ligand–receptor interactions. The MEDock Web server also includes these two Web servers for automatically converting the PDB format to the PDBQ format, just in case users do not have other preferred procedures for the required calculations and conversions.

Laurie and Jackson (2005) described a method of ligand-binding site prediction called Q-SiteFinder. It uses the interaction energy between the protein and a simple van der Waals probe to locate energetically favourable binding sites. Energetically favourable probe sites are clustered according to their spatial proximity, and clusters are then ranked according to the sum of interaction energies for sites within each cluster. There is at least one successful prediction in the top three predicted sites in 90% of proteins tested when using Q-SiteFinder. This success rate is higher than that of a commonly used pocket detection algorithm (Pocket-Finder), which uses geometric criteria. In addition, Q-SiteFinder is twice as effective as Pocket-Finder in generating predicted sites that map accurately onto ligand coordinates. It also generates predicted sites with the lowest average volumes of the methods examined in this study. Unlike pocket detection, the volumes of the predicted sites appear to show relatively low dependence on protein volume and are similar in volume to the ligands they contain. Restricting the size of the pocket is important for reducing the search space required for docking and de novo drug design or site comparison. Both Q-SiteFinder and Pocket-Finder have been made available online, respectively, at http://www.bioinformatics.leeds.ac.uk/qsitefinder and http://www.bioinformatics.leeds.ac.uk/pocketfinder.

Other programs to predict the ligand binding sites in proteins include (i) Fuzzy-oil-drop (http://www.bioinformatics.cm-uj.krakow.pl/activesite; Brylinski et al. 2007), (ii) LigProf (http://www.cropnet.pl/ligprof; Koczyk et al. 2007), (iii) Protemot (http://protemot.csbb.ntu.edu.tw/; Chang et al. 2006), (iv) CASTp (http://sts.bioengr.uic.edu/castp/; Dundas et al. 2006), (v) PASS (http://www.ccl. net/cca/software/UNIX/pass/overview.shtml; Brady and Stouten, 2000), (vi) SURFNET-ConSurf (http://consurf-hssp.tau.ac.il; Glaser et al. 2006), (vii) eHiTS (http://www.simbiosys.ca/ehits/; Zsoldos et al. 2006), and (viii) MolDock (Molegro Virtual Docker, MVD) (http://www.molegro.com/; Thomsen and Christensen, 2006).

7.4.2 Protein–ligand docking

Vaque et al. (2008) presented an excellent review on the advances of protein–ligand docking. It covers the fundamental algorithms used in docking programs, comparison of different docking methods with their strengths and weakness, and the challenges for docking techniques. The procedure for protein–ligand docking involves the following steps: (i) finding the structures of proteins (receptors) and ligands, (ii) identifying ligand binding sites, (iii) considering receptor/ligand flexibility, and (iv) computing interaction energy between the receptor and the ligand based on force fields (see **Section 3.9.5**), which is used as scoring function. Several methods have been developed for docking proteins with ligands, which have their own advantages and limitations.

The main features of five selected docking programs, eHiTs (Zsoldos et al. 2006), GOLD (Jones et al. 1997), MVD (Thomsen and Christensen, 2006), AutoDock (Morris et al. 1996), and Glide (Friesner et al. 2004; Halgren et al. 2004) are summarized below (Vaque et al. 2008):

eHiTS

The eHiTS (electronic High Throughput Screening; Zsoldos et al. 2006) program is (i) easy to use, (ii) performs well both in speed and accuracy, and (c) has a lot of automated features that simplify the drug design workflow and provide innovative solutions to common docking problems. It is available at http://www.simbiosys.ca/ehits/.

GOLD

GOLD (Genetic Optimization for Ligand Docking; Jones et al. 1997) has several strengths that include (i) backbone and side chain flexibility, (ii) user-defined scoring functions, (iii) energy functions based on conformational and nonbonded contacts, (iv) variety of constraint options with the inclusion of water molecules and metal atoms, and (v) optimized for parallel execution on processor networks as well as use on grid systems. It is available at http://www.ccdc.cam.ac.uk/products/life_sciences/gold/.

MVD

The main features of MVD (Molegro Virtual Docker; Thomsen and Christensen, 2006) are as follows: (i) It can automatically set up the input structures by assigning charges, bond orders, and hybridization; (ii) it predicts automatically the potential binding sites in the receptor; (iii) it can deal with receptor side chain flexibility with user-defined constraints; (iv) it has the options to dock in precalculated energy grids and with templates; and (v) it is able to distribute the calculations on multiple computers. It is available at http://www.molegro.com/.

AutoDock

AutoDock (Morris et al. 1996) is the first docking package to model the ligand with full conformational flexibility. The package consists of two sequentially applied programs, AutoGrid and AutoDock. AutoGrid is initially used to calculate the noncovalent energy of interaction between the rigid part of the receptor and a probe atom that is located at various grid points of the lattice. Furthermore, AutoGrid generates an electrostatic potential grid map and a desolvation map. The full set of grid maps and the flexible part of the receptor are used by AutoDock to guide the docking process of the selected ligands.

AutoDock's main strengths are (i) receptor flexibility; (ii) blind-docking; (iii) precalculated grid maps on a binding site; (iv) free-energy scoring function based on linear regression analysis, the AMBER force field, and a large set of protein-ligand complexes with known inhibition constants; and (v) good correlation between predicted inhibition constants and experimental data. It is available at http://autodock.scripps.edu/.

Glide

Glide (Grid-based Ligand Docking with Energetics; Friesner et al. 2004; Halgren et al. 2004) was developed by Schrödinger (Portland, Oregon, USA) and it has the following features: (i) It considers the receptor as flexible, and the receptor can change its conformation upon ligand binding, (ii) it allows user-defined constraints for restricting a ligand atom to lie within a specific region, (iii) it can be used together with two other programs, Liaison and Qsite, to obtain accurate binding energies for ligand–receptor pairs, (iv) it can consider water molecules in the receptor's active site during docking, and (v) it has parallel processing or distribution options. The details are available at http://www.schrodinger.com/.

7.4.3   Estimation of protein–ligand binding free energy

Pei et al. (2004) developed a method for estimating the solvation energy for protein–ligand binding based on atomic solvation parameters (ASPs). Two sets of ASPs have been derived from experimental n-octanol/water partition coefficient (log P) data, which contains (i) 100 atom types for a united model that treats hydrogen atoms implicitly and (ii) 119 atom types for all-atom model that treats hydrogen atoms explicitly. Utilizing the unified ASP set, an algorithm was developed for solvation energy calculation and was further integrated into a score function for predicting protein–ligand binding affinity. This score function could reasonably predict the binding free energies of protein–ligand interactions.

Jayaram and his co-workers developed a series of algorithms for predicting the binding affinity of protein–ligand complexes. Jain and Jayaram (2005, 2007) reported a computationally fast protocol for predicting binding affinities of nonmetallo protein–ligand complexes and zinc containing metalloprotein–ligand complexes. The protocol builds in an all atom energy-based empirical scoring function, comprising electrostatics, van der Waals, hydrophobicity, and loss of

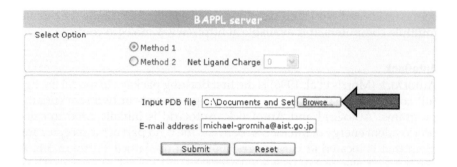

The Predicted Binding Free Energy of the Protein-Ligand Complex 1A30.pdb is   -5.78
kcal/mol

FIGURE 7.20   Computation of protein–ligand interaction energy using BAPPL server.

conformational entropy of protein side chains upon ligand binding. A Web server, BAPPL, has been developed to compute the interaction energy and it is available at http://www.scfbio-iitd.res.in/software/drugdesign/bappl.jsp. The server requires a specific format of PDB coordinates to compute the protein–ligand interaction energy, and the details have been provided on the Web with examples. **Figure 7.20** shows the utility of BAPPL. It takes the atomic coordinates of protein and ligand and sends the protein–ligand binding free energy by e-mail. Furthermore, Gupta et al. (2007) developed an all atom energy-based Monte Carlo docking protocol, ParDock, for protein–ligand complexes, and the software is available at www.scfbio-iitd.res.in/dock.

7.4.4   Validating ligand and active site residues in protein structures

Kleywegt et al (2004) developed a protocol to compare and validate ligand structures in PDB with the electron density map. A detailed tutorial is available at http://xray.bmc.uu.se/embo/ligandtut/index.html. **Figure 7.21** illustrates the procedure for the validation of ligand structures in the PDB. The information on ligand can be found in PDBsum (http://www.ebi.ac.uk/pdbsum/) by providing the PDB code (e.g., 1HW8; **Figure 7.21a**). The link "EDS" available in PDBsum (**Figure 7.21b**) connects to Uppsala Electron Density Server and provides the summary and plots for the structure (**Figure 7.21c**). The links to real space R-value shows the map of R-value as well as the hetero groups (**Figure 7.21d**). By clicking on the ligand (for example, 114 in 1HW8), one can obtain the electron density around the residue (position 1 for the ligand 114 in 1AZM). The final view can be obtained by following the below mentioned steps: (i) **Show -> Solvent** (this will turn off the water molecules), (ii) **Protein -> Line** (this will turn off the protein wireframe), (iii) **Ligand -> Line** (this will turn off the ligand wireframe), (iv) click **Select**, (v) click on the **(+)** sign in **()** chain, (vi) click on **1 : 114** to select this ligand (it will appear in red to indicate that it is selected), (vii) **Ligand -> Ball & Stick** (this will make **1 : 114** appear in the visualization screen), (viii) go to the screen and push right button in

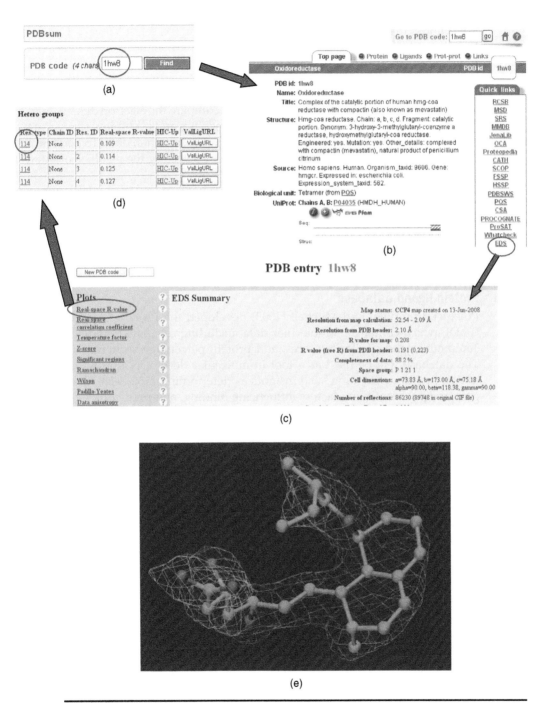

FIGURE 7.21 Procedure for validating ligand structures in PDB: (a) searching PDBsum, (b) output obtained with PDBsum and a link to EDS, (c) summary of the given structure at EDS, (d) Hetero groups in EDS, and (e) ligand structure with electron density map.

your mouse to make the contextual menu to appear, (ix) select **View -> Center on Selection** (this will make the view to become centered on the ligand **1 : 114**), (x) go to the screen and push right button of your mouse to make the contextual menu to appear, and (xi) select **View -> Clip maps on Selection** (this will delete the electron density map that is not around ligand **1 : 114**). The final selected image is shown in **Figure 7.21e**. This protocol can also be used to compare the electron density map of any region in the PDB file (e.g., the active site residues and prohibited regions of the Ramachandran map) against that derived from experimental data. To perform the comparison, repeat all steps described above by selecting the residues of interest instead of the ligand.

Kleywegt and Harris (2007) developed a Web-based tool, ValLigURL, for validating the geometry of a ligand and comparing the conformation of a ligand with all instances of that ligand in PDB. In addition, it can be used to survey the quality or conformational diversity of any ligand across the entire PDB. The server is freely accessible at http://eds.bmc.uu.se/eds/valligurl.php.

7.4.5 Protein-ligand database

Puvanendrampillai and Mitchell (2003) developed the Protein Ligand Database (PLD), which contains biomolecular data, including calculated binding energies, Tanimoto ligand similarity scores, and protein percentage sequence similarities. The database will be used as a tool in molecular design, primarily in the drug design process, providing such insights as determining molecular function from genetic data, identifying factors influencing binding, and predicting biological effects. It can be searched with PDB code, protein and ligand names, protein function, interaction types, binding energies, authors, etc. It is freely available at http://www-mitchell.ch.cam.ac.uk/pld/.

Shin and Cho (2005) developed a ligand database based on PDB, PDB-Ligand (http://www.idrtech.com/PDB-Ligand/), which is a three-dimensional structure database of small molecular ligands that are bound to larger biomolecules deposited in the PDB. It is also a database tool that allows one to browse, classify, superimpose, and visualize these structures. The proteins that a given ligand binds are often homologous and present the same binding structure to the ligand. However, there are also many instances wherein a given ligand binds to two or more unrelated proteins, or to the same or homologous protein in different binding environments. PDB-Ligand serves as an interactive structural analysis and clustering tool for all the ligand-binding structures in the PDB. PDB-Ligand also provides an easy way to obtain a number of different structure alignments of many related ligand-binding structures based on a simple and flexible ligand clustering method. PDB-Ligand will be a good resource for both a better interpretation of ligand-binding structures and the development of better scoring functions to be used in many drug discovery applications.

Kellenberger et al. (2006) developed an annotated database (sc-PDB) for druggable binding sites from the PDB. It is a collection of 5952 three-dimensional structures of binding sites (as of 15 June 2009) found in the PDB. Binding sites were extracted from all high-resolution crystal structures in which a complex between a protein cavity and a small-molecular-weight ligand could be identified. Ligands are classified into four main categories: nucleotides (< 4-mer), peptides (< 9-mer), cofactors, and organic compounds. The corresponding binding site is formed by all protein residues (including amino acids, cofactors, and important metal ions) with at least one atom within 6.5 Å of any ligand atom. The database was carefully annotated by browsing several protein databases (PDB, UniProt, and GO) and storing, for every sc-PDB entry with the following features: protein name, function, source, domain and mutations, ligand name, and structure. The repository of ligands has also been archived by diversity analysis of molecular scaffolds, and several chemoinformatics descriptors were computed to better understand the chemical space covered by stored ligands. The sc-PDB may be used for several purposes: (i) screening a collection of binding sites for predicting the most likely target(s) of any ligand, (ii) analyzing the molecular similarity between different cavities, and (iii) deriving rules that describe the relationship between ligand pharmacophoric points and active-site properties. The database is accessible on the Web at http://bioinfo-pharma.u-strasbg.fr/scPDB/.

Furthermore, several other databases have been created from the knowledge of the three-dimensional structures available in the PDB and binding affinities. This includes SuperLigands (http://bioinf.charite.de/superligands/; Michalsky et al. 2005), LigBase (http://modbase.compbio.ucsf.edu/ligbase/; Stuart et al. 2002), Het-PDB Navi (http://daisy.nagahama-i-bio.ac.jp/golab/hetpdbnavi.html; Yamaguchi et al. 2004), Ligand-protein database (http://lpdb.chem.lsa.umich.edu/; Roche et al. 2001), PDBSite (http://wwwmgs.bionet.nsc.ru/mgs/gnw/pdbsite/; Ivanisenko et al. 2005), SitesBase (http://www.bioinformatics.leeds.ac.uk/sb; Gold and Jackson, 2006), AffinDB (http://www.agklebe.de/affinity; Block et al. 2006), SuperStar (http://www.ccdc.cam.ac.uk/products/life_sciences/superstar/; Boer et al. 2001), PLASS (Ozrin et al. 2004), PDBbind (http://www.pdbbind.org/; Wang et al. 2005), PDBLIG (Chalk et al. 2004), EzCatDB (http://mbs.cbrc.jp/EzCatDB/; Nagano, 2005), MSDsite (http://www.ebi.ac.uk/msd-srv/msdsite; Golovin et al. 2005), BindingDB (http://www.bindingdb.org; Liu et al. 2007), protein–ligand interaction database, PLID (http://203.199.182.73/gnsmmg/databases/plid/; Reddy et al. 2008), and so on.

7.5   Quantitative structure activity relationship in protein–ligand interactions

An essential feature of drug discovery is to synthesize analogs of lead molecules and to test their biological activity for obtaining better analogs. It is mainly based on the hypothesis that any change in the chemical structure produces a positive or negative change in the bioactivity, which is termed as structure–activity relationship. The concept of relating physicochemical properties of molecules with their biological activities is known as "quantitative structure–activity relationship, QSAR" (Hansch et al. 1962). The QSAR studies have been described in different aspects, including rational drug design (Fujita, 1995; Parrill and Reddy, 1999; Debnath, 2006) and applications in chemistry and biology (Hansch and Leo, 1995). Garg and Bhhatarai (2006) extensively reviewed QSAR and modeling studies of HIV protease inhibitors.

In QSAR, the biological (enzyme) activity of compounds is generally expressed with IC50 or EC50. The IC50 refers to the minimal concentration of the compound leading to 50% inhibition of the enzyme. EC50 is the concentration (mol/L or mol/g) of the compound required to achieve 50% protection of cells against the effect of the virus. It also refers to the molar concentration of an agonist, which produces 50% of the maximum possible response for that agonist. For the selectivity of the compounds, their cytotoxic effect has been measured in terms of CC50, the concentration of the compound required to reduce by 50% the number of mock-infected cells. Another term is K_i, which is the enzyme inhibition constant. The logarithms of inverse of these parameters [log(1/C)] refer to the biological parameters (endpoints) in QSAR studies (Garg and Bhhatarai, 2006).

The biological endpoints are correlated with various physicochemical parameters, which include hydrophobic, electronic, sterical, and topological parameters. Hydrophobic parameters are mostly experimentally obtained logP or calculated logP, where P is the octanol–water partition coefficient. The electronic parameters are σ, σ^-, and σ^+ for aromatic systems and Taft's σ^* for aliphatic systems. Steric parameters include Tafts steric parameter (Es), McGowan volume (MgVol), van der Waals volume (Vw), and molecular weight (Mw). Verloop's sterimol parameter, B1, is a measure of the width of the first atom of a substituent, B5 is the overall volume, and L is the substituent length. Molar refractivity MR is a measure of volume with a small correction for polarizability. These parameters are widely used in QSAR studies.

2D QSAR

The 2D-QSAR models are generally developed with multiple regression analysis. The general equation for 2D-QSAR is

$$\text{Activity} = A^*P_1 + B^*P_2 + C, \tag{7.12}$$

where P_1 and P_2 are physicochemical properties; A and B are the fitted coefficients with activity for the properties P_1 and P_2, respectively. C is a constant. The data are analyzed with total number of points, correlation coefficient, standard deviation, etc. A typical equation obtained for the activity of protease inhibitor with cyclic cyanoguanidines is (Garg and Bhhatarai, 2006):

$$\log(1/K_i) = -1.77\,\text{MgVol} - 1.26\,\sigma_{\text{sum}} + 16.21 \tag{7.13}$$

The correlation coefficient (r) between activity data and combination of physicochemical properties is reported to be 0.92 (Garg and Bhhatarai, 2006).

Several compounds have been used to perform QSAR studies on protease inhibitors. Nugiel et al. (1996) reported the enzyme inhibitory activity of cyclic urea derivatives by replacing benzyl moiety into CH2-cyclopropyl groups. Wilkerson et al. (1996) reported QSAR for a series of N, N' di-substituted cyclic urea 3-benzamides for HIV protease inhibition and antiviral activity. Jadhav et al. (1998) observed a significant improvement in inhibitor potency for cyclic cyanoguanidine derivatives. Other compounds such as aminoindazole, tetrahydropyrimidinones, pyranones and dihydropyranones have been used to understand the relationship between the activity of protease inhibitors and physicochemical properties.

Furthermore, similar studies have been carried out for different inhibitors with several compounds.

3D QSAR

The 3D-QSAR studies are carried out with different techniques such as Computational Molecular Field Analysis (CoMFA), Comparative Molecular Indices Analysis (CoMSIA), pharamacophore generation, and free-energy binding analysis. The applications of 3D-QSAR are discussed with the latest work reported for the activity of choline kinase inhibitors (Srivani and Sastry, 2009). They have used both CoMFA and CoMSIA to relate the activity and 3D structures of the molecules. The steric and electrostatic potential fields for CoMFA were calculated using Lennard-Jones 6–12 and columbic terms, respectively with a distance dependent dielectric constant of 1.0. Five CoMSIA fields were calculated for steric, electrostatic, hydrophobic, hydrogen bond donor, and hydrogen bond acceptor using the probe atom. The steric indices are related to the third power of the atomic radii, the electrostatic descriptors are derived from atomic partial charges, the hydrophobic fields are derived from atom-based parameters developed by Viswanadhan et al. (1989), and the hydrogen bond donor and acceptor indices are obtained by a rule-based method derived from experimental values (Bohacek and McMartin, 1992). **Figure 7.22** shows the relationship between experimental and predicted

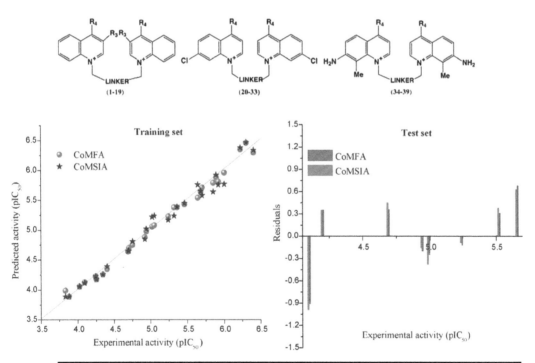

FIGURE 7.22 Plot of actual versus predicted biological activities for training and test datasets obtained with CoMFA and CoMSIA models. The ligands were changed upon the substitution of atoms (groups) in R3 and R4. Figure was taken from Srivani and Sastry (2009) with permission from Elsevier.

activities, obtained with CoMFA and CoMSIA for training and test sets of 29 and 10 compounds, respectively. The compounds were obtained with different atoms (groups) at positions R3 and R4 shown in **Figure 7.22**. The correlation between experimental and predicted activities is 0.99 and 0.98 with CoMFA and CoMSIA, respectively (Srivani and Sastry, 2009). Furthermore, Sala et al. (2009) reported the applications of pyridine derivatives to develop a 3D-QSAR model for rationalizing the structure–activity relationship studies on IkB kinase β (IKK-2) inhibition by these molecules. The QSAR model has been used to predict the IC50 values of other pyridine derivatives of IKK-2 inhibitors, whose activity is unknown. The results showed that 3D-QSAR models can be used to predict the activity of IKK-2 inhibitors by pyridine derivatives reliably.

7.6   Exercises

1. Find the protein–protein interactions in IntAct database for yeast obtained with biochemical method.
   Hint: Search with species (yeast) and detection method.
2. Identify the interacting proteins with the motif CXXC.
   Hint: Search HPRD using the given motif.
3. Obtain the kinetic data deposited for the complex, barnase-barstar.
   Hint: Use KDBI with the names of the molecules.
4. Retrieve the thermodynamic data for barnase–barstar complex.
   Hint: search PINT with protein, barnase and peptide, barstar.
5. Obtain the single mutants that changed the binding free energy of 2–10 kcal/mol.
   Hint: Search PINT with single mutants (wild) in protein and peptide and limit the DDG as 2–10 kcal/mol.
6. Compare the binding data for the alanine mutations with the binding free energy change of 2–5 kcal/mol.
   Hint: Search ASEdb and find the mutation with the free energy range.
7. Predict the protein–protein binding sites for the following sequence using Firestar

   ```
   >1IAR:A|PDBID|CHAIN|
   SEQUENCEHKCDITLQEIIKTLNSLTEQKTLCTELTVTDIFAASKNTTEKETFCRAATVL
   RQFYSHHEKDTRCLGATAQQFHRHKQLIRFLKRLDRNLWGLAGLNSCPVKEANQSTLENF
   LERLKTIMREKYSKCSS
   ```

   Hint: Input the sequence and find the probability score for binding sites.
8. Analyze the binding site prediction using different methods.
   Hint: Submit the sequence in different sequence based prediction servers.
9. Predict the binding sites of 1C4Z-D chain using cons-PPISP.
   Hint: Obtain the coordinates from PDB and submit with chain information.
10. Analyze the binding site prediction using different methods.
    Hint: Submit the sequence in different structure based prediction servers.
11. Compute the direct and indirect readout Z-scores for 1a02.
    Hint: Run ReadOut program using the PDB code.

12. Compute the direct and indirect readout Z-scores in a set of protein–DNA complexes.
Hint: Develop the dataset from PDB and run ReadOut program.

13. Check the following sequence as a DNA-binding protein using DBS-PRED.

```
>  |CHAIN|SEQUENCE
MALTNAQILAVIDSWEETVGQFPVITHHVPLGGGLQGTLHCYEIPLAAPYGVGFAKNGPT
RWQYKRTINQVVHRWGSHTVPFLLEPDNINGKTCTASHLCHNTRCHNPLHLCWESLDDNK
GRNWCPGPNGGCVHAVVCLRQGPLYGPGATVAGPQQRGSHFVV
```

Hint: Input the sequence in DBS-PRED and check for the probability.

14. Predict the binding sites in the following sequence

```
>  |CHAIN|SEQUENCE
MALTNAQILAVIDSWEETVGQFPVITHHVPLGGGLQGTLHCYEIPLAAPYGVGFAKNGPT
RWQYKRTINQVVHRWGSHTVPFLLEPDNINGKTCTASHLCHNTRCHNPLHLCWESLDDNK
GRNWCPGPNGGCVHAVVCLRQGPLYGPGATVAGPQQRGSHFVV
```

Hint: Use DBS-PRED

15. Predict the binding sites using different methods.
Hint: Get the binding sites with different servers.

16. Discriminate the following sequence using DNAbinder
```
>  |CHAIN|SEQUENCE
MALTNAQILAVIDSWEETVGQFPVITHHVPLGGGLQGTLHCYEIPLAAPYGVGFAKNGPT
RWQYKRTINQVVHRWGSHTVPFLLEPDNINGKTCTASHLCHNTRCHNPLHLCWESLDDNK
GRNWCPGPNGGCVHAVVCLRQGPLYGPGATVAGPQQRGSHFVV
```

Hint: Input the sequence in FASTA format.

17. Identify the RNA-binding residues in the following sequence

```
>SEQUENCE
AVPETRPNHTIYINNLNEKIKKDELKKSLHAIFSRFGQILDILVSRSLKMRGQAFVIFKE
VSSATNALRSMQGFPFYDKPMRIQYAKTDSDIIAKMK
```

Hint: submit the sequence in Pprint server.

18. Compute the protein–ligand binding free energy of 1a4k.
Hint: Prepare the input file for the PDB, 1a4k as per the prescribed format and submit the coordinates at BAPPL server. Alternatively, the formatted file is available at http://www.scfbio-iitd.res.in/software/drugdesign/proteinliganddataset.htm, which can be obtained by clicking on the PDB code.

19. Validate the geometry of the ligand AZM in the structure, 1AZM.
Hint: Follow the procedure described in 7.4.3.

References

Adachi M, Kurihara Y, Nojima H, Takeda-Shitaka M, Kamiya K, Umeyama H. Interaction between the antigen and antibody is controlled by the constant domains: normal mode dynamics of the HEL-HyHEL-10 complex. Protein Sci. 2003;12:2125–2131.

Ahmad S, Sarai A. Moment-based prediction of DNA-binding proteins. J Mol Biol. 2004;341(1):65–71.

Ahmad S, Sarai A. PSSM-based prediction of DNA binding sites in proteins. BMC Bioinformatics. 2005;6:33.

Ahmad S, Gromiha MM, Sarai A. Analysis and prediction of DNA-binding proteins and their binding residues based on composition, sequence and structural information. Bioinformatics. 2004;20(4):477–486.

Ahmad S, Keskin O, Sarai A, Nussinov R. Protein-DNA interactions: structural, thermodynamic and clustering patterns of conserved residues in DNA-binding proteins. Nucleic Acids Res. 2008;36(18):5922–5932.

Ahmad S, Kono H, Araúzo-Bravo MJ, Sarai A. ReadOut: structure-based calculation of direct and indirect readout energies and specificities for protein-DNA recognition. Nucleic Acids Res. 2006;34(Web Server issue):W124–W127.

Alfarano C, Andrade CE, Anthony K, Bahroos N, Bajec M, Bantoft K, Betel D, Bobechko B, Boutilier K, Burgess E, Buzadzija K, Cavero R, D'Abreo C, Donaldson I, Dorairajoo D, Dumontier MJ, Dumontier MR, Earles V, Farrall R, Feldman H, Garderman E, Gong Y, Gonzaga R, Grytsan V, Gryz E, Gu V, Haldorsen E, Halupa A, Haw R, Hrvojic A, Hurrell L, Isserlin R, Jack F, Juma F, Khan A, Kon T, Konopinsky S, Le V, Lee E, Ling S, Magidin M, Moniakis J, Montojo J, Moore S, Muskat B, Ng I, Paraiso JP, Parker B, Pintilie G, Pirone R, Salama JJ, Sgro S, Shan T, Shu Y, Siew J, Skinner D, Snyder K, Stasiuk R, Strumpf D, Tuekam B, Tao S, Wang Z, White M, Willis R, Wolting C, Wong S, Wrong A, Xin C, Yao R, Yates B, Zhang S, Zheng K, Pawson T, Ouellette BF, Hogue CW. The Biomolecular Interaction Network Database and related tools 2005 update. Nucleic Acids Res. 2005;33(Database issue):D418–D424.

Allers J, Shamoo Y. Structure-based analysis of protein-RNA interactions using the program ENTANGLE. J Mol Biol. 2001;311(1):75–86.

Aloy P, Pichaud M, Russell RB. Protein complexes: structure prediction challenges for the 21st century. Curr Opin Struct Biol. 2005;15(1):15–22.

Andrabi M, Mizuguchi K, Sarai A, Ahmad S. Prediction of mono-and di- nucleotide specific DNA-binding sites in proteins using neural networks. BMC Struct Biol. 2009;9(1):30.

Andrusier N, Nussinov R, Wolfson HJ. FireDock: fast interaction refinement in molecular docking. Proteins. 2007;69:139–159.

Audie J. Development and validation of an empirical free energy function for calculating protein–protein binding free energy surfaces. Biophys Chem. 2009;139(2–3):84–91.

Bahadur RP, Chakrabarti P, Rodier F, Janin J. A dissection of specific and non-specific protein-protein interfaces. J Mol Biol. 2004;336(4):943–955.

Baker CM, Grant GH. The role of aromatic amino acids in protein-nucleic acid recognition. Biopolymers. 2007;85(5–6):456–470.

Bhardwaj N, Lu H. Residue-level prediction of DNA-binding sites and its application on DNA-binding protein predictions. FEBS Lett. 2007;581(5):1058–1066.

Block P, Sotriffer CA, Dramburg I, Klebe G. AffinDB: a freely accessible database of affinities for protein-ligand complexes from the PDB. Nucleic Acids Res. 2006;34(Database issue):D522–D526.

Boer DR, Kroon J, Cole JC, Smith B, Verdonk ML. SuperStar: comparison of CSD and PDB-based interaction fields as a basis for the prediction of protein-ligand interactions. J Mol Biol. 2001;312(1):275–287.

Bohacek R, McMartin S. Definition and display of steric, hydrophobic, and hydrogen bonding properties of ligand binding sites in proteins using Lee and Richards accessible surface: validation of a high-resolution graphical tool for drug design. J Med Chem. 1992;35:1671–1684.

Bradford JR, Westhead DR. Improved prediction of protein–protein binding sites using a support vector machines approach. Bioinformatics. 2005;21:1487–1494.

Brady GP Jr, Stouten PF. Fast prediction and visualization of protein binding pockets with PASS. J Comput Aided Mol Des. 2000;14(4):383–401.

Breitkreutz BJ, Stark C, Reguly T, Boucher L, Breitkreutz A, Livstone M, Oughtred R, Lackner DH, Bähler J, Wood V, Dolinski K, Tyers M. The BioGRID Interaction Database: 2008 update. Nucleic Acids Res. 2008;36(Database issue):D637–D640.

Brylinski M, Kochanczyk M, Broniatowska E, Roterman I. Localization of ligand binding site in proteins identified in silico. J Mol Model. 2007;13(6–7):665–675.

Camacho CJ. Modeling side-chains using molecular dynamics improve recognition of binding region in CAPRI targets. Proteins. 2005;60(2):245–251.

Chalk AJ, Worth CL, Overington JP, Chan AW. PDBLIG: classification of small molecular protein binding in the Protein Data Bank. J Med Chem. 2004;47(15):3807–3816.

Chang DT, Oyang YJ, Lin JH. MEDock: a web server for efficient prediction of ligand binding sites based on a novel optimization algorithm. Nucleic Acids Res. 2005;33(Web Server issue):W233–W238.

Chang DT, Weng YZ, Lin JH, Hwang MJ, Oyang YJ. Protemot: prediction of protein binding sites with automatically extracted geometrical templates. Nucleic Acids Res. 2006;34(Web Server issue):W303–W309.

Chatr-aryamontri A, Ceol A, Palazzi LM, Nardelli G, Schneider MV, Castagnoli L, Cesareni G. MINT: the Molecular INTeraction database. Nucleic Acids Res. 2007;35(Database issue):D572–D574.

Cheng TM, Blundell TL, Fernandez-Recio J. pyDock: electrostatics and desolvation for effective scoring of rigid-body protein-protein docking. Proteins. 2007;68:503–515.

Comeau SR, Gatchell DW, Vajda S, Camacho CJ. ClusPro: a fully automated algorithm for protein-protein docking. Nucleic Acids Res. 2004;32(Web server issue):W96–W99.

Debnath AK. Quantitative structure-activity relationship (QSAR): a versatile tool in drug design. In Ghose AK, Viswanathan VN, eds. Combinatorial library design and evaluation: principles, software tools and applications. New York: Marcel Dekker Inc; 2006.

Dickerson RE. Definitions and nomenclature of nucleic acid structure components. Nucleic Acids Res. 1989;17(5):1797–1803.

Dolinsky TJ, Nielsen JE, McCammon JA, Baker NA. PDB2PQR: an automated pipeline for the setup of Poisson-Boltzmann electrostatics calculations. Nucleic Acids Res. 2004;32:W665–W667.

Dominguez C, Boelens R, Bonvin AM. HADDOCK: a protein–protein docking approach based on biochemical or biophysical information. J Am Chem Soc. 2003;125:1731–1737.

Dundas J, Ouyang Z, Tseng J, Binkowski A, Turpaz Y, Liang J. CASTp: computed atlas of surface topography of proteins with structural and topographical mapping of functionally annotated residues. Nucleic Acids Res. 2006;34(Web Server issue):W116–W118.

Ellis JJ, Broom M, Jones S. Protein-RNA interactions: structural analysis and functional classes. Proteins. 2007;66(4):903–911.

Ertekin A, Nussinov R, Haliloglu T. Association of putative concave protein-binding sites with the fluctuation behavior of residues. Protein Sci. 2006;15(10):2265–2277.

Espadaler J, Romero-Isart O, Jackson RM, Oliva B. Prediction of protein-protein interactions using distant conservation of sequence patterns and structure relationships. Bioinformatics. 2005;21(16):3360–3368.

Ezkurdia I, Bartoli L, Fariselli P, Casadio R, Valencia A, Tress ML. Progress and challenges in predicting protein-protein interaction sites. Brief Bioinform. 2009;10(3):233–246.

Fariselli P, Pazos F, Valencia A, Casadio R. Prediction of protein–protein interaction sites in heterocomplexes with neural networks. Eur J Biochem. 2002;269(5):1356–1361.

Friesner RA, Banks JL, Murphy RB, Halgren TA, Klicic JJ, Mainz DT, Repasky MP, Knoll EH, Shelley M, Perry JK, Shaw DE, Francis P, Shenkin PS. Glide: a new approach for rapid, accurate docking and scoring. 1. Method and assessment of docking accuracy. J Med Chem. 2004;47(7):1739–1749.

Fujita T, ed. QSAR and Drug Design: New Developments and Applications. Amsterdam: Elsevier; 1995.

Gabb HA, Jackson RM, Sternberg MJ. Modelling protein docking using shape complementarity, electrostatics and biochemical information. J Mol Biol. 1997;272:106–120.

Garg R, Bhatarai B. QSAR and molecular modeling studies of HIV protease inhibitors. Topics. Heterocycl Chem. 2006;3:181–271.

Glaser F, Steinberg DM, Vakser IA, Ben-Tal N. Residue frequencies and pairing preferences at protein-protein interfaces. Proteins. 2001;43(2):89–102.

Glaser F, Morris RJ, Najmanovich RJ, Laskowski RA, Thornton JM. A method for localizing ligand binding pockets in protein structures. Proteins. 2006;62(2):479–488.

Gold ND, Jackson RM. SitesBase: a database for structure-based protein-ligand binding site comparisons. Nucleic Acids Res. 2006;34(Database issue):D231–D234.

Golovin A, Dimitropoulos D, Oldfield T, Rachedi A, Henrick K. MSDsite: a database search and retrieval system for the analysis and viewing of bound ligands and active sites. Proteins. 2005;58(1):190–199.

Gromiha MM, Munteanu MG, Gabrielian A, Pongor S. Anisotropic elastic bending models of DNA. J Biol Phys. 1996;22:227–243.

Gromiha MM, Munteanu MG, Simon I, Pongor S. The role of DNA bending in Cro protein–DNA interactions. Biophys Chem. 1997;69(2–3):153–160.

Gromiha MM. Structure based sequence dependent stiffness scale for trinucleotides: a direct method. J Biol Phys. 2000;26:43–50.

Gromiha MM. Influence of DNA stiffness in protein-DNA recognition. J Biotech. 2005; 117:137–145.

Gromiha MM, Santhosh C, Ahmad S. Structural analysis of cation-pi interactions in DNA binding proteins. Int J Biol Macromol. 2004a;34(3):203–211.

Gromiha MM, Siebers JG, Selvaraj S, Kono H, Sarai A. Intermolecular and intramolecular readout mechanisms in protein-DNA recognition. J Mol Biol. 2004b;337:285–294.

Gromiha MM, Siebers JG, Selvaraj S, Kono H, Sarai A. Role of inter and intramolecular interactions in protein-DNA recognition. Gene. 2005;364:108–13.

Gromiha MM, Yokota K, Fukui K. Energy based approach for understanding the recognition mechanism in protein-protein complexes. Mol Biosystems, 2009 (in press). DOI:10.1039/B904161N.

Gupta A, Gandhimathi A, Sharma P, Jayaram B. ParDOCK: an all atom energy based Monte Carlo docking protocol for protein-ligand complexes. Protein Pept Lett. 2007;14(7):632–646.

Halgren TA, Murphy RB, Friesner RA, Beard HS, Frye LL, Pollard WT, Banks JL. Glide: a new approach for rapid, accurate docking and scoring. 2. Enrichment factors in database screening. J Med Chem. 2004;47(7):1750–1759.

Hansch C, Leo A. Exploring QSAR: Fundamentals and Applications in Chemistry and Biology. Washington, DC: American Chemical Society; 1995.

Hansch C, Maloney PP, Fujita T, Muir RM. Correlation of biological activity of phenoxyacetic acids with Hammett substituent constants and partition coefficients. Nature. 1962;194:178–180.

Harrington RE, Winicov I. New concepts in protein-DNA recognition: sequence-directed DNA bending and flexibility. Prog Nucleic Acid Res. Mol Biol. 1994;47:195–270.

Heifetz A, Katchalski-Katzir E, Eisenstein M. Electrostatics in protein–protein docking. Protein Sci. 2002;11:571–587.

Henrick K, Thornton JM. PQS: a protein quaternary structure file server. Trends Biochem Sci. 1998;23(9):358–361.

Hoffman MM, Khrapov MA, Cox JC, Yao J, Tong L, Ellington AD. AANT: the amino acid-nucleotide interaction database. Nucleic Acids Res. 2004;32(Database issue):D174–D181.

Hogan ME, Austin RH. Importance of DNA stiffness in protein-DNA binding specificity. Nature. 1987;329:263–266.

Huang B, Schroeder M. Using protein binding site prediction to improve protein docking. Gene. 2008;422(1–2):14–21.

Hwang S, Gou Z, Kuznetsov IB. DP-Bind: a web server for sequence-based prediction of DNA-binding residues in DNA-binding proteins. Bioinformatics. 2007;23(5):634–636.

Ivanisenko VA, Pintus SS, Grigorovich DA, Kolchanov NA. PDBSite: a database of the 3D structure of protein functional sites. Nucleic Acids Res. 2005;33(Database issue):D183–D187.

Jadhav PK, Woerner FJ, Lam PY, Hodge CN, Eyermann CJ, Man HW, Daneker WF, Bacheler LT, Rayner MM, Meek JL, Erickson-Viitanen S, Jackson DA, Calabrese JC, Schadt M, Chang CH. Nonpeptide cyclic cyanoguanidines as HIV-1 protease inhibitors: synthesis, structure-activity relationships, and X-ray crystal structure studies. J Med Chem. 1998;41(9):1446–1455.

Jain T, Jayaram B. An all atom energy based computational protocol for predicting binding affinities of protein-ligand complexes. FEBS Lett. 2005;579(29):6659–6666.

Jain T, Jayaram B. Computational protocol for predicting the binding affinities of zinc containing metalloprotein-ligand complexes. Proteins. 2007;67(4):1167–1178.

Janin J, Miller S, Chothia C. Surface, subunit interfaces and interior of oligomeric proteins. J Mol Biol. 1988;204(1):155–164.

Jeong E, Miyano S. A weighted profile method for protein–RNA interacting residue prediction. Trans Comput Syst Biol. 2006;4:123–139.

Jeong E, Chung I, Miyano S. A neural network method for identification of RNA-interacting residues in protein. Genome Inform Ser Workshop Genome Inform. 2004;15:105–116.

Ji ZL, Chen X, Zhen CJ, Yao LX, Han LY, Yeo WK, Chung PC, Puy HS, Tay YT, Muhammad A, Chen YZ. KDBI: kinetic data of bio-molecular interactions database. Nucleic Acids Res. 2003;31(1):255–257.

Jones G, Willett P, Glen RC, Leach AR, Taylor R. Development and validation of a genetic algorithm for flexible docking. J Mol Biol. 1997;267(3):727–748.

Jones S, Daley DT, Luscombe NM, Berman HM, Thornton JM. Protein-RNA interactions: a structural analysis. Nucleic Acids Res. 2001;29(4):943–954.

Jones S, Thornton JM. Prediction of protein-protein interaction sites using patch analysis. J Mol Biol. 1997;272(1):133–143.

Jones S, van Heyningen P, Berman HM, Thornton JM. Protein–DNA interactions: a structural analysis. J Mol Biol. 1999;287:877–896.

Kamiya K, Sugawara Y, Umeyama H. Algorithm for normal mode analysis with general internal coordinates. J Comput Chem. 2003;24:826–841.

Kellenberger E, Muller P, Schalon C, Bret G, Foata N, Rognan D. sc-PDB: an annotated database of druggable binding sites from the Protein Data Bank. J Chem Inf Model. 2006;46(2):717–727.

Kerrien S, Alam-Faruque Y, Aranda B, Bancarz I, Bridge A, Derow C, Dimmer E, Feuermann M, Friedrichsen A, Huntley R, Kohler C, Khadake J, Leroy C, Liban A, Lieftink C, Montecchi-Palazzi L, Orchard S, Risse J, Robbe K, Roechert B, Thorneycroft D, Zhang Y, Apweiler R, Hermjakob H. IntAct–open source resource for molecular interaction data. Nucleic Acids Res. 2007;35(Database issue):D561–D565.

Keshava Prasad TS, Goel R, Kandasamy K, Keerthikumar S, Kumar S, Mathivanan S, Telikicherla D, Raju R, Shafreen B, Venugopal A, Balakrishnan L, Marimuthu A, Banerjee S, Somanathan DS, Sebastian A, Rani S, Ray S, Harrys Kishore CJ, Kanth S, Ahmed M, Kashyap MK, Mohmood R, Ramachandra YL, Krishna V, Rahiman BA, Mohan S, Ranganathan P, Ramabadran S, Chaerkady R, Pandey A. Human Protein Reference Database–2009 update. Nucleic Acids Res. 2009;37(Database issue):D767–D772.

Keskin O, Gursoy A, Ma B, Nussinov R. Principles of protein-protein interactions: what are the preferred ways for proteins to interact? Chem Rev. 2008;108:1225–1244.

Keskin O, Ma B, Rogale K, Gunasekaran K, Nussinov R. Protein-protein interactions: organization, cooperativity and mapping in a bottom-up systems biology approach. Phys Biol. 2005;2:S24–S35.

Keskin O, Tsai CJ, Wolfson H, Nussinov R. A new, structurally nonredundant, diverse data set of protein-protein interfaces and its implications. Protein Sci. 2004;13(4):1043–1055.

Kleywegt GJ, Harris MR, Zou J, Taylor TC, Wåhlby A, Jones TA. The uppsala electron-density server. Acta Crystallographica. 2004;D60:2240–2249.

Kleywegt GJ, Harris MR. ValLigURL: a server for ligand-structure comparison and validation. Acta Crystallographica. 2007;D63:935–938.

Koczyk G, Wyrwicz LS, Rychlewski L. LigProf: a simple tool for in silico prediction of ligand-binding sites. J Mol Model. 2007;13(3):445–455.

Koike A, Takagi T. Prediction of protein-protein interaction sites using support vector machines. Protein Eng Des Sel. 2004;17(2):165–173.

Komatsu K, Kurihara Y, Iwadate M, Takeda-Shitaka M, Umeyama H. Evaluation of the third solvent clusters fitting procedure for the prediction of protein-protein interactions based on the results at the CAPRI blind docking study. Proteins. 2003;52(1):15–18.

Kono H, Sarai A. Structure-based prediction of DNA target sites by regulatory proteins. Proteins. 1999;35:114–131.

Kumar M, Gromiha MM, Raghava GP. Identification of DNA-binding proteins using support vector machines and evolutionary profiles. BMC Bioinformatics. 2007;8:463.

Kumar M, Gromiha MM, Raghava GP. Prediction of RNA binding sites in a protein using SVM and PSSM profile. Proteins. 2008;71(1):189–194.

Kumar MDS, Gromiha MM. PINT: Protein-protein interactions thermodynamic database. Nucleic Acids Res. 2006;34(Database issue):D195–D198.

Laurie AT, Jackson RM. Q-SiteFinder: an energy-based method for the prediction of protein-ligand binding sites. Bioinformatics. 2005;21(9):1908–1916.

Lee K, Sim J, Lee J. Study of protein-protein interaction using conformational space annealing. Proteins. 2005;60(2):257–262.

Li W, Keeble AH, Giffard C, James R, Moore GR, Kleanthous C. Highly discriminating protein-protein interaction specificities in the context of a conserved binding energy hotspot. J Mol Biol. 2004;337(3):743–759.

Liang S, Zhang C, Liu S, Zhou Y. Protein binding site prediction using an empirical scoring function. Nucleic Acids Res. 2006;34:3698–3707.

Lin JH, Perryman AL, Schames JR, McCammon JA. The relaxed complex method: accommodating receptor flexibility for drug design with an improved scoring scheme. Biopolymers. 2003;68:47–62.

Liu T, Lin Y, Wen X, Jorissen RN, Gilson MK. BindingDB: a web-accessible database of experimentally determined protein-ligand binding affinities. Nucleic Acids Res. 2007;35(Database issue):D198–D201.

Lopez G, Valencia A, Tress ML. Firestar—prediction of functionally important residues using structural templates and alignment reliability. Nucleic Acids Res. 2007;35(Web server issue):W573–W577.

Mandel-Gutfreund Y, Margalit H, Jernigan RL, Zhurkin VB. A role for CH. . .O interactions in protein–DNA recognition. J Mol Biol. 1998;277:1129–1140.

Mandell JG, Roberts VA, Pique ME, Kotlovyi V, Mitchell JC, Nelson E, Tsigelny I, Ten Eyck LF. Protein docking using continuum electrostatics and geometric fit. Protein Eng. 2001;14:105–113.

Matthews BW. Protein–DNA interaction. No code for recognition. Nature. 1988;335: 294–295.

Michalsky E, Dunkel M, Goede A, Preissner R. SuperLigands—a database of ligand structures derived from the Protein Data Bank. BMC Bioinformatics. 2005;6:122.

Morris GM, Goodsell DS, Huey R, Olson AJ. Distributed automated docking of flexible ligands to proteins: parallel applications of AutoDock 2.4. J Comput Aided Mol Des. 1996;10(4):293–304.

Murakami Y, Jones S. SHARP2: protein-protein interaction predictions using patch analysis. Bioinformatics. 2006;22:1794–1795.

Nadassy K, Wodak SJ, Janin J. Structural features of protein-nucleic acid recognition sites. Biochemistry. 1999;38:1999–2017.

Nagano N. EzCatDB: the Enzyme Catalytic-mechanism Database. Nucleic Acids Res. 2005;33(Database issue):D407–D412.

Negi SS, Schein CH, Oezguen N, Power TD, Braun W. InterProSurf: a web server for predicting interacting sites on protein surfaces. Bioinformatics. 2007;23:3397–3399.

Neuvirth H, Raz R, Schreiber G. ProMate: a structure based prediction program to identify the location of protein–protein binding sites. J Mol Biol. 2004;338:181–199.

Nugiel DA, Jacobs K, Kaltenbach RF, Worley T, Patel M, Meyer DT, Jadhav PK, De Lucca GV, Smyser TE, Klabe RM, Bacheler LT, Rayner MM, Seitz SP. Preparation and structure-activity relationship of novel P1/P1′-substituted cyclic urea-based human immunodeficiency virus type-1 protease inhibitors. J Med Chem. 1996;39(11):2156–2169.

Ofran Y, Rost B. ISIS: interaction sites identified from sequence. Bioinformatics. 2007;23(2):e13–e16.

Olson WK, Gorin AA, Lu XJ, Hock LM, Zhurkin VB. DNA sequence-dependent deformability deduced from protein-DNA crystal complexes. Proc Natl Acad Sci USA. 1998;95;11163–11168.

Ozrin VD, Subbotin MV, Nikitin SM. PLASS: protein-ligand affinity statistical score–a knowledge-based force-field model of interaction derived from the PDB. J Comput Aided Mol Des. 2004;18(4):261–270.

Pabo CO, Nekludova L. Geometric analysis and comparison of protein-DNA interfaces: why is there no simple code for recognition? J Mol Biol. 2000;301:597–624.

Parrill AL, Reddy MR. Rational Drug Design: Novel Methods and Practical Applications. Washington, DC: American Chemical Society; 1999.

Pazos F, Helmer-Citterich M, Ausiello G, Valencia A. Correlated mutations contain information about protein-protein interaction. J Mol Biol. 1997;271(4):511–523.

Pei J, Wang Q, Zhou J, Lai L. Estimating protein-ligand binding free energy: atomic solvation parameters for partition coefficient and solvation free energy calculation. Proteins. 2004;57(4):651–664.

Phipps KR, Li H. Protein-RNA contacts at crystal packing surfaces. Proteins. 2007;67(1): 121–127.

Porollo A, Meller J. Prediction-based fingerprints of protein–protein interactions. Proteins. 2007;66:630–645.

Potapov V, Reichmann D, Abramovich R, Filchtinski D, Zohar N, Ben Halevy D, Edelman M, Sobolev V, Schreiber G. Computational redesign of a protein-protein interface for high affinity and binding specificity using modular architecture and naturally occurring template fragments. J Mol Biol. 2008;384(1):109–119.

Prabakaran P, An J, Gromiha MM, Selvaraj S, Uedaira H, Kono H, Sarai A. Thermodynamic database for protein-nucleic acid interactions (ProNIT). Bioinformatics. 2001;17(11):1027–1034.

Prabakaran P, Siebers JG, Ahmad S, Gromiha MM, Singarayan MG, Sarai A. Classification of protein-DNA complexes based on structural descriptors. Structure. 2006;14(9):1355–1367.

Puvanendrampillai D, Mitchell JB. Protein Ligand Database (PLD): additional understanding of the nature and specificity of protein-ligand complexes. Bioinformatics. 2003;19(14):1856–1857.

Reddy AS, Amarnath HS, Bapi RS, Sastry GM, Sastry GN. Protein ligand interaction database (PLID). Comput Biol Chem. 2008;32(5):387–390.

Res I, Mihalek I, Lichtarge O. An evolution based classifier for prediction of protein interfaces without using protein structures. Bioinformatics. 2005;21(10):2496–2501.

Ritchie DW, Kemp GJ. Protein docking using spherical polar Fourier correlations. Proteins. 2000;39:178–194.

Roche O, Kiyama R, Brooks CL III. Ligand-protein database: linking protein-ligand complex structures to binding data. J Med Chem. 2001;44(22):3592–3598.

Sala E, Guasch L, Vaque M, Mateo-Sanz, JM, Blay M, Blade C, Garcia-Vallve S, Pujadas G. 3D-QSAR study of pyridine derivatives as IKK-2 inhibitors. QSAR & Combinatorial Sci. 2009 (in press). DOI:10.1002/qsar.200860167.

Salwinski L, Eisenberg D. Computational methods of analysis of protein-protein interactions. Curr Opin Struct Biol. 2003;13(3):377–382.

Salwinski L, Miller CS, Smith AJ, Pettit FK, Bowie JU, Eisenberg D. The Database of Interacting Proteins: 2004 update. Nucleic Acids Res. 2004;32(Database issue):D449–D451.

Sarai A, Mazur J, Nussinov R, Jernigan RL. Sequence dependence of DNA conformational flexibility. Biochemistry. 1989;28:7842–7849.

Sayle RA, Milner-White EJ. RASMOL: biomolecular graphics for all. Trends Biochem Sci. 1995;20(9):374.

Schneidman-Duhovny D, Inbar Y, Nussinov R, Wolfson HJ. PatchDock and SymmDock: servers for rigid and symmetric docking. Nucleic Acids Res. 2005;33(Web server issue):W363–W367.

Schueler-Furman O, Wang C, Baker D. Progress in protein-protein docking: atomic resolution predictions in the CAPRI experiment using RosettaDock with an improved treatment of side-chain flexibility. Proteins. 2005;60(2):187–194.

Schuttelkopf AW, van Aalten DMF. PRODRG: a tool for high-throughput crystallography of protein-ligand complexes. Acta Crystallogr. D Biol Crystallogr. 2004;60:1355–1363.

Schwabe JW. The role of water in protein-DNA interactions. Curr Opin Struct Biol. 1997;7:126–134.

Shin JM, Cho DH. PDB-Ligand: a ligand database based on PDB for the automated and customized classification of ligand-binding structures. Nucleic Acids Res. 2005;33(Database issue):D238–D241.

Shoemaker BA, Panchenko AR. Deciphering protein-protein interactions. Part I. Experimental techniques and databases. PLoS Comput Biol. 2007;3:e42.

Shoemaker BA, Panchenko AR. Deciphering protein-protein interactions. Part II. Computational methods to predict protein and domain interaction partners. PLoS Comput Biol. 2007;3:e43.

Shulman-Peleg A, Shatsky M, Nussinov R, Wolfson HJ. MultiBind and MAPPIS: webservers for multiple alignment of protein 3D-binding sites and their interactions. Nucleic Acids Res. 2008;36(Web Server issue):W260–W264.

Sikić M, Tomić S, Vlahovicek K. Prediction of protein-protein interaction sites in sequences and 3D structures by random forests. PLoS Comput Biol. 2009;5(1): e1000278.

Sippl M. Calculation of conformational ensembles for potentials of mean force: an approach to the knowledge-based prediction of local structures in globular proteins. J Mol Biol. 1990;213:859–883.

Srivani P, Sastry GN. Potential choline kinase inhibitors: a molecular modeling study of bis-quinolinium compounds. J Mol Graph Model. 2009;27(6):676–688.

Stuart AC, Ilyin VA, Sali A. LigBase: a database of families of aligned ligand binding sites in known protein sequences and structures. Bioinformatics. 2002;18(1):200–201.

Takeda Y, Sarai A, Rivera VM. Analysis of the sequence-specific interactions between Cro repressor and operator DNA by systematic base substitution experiments. Proc Natl Acad Sci U S A. 1989;86(2):439–443.

Terashi G, Takeda-Shitaka M, Takaya D, Komatsu K, Umeyama H. Searching for protein-protein interaction sites and docking by the methods of molecular dynamics, grid scoring, and the pairwise interaction potential of amino acid residues. Proteins. 2005;60(2):289–295.

Terribilini M, Lee JH, Yan C, Jernigan RL, Honavar V, Dobbs D. Prediction of RNA binding sites in proteins from amino acid sequence. RNA. 2006;12(8):1450–1462.

Thomsen R, Christensen MH. MolDock: a new technique for high-accuracy molecular docking. J Med Chem. 2006;49(11):3315–3321.

Thorn KS, Bogan AA. ASEdb: a database of alanine mutations and their effects on the free energy of binding in protein interactions. Bioinformatics. 2001;17(3):284–285.

Tovchigrechko A, Vakser IA. GRAMM-X public web server for protein-protein docking. Nucleic Acids Res. 2006;34(Web Server issue):W310–W314.

Treger M, Westhof E. Statistical analysis of atomic contacts at RNA-protein interfaces. J Mol Recognit. 2001;14(4):199–214.

Tuncbag N, Kar G, Keskin O, Gursoy A, Nussinov R. A survey of available tools and web servers for analysis of protein-protein interactions and interfaces. Brief Bioinform. 2009;10(3):217–232.

Valencia A, Pazos F. Computational methods for the prediction of protein interactions. Curr Opin Struct Biol. 2002;12(3):368–373.

Vaqué M, Ardévol A, Bladé C, Salvadó MJ, Blay M, Fernández-Larrea J, Arola L, Pujadas G. Protein-ligand docking: a review of recent advances and future perspectives. Curr Pharm Anal. 2008;4:1–19.

Viswanadhan VN, Ghose AK, Revenkar R, Robins, RN. Atomic physicochemical parameters for three-dimensional structure directed quantitative structure-activity relationships. 4. Additional parameters for hydrophobic and dispersive interactions and their application for an automated superposition of certain naturally occurring nucleoside antibiotics. J Chem Inf Comput Sci. 1989;29:163–172.

Wang L, Brown SJ. BindN: a web-based tool for efficient prediction of DNA and RNA binding sites in amino acid sequences. Nucleic Acids Res. 2006;34(Web Server issue):W243–W248.

Wang R, Fang X, Lu Y, Yang CY, Wang S. The PDBbind database: methodologies and updates. J Med Chem. 2005;48(12):4111–4119.

Weiner SJ, Kollman PA, Case DA, Singh UC, Ghio C, Alagona G, Profeta S Jr, Weiner P. A new force field for molecular mechanical simulation of nucleic acids and proteins. J Am Chem Soc. 1984;106:765–784.

Wiehe K, Pierce B, Mintseris J, Tong WW, Anderson R, Chen R, Weng Z. ZDOCK and RDOCK performance in CAPRI rounds 3, 4, and 5. Proteins. 2005;60(2):207–213.

Wilkerson WW, Akamike E, Cheatham WW, Hollis AY, Collins RD, DeLucca I, Lam PY, Ru Y. HIV protease inhibitory bis-benzamide cyclic ureas: a quantitative structure-activity relationship analysis J Med Chem. 1996;39(21):4299–4312.

Wodak SJ, Mendez R. Prediction of protein-protein interactions: the CAPRI experiment, its evaluation and implications. Curr Opin Struct Biol. 2004;14(2):242–249.

Xu D, Tsai CJ, Nussinov R. Hydrogen bonds and salt bridges across protein-protein interfaces. Protein Eng. 1997;10(9):999–1012.

Yamaguchi A, Iida K, Matsui N, Tomoda S, Yura K, Go M. Het-PDB Navi.: a database for protein-small molecule interactions. J Biochem (Tokyo). 2004;135(1):79–84.

Yan C, Terribilini M, Wu F, Jernigan RL, Dobbs D, Honavar V. Predicting DNA-binding sites of proteins from amino acid sequence. BMC Bioinformatics. 2006;7:262.

Yoshikawa T, Tsukamoto K, Hourai Y, Fukui K. Improving the accuracy of an affinity prediction method by using statistics on shape complementarity between proteins. J Chem Inf Model. 2009;49(3):693–703.

Yu X, Cao J, Cai Y, Shi T, Li Y. Predicting rRNA-, RNA-, and DNA-binding proteins from primary structure with support vector machines. J Theor Biol. 2006;240(2):175–184.

Zanzoni A, Montecchi-Palazzi L, Quondam M, Ausiello G, Helmer-Citterich M, Cesareni G. MINT: a Molecular INTeraction database. FEBS Lett. 2002;513(1): 135–140.

Zhou HX, Shan Y. Prediction of protein interaction sites from sequence profile and residue neighbor list. Proteins. 2001;44:336–343.

Zsoldos Z, Reid D, Simon A, Sadjad BS, Johnson AP. eHiTS: an innovative approach to the docking and scoring function problems. Curr Protein Pept Sci. 2006;7(5):421–435.

Appendix A

List of protein databases

Protein sequence

PIR	http://pir.georgetown.edu/
SWISS-PROT	http://www.expasy.org/sprot/
	http://www.ebi.ac.uk/swissprot/
UniProt	http://www.uniprot.org/
Export	http://www.cmbi.kun.nl/EXProt/
NCBI Protein database	http://www.ncbi.nlm.nih.gov/entrez
TCDB	http://www.tcdb.org/
SBASE	http://www.icgeb.trieste.it/sbase

Protein structure

PDB, Protein Data Bank	http://www.rcsb.org/
PDB ftp access	ftp://ftp.wwpdb.org/pub/pdb
PDBsum	http://www.ebi.ac.uk/pdbsum/
ASTRAL	http://astral.berkeley.edu/
PISCES	http://dunbrack.fccc.edu/pisces/
PDB-REPRDB	http://www.cbrc.jp/pdbreprdb/
MPtopo	http://blanco.biomol.uci.edu/mptopo
TMPDB	http://bioinfo.si.hirosaki-u.ac.jp/~TMPDB/
PDBTM	http://pdbtm.enzim.hu/
SCOP	http://scop.mrc-lmb.cam.ac.uk/scop/
CATH	http://www.biochem.ucl.ac.uk/bsm/cath/cath.html
	http://www.cathdb.info/

3D Graphics

Pymol	http://www.pymol.org
Rasmol	http://www.umass.edu/microbio/rasmol/
Jmol	http://jmol.sourceforge.net/
KING	http://kinemage.biochem.duke.edu/software/king.php
Webmol	http://www.cmpharm.ucsf.edu/~walther/webmol/readme.html
SWISS-PDB viewer	http://spdbv.vital-it.ch/

Amino acid properties

AAindex   http://www.genome.ad.jp/aaindex/
Numerical/normalized values   http://www.cbrc.jp/~gromiha/fold_rate/
         property.html

Solvent accessibility representation

ASAview   http://www.netasa.org/asaview/
POLYVIEW   http://polyview.cchmc.org/

Protein folding

PFD   http://pfd.med.monash.edu.au/
     http://www.foldeomics.org/pfd/
KineticDB   http://kineticdb.protres.ru/db/index.pl

Thermodynamic

ProTherm   http://gibk26.bse.kyutech.ac.jp/jouhou/
         protherm/protherm.html

Prokaryotic growth temperature

PGTdb   http://pgtdb.csie.ncu.edu.tw

Protein–protein interactions

DIP   http://dip.doe-mbi.ucla.edu
MINT   http://mint.bio.uniroma2.it/mint/
BIND   http://bind.ca
IntAct   http://www.ebi.ac.uk/intact
BioGRID   http://www.thebiogrid.org
HPRD   http://www.hprd.org

Thermodynamic/kinetic

ASEdb   http://nic.ucsf.edu/asedb/
PINT   http://www.bioinfodatabase.com/pint/
         index.html
KDBI   http://xin.cz3.nus.edu.sg/group/kdbi/
         kdbi.asp

Protein–nucleic acid complexes

ProNuc   http://gibk26.bse.kyutech.ac.jp/jouhou/
         pronuc/pronuc.html
BAInt   http://gibk26.bse.kyutech.ac.jp/jouhou/
         baint/baint.html
AANT   http://aant.icmb.utexas.edu/
ENTANGLE   http://www.bioc.rice.edu/~shamoo/
         resources.html

Thermodynamic

ProNIT
http://gibk26.bse.kyutech.ac.jp/jouhou/pronit/
pronit.html

Protein–ligand complexes

PLD
http://www-mitchell.ch.cam.ac.uk/pld/
PDB-Ligand
http://www.idrtech.com/PDB-Ligand/
sc-PDB
http://bioinfo-pharma.u-strasbg.fr/scPDB/
Het-PDB Navi
http://daisy.nagahama-i-bio.ac.jp/golab/
hetpdbnavi.html

SuperLigands
http://bioinf.charite.de/superligands/
PDBSite
http://wwwmgs.bionet.nsc.ru/mgs/gnw/
pdbsite/

LigBase
http://modbase.compbio.ucsf.edu/ligbase/
SitesBase
http://www.bioinformatics.leeds.ac.uk/sb
Ligand-protein database
http://lpdb.chem.lsa.umich.edu/
SuperStar
http://www.ccdc.cam.ac.uk/products/
life_sciences/superstar/

AffinDB
http://www.agklebe.de/affinity
PDBbind
http:// www.pdbbind.org/
EzCatDB
http://mbs.cbrc.jp/EzCatDB/
MSDsite
http://www.ebi.ac.uk/msd-srv/msdsite
BindingDB
http://www.bindingdb.org
PLID
http://203.199.182.73/gnsmmg/databases/plid/

Literature

PubMed
http://www.ncbi.nlm.nih.gov/pubmed/
Google scholar
http://scholar.google.co.jp/
Scopus
http://www.scopus.com/

Database collection
http://www.oxfordjournals.org/nar/database/
cap/

List of protein Web servers

Sequence alignment

BLAST
http://www.ncbi.nlm.nih.gov/BLAST/
FASTA
http://www.ebi.ac.uk/fasta/
ClustalW
http://www.ebi.ac.uk/clustalw/
PRIME
http://prime.cbrc.jp/
PROMALS
http://prodata.swmed.edu/promals/
PSI-BLAST
http://www.ebi.ac.uk/Tools/psiblast/
http://www.ncbi.nlm.nih.gov/Education/
BLASTinfo/psi1.html

Protein structure comparison

DALI	http://ekhidna.biocenter.helsinki.fi/dali/start
CE	http://cl.sdsc.edu/ce.html
PRIDE	http://hydra.icgeb.trieste.it/pride/
MATRAS	http://biunit.aist-nara.ac.jp/matras/
TOPS	http://www.tops.leeds.ac.uk/

Retrieval of nonredundant sequences/structures

CD-HIT	http://cd-hit.org/
ASTRAL	http://astral.berkeley.edu/
PISCES	http://dunbrack.fccc.edu/pisces/

Secondary structure content and class prediction

SSCP	http://www.bork.embl-heidelberg.de/SSCP/
SOV measure	http://proteinmodel.org/AS2TS/SOV/sov.html
Chou-Fasman	http://fasta.bioch.virginia.edu/fasta/chofas.htm
Garnier	http://npsa-pbil.ibcp.fr/cgi-bin/npsa_automat.pl?page=npsa_gor4.html
Jpred	http://www.compbio.dundee.ac.uk/www-jpred/
SSpro	http://promoter.ics.uci.edu/BRNN-PRED/
PHD	http://cubic.bioc.columbia.edu/predictprotein/
Jnet	http://www.compbio.dundee.ac.uk/www-jpred/
PSIPRED	http://bioinf.cs.ucl.ac.uk/psipred/
	Source code: http://bioinf.cs.ucl.ac.uk/memsat/memsat-svm/
YASPIN	http://ibivu.cs.vu.nl/programs/yaspinwww/
SVM package	http://svmlight.joachims.org/

Solvent accessibility representation

ASAview	http://www.netasa.org/asaview/
POLYVIEW	http://polyview.cchmc.org/

Solvent accessibility computation

ACCESS	http://www.csb.yale.edu/
ASC	http://mendel.imp.univie.ac.at/mendeljsp/studies/asc.jsp
DSSP	http://www.cmbi.kun.nl/gv/dssp/
	ftp://ftp.cmbi.kun.nl/pub/molbio/data/dssp/
GETAREA	http://www.scsb.utmb.edu/getarea/
NACCESS	http://wolf.bms.umist.ac.uk/naccess/
POPS	http://mathbio.nimr.mrc.ac.uk/wiki/POPS
	http://www.cs.vu.nl/~ibivu/programs/popswww

Solvent accessibility prediction (state-wise)

PHDacc	http://cubic.bioc.columbia.edu/predictprotein/
PredAcc	http://bioserv.rpbs.jussieu.fr/RPBS/cgi-bin/Ressource .cgi?chzn_lg=an&chzn_rsrc=PredAcc
Jnet	http://www.compbio.dundee.ac.uk/
NETASA	http://www.netasa.org/netasa/
RCNPRED	http://gpcr.biocomp.unibo.it/cgi/predictors/rcn/ pred_rcncgi.cgi

Prediction of real value solvent accessibility

RVP-net	http://www.netasa.org/rvp-net/
SABLE	http://sable.cchmc.org
SARpred	http://www.imtech.res.in/raghava/sarpred/
Two-stage SVM	http://birc.ntu.edu.sg/~pas0186457/asa.html

Transmembrane helix prediction

ALOM	http://psort.nibb.ac.jp/form.html
MPEx	http://blanco.biomol.uci.edu/mpex/
KD plot	http://fasta.bioch.virginia.edu/fasta/grease.htm
TMpred	http://www.ch.embnet.org/software/TMPRED_form.html
SPLIT	http://split.pmfst.hr/split/
TopPred	http://www.sbc.su.se/~erikw/toppred2/
PRED-TMR	http://o2.biol.uoa.gr/PRED-TMR/
SOSUI	http://sosui.proteome.bio.tuat.ac.jp/sosuiframe0.html
MEMSAT	http://bioinf.cs.ucl.ac.uk/psipred/psiform.html
PHDhtm	http://cubic.bioc.columbia.edu/predictprotein/
TMAP	http://bioinfo.limbo.ifm.liu.se/tmap/
DAS	http://www.sbc.su.se/~miklos/DAS/
HMMTOP	http://www.enzim.hu/hmmtop/
TMHMM	http://www.cbs.dtu.dk/services/TMHMM/
CoPreTHi	http://athina.biol.uoa.gr/CoPreTHi/

Transmembrane strand prediction

PRED-TMBB	http://bioinformatics.biol.uoa.gr/PRED-TMBB
TMB-HUNT	http://www.bioinformatics.leeds.ac.uk/betaBarrel
TMBETA-DISC	http://psfs.cbrc.jp/tmbetadisc/
TMBETA-SVM	http://tmbeta-svm.cbrc.jp/
TMETADISC-RBF	http://rbf.bioinfo.tw/~sachen/OMP.html
ProfTMB	http://cubic.bioc.columbia.edu/services/proftmb/
TBBpred	http://www.imtech.res.in/raghava/tbbpred/
TMBETA-NET	http://psfs.cbrc.jp/tmbeta-net/
ConBBPRED	http://bioinformatics.biol.uoa.gr/ConBBPRED/
TMBpro	http://www.igb.uci.edu/servers/psss.html
TMBETAPRED-RBF	http://rbf.bioinfo.tw/~sachen/tmrbf.html

Prediction of disordered proteins and domains

FoldUnfold	http://skuld.protres.ru/~mlobanov/ogu/ogu.cgi
DPROT	http://www.imtech.res.in/raghava/dprot/
MD	http://www.rostlab.org/services/md/index.php
DISPROT (VSL2)	http://www.ist.temple.edu/disprot/Predictors.html
	http://www.ist.temple.edu/disprot/predictorVSL2.php
POODLE	http://mbs.cbrc.jp/poodle/
IUPRED	http://iupred.enzim.hu/
DISOPRED	ftp://bioinf.cs.ucl.ac.uk/pub/DISOPRED
GlobPlot	http://globplot.embl.de
DisEMBL	http://dis.embl.de
FoldIndex	http://bioportal.weizmann.ac.il/fldbin/findex

Contact prediction

CORNET	http://gpcr.biocomp.unibo.it/cgi/predictors/cornet/pred_cmapcgi.cgi
CMAPpro	http://www.ics.uci.edu/~baldig/
GPCPRED	http://www.sbc.su.se/~maccallr/contactmaps
PoCM	http://foo.acmc.uq.edu.au/~nick/Protein/contact_casp6.html
PROFcon	http://www.predictprotein.org/submit_profcon.html

Protein three-dimensional structure prediction

FRankenstein	http://genesilico.pl/meta/
PROSPECTOR	http://128.205.242.1/current_buffalo/skolnick/prospector.html
MODELLER	http://salilab.org/modeller/modeller.html
GenTHREADER	http://bioinf.cs.ucl.ac.uk/psipred/psiform.html
ROBETTA	http://robetta.bakerlab.org/
FORTE	http://www.cbrc.jp/htbin/forte-cgi/forte_form.pl
Bhageerath	http://www.scfbio-iitd.res.in/bhageerath
SWISS-MODEL	http://swissmodel.expasy.org

Folding rates/folding nuclei

Folding nuclei	http://www.cs.ubc.ca/~oshmygel/foldingnuclei.html
FOLD-RATE	http://psfs.cbrc.jp/fold-rate/
K-fold	http://gpcr.biocomp.unibo.it/cgi/predictors/K-Fold/K-Fold.cgi
FOLD-RATE Q	http://bioinformatics.myweb.hinet.net/foldrate.htm

Stabilizing residues in protein structures

SRide	http://sride.enzim.hu

Prediction of protein mutant stability

FOLD-X (FOLDEF)	http://fold-x.embl-heidelberg.de
CUPSAT	http://cupsat.uni-koeln.de/
I-Mutant2.0	http://gpcr.biocomp.unibo.it/cgi/predictors/ I-Mutant2.0/I-Mutant2.0.cgi.
MUpro	http://www.igb.uci.edu/servers/servers.html
PoPMuSiC	http://babylone.ulb.ac.be/popmusic/
iPTREE-STAB	http://bioinformatics.myweb.hinet.net/ iptree.htm
Automute	http://proteins.gmu.edu/automute
iROBOT	http://bioinformatics.myweb.hinet.net/irobot. htm

Binding site prediction in protein–protein complexes

cons-PPISP	http://pipe.scs.fsu.edu/ppisp.html
Firestar	http://firedb.bioinfo.cnio.es/Php/FireStar.php
InterProSurf	http://curie.utmb.edu/
PINUP	http://sparks.informatics.iupui.edu/PINUP/
PPI-Pred	http://bioinformatics.leeds.ac.uk/ppi-pred
ProMate	http://bioportal.weizmann.ac.il/promate/
SHARP2	http://www.bioinformatics.sussex.ac.uk/SHARP2/ sharp2.html Patch
SPPIDER	http://sppider.cchmc.org/

Protein–protein docking

PQS	http://pqs.ebi.ac.uk/
BDOCK	http://www.biotec.tudresden.de/~bhuang/bdock/ bdock.html
ClusPro	http://nrc.bu.edu/cluster/
DOT	http://www.sdsc.edu/CCMS/DOT/
FireDock	http://bioinfo3d.cs.tau.ac.il/FireDock/
GRAMMX	http://vakser.bioinformatics.ku.edu/resources/ gramm/grammx
HEX	http://www.csd.abdn.ac.uk/hex/
MolFit	http://www.weizmann.ac.il/Chemical_Research_ Support//molfit/home.html
PatchDock	http://bioinfo3d.cs.tau.ac.il/PatchDock/
PyDock	http://mmb.pcb.ub.es/PyDock/
RosettaDock	http://rosettadock.graylab.jhu.edu
ZDOCK	http://zlab.bu.edu/zdock/index.shtml
HADDOCK	http://www.nmr.chem.uu.nl/haddock/
3D-Dock	http://www.sbg.bio.ic.ac.uk/docking/

Protein–DNA interactions (discrimination/prediction)

DBS-PRED	http://www.netasa.org/dbs-pred
	http://gibk26.bse.kyutech.ac.jp/jouhou/shandar/
	netasa/dbs-pred/
DP-Bind	http://lcg.rit.albany.edu/dp-bind
BindN	http://bioinformatics.ksu.edu/bindn/
DNAbinder	http://www.imtech.res.in/raghava/dnabinder/

Protein–RNA interactions (discrimination/prediction)

RNABindR	http://bindr.gdcb.iastate.edu
Pprint	http://www.imtech.res.in/raghava/pprint/

Protein–ligand interactions (bindig site prediction/docking)

MEDock	http://bioinfo.mc.ntu.edu.tw/medock/
PRODRG	http://davapc1.bioch.dundee.ac.uk/programs/prodrg/
PDB2PQR	http://agave.wustl.edu/pdb2pqr/
Q-SiteFinder	http://www.bioinformatics.leeds.ac.uk/qsitefinder
Pocket-Finder	http://www.bioinformatics.leeds.ac.uk/pocketfinder
ParDOCK	http://www.scfbio-iitd.res.in/dock
Fuzzy-oil-drop	http://www.bioinformatics.cm-uj.krakow.pl/activesite
LigProf	http://www.cropnet.pl/ligprof
Protemot	http://protemot.csbb.ntu.edu.tw/
CASTp	http://sts.bioengr.uic.edu/castp/
PASS	http://www.ccl.net/cca/software/UNIX/pass/
	overview.shtml
SURFNET-ConSurf	http://consurf-hssp.tau.ac.il
eHiTS	http://www.simbiosys.ca/ehits/
MolDock (Molegro)	http://www.molegro.com/
GOLD	http://www.ccdc.cam.ac.uk/products/life_sciences/
	gold/
AutoDock	http://autodock.scripps.edu/
Glide	http://www.schrodinger.com/

Binding free energy

BAPPL	http://www.scfbio-iitd.res.in/software/drugdesign/
	bappl.jsp

Ligand structure validation/comparison

ValLigURL	http://eds.bmc.uu.se/eds/valligurl.php
EDS Tutorial	http://xray.bmc.uu.se/embo/ligandtut/index.html

Amino acid properties

Amino acid composition   http://www.expasy.ch/tools/protparam.html
                                http://pir.georgetown.edu/pirwww/search/
                                    comp_mw.shtml
Amino acid properties   http://gibk26.bse.kyutech.ac.jp/cgi-bin/jouhou/
                                    3D/bin/pdb_amino.sh?FILENAME=pdb1prc.
                                    ent&SUB=DATA&

Conservation score

ConSurf   http://consurf.tau.ac.il/
AL2CO   http://prodata.swmed.edu/al2co/al2co.php

ROC analysis

Web based calculator   http://www.jrocfit.org

Noncanonical interactions

CAPTURE   http://capture.caltech.edu/
NCI   http://www.mrc-lmb.cam.ac.uk/genomes/nci/

Hydrogen bonds

HBPLUS   http://www.biochem.ucl.ac.uk/bsm/hbplus/
                                    home.html

Parameters for proteins

Stabilization center (SCide)   http://www.enzim.hu/scide
Atom depth (DPX)   http://hydra.icgeb.trieste.it/dpx/

Membrane protein function

DISC-FUNCTION   http://tmbeta-genome.cbrc.jp/disc-function/

Direct and indirect readout in protein–DNA complexes

Readout   http://gibk26.bse.kyutech.ac.jp/jouhou/
                                    readout/

Compilation of Web servers   http://www.oxfordjournals.org/nar/
                                    webserver/cap/

Index

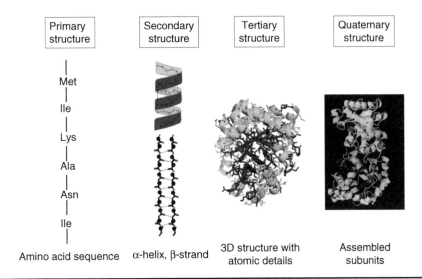

COLOR INSERT 1 Structural organization of proteins.

(b)

COLOR INSERT 2 Multiple sequence alignment using ClustalW: (b) the alignment of sequences.

Variable   Average   Conserved

(b)

COLOR INSERT 3   Computation of conservation score using ConSurf server: (b) Jmol view of conservation score.

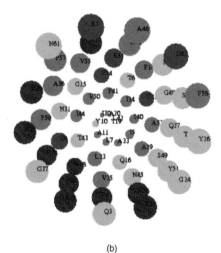

(b)

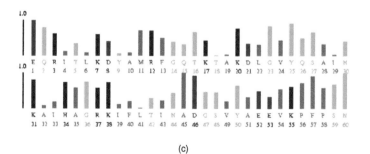

(c)

COLOR INSERT 4   ASAView of a DNA-binding protein (PDB code: 6CROA): (b) spiral view—blue, red, green, gray, and yellow colors—indicates the positively charged, negatively charged, polar, nonpolar, and Cys residues, respectively. The size of the sphere shows the relative ASA, and (c) bar diagram: length of the bar represents the ASA.

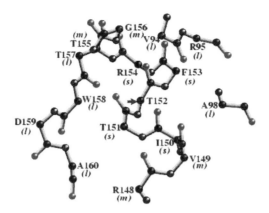

Color Insert 5 Representation of short-, medium-, and long-range contacts in protein structures. A typical example of surrounding residues around T152 of T4 lysozyme within 8 Å is shown): s: short-range contacts, m: medium-range contacts, and l: long-range contacts. Figure was adapted from Gromiha and Selvaraj (2004).

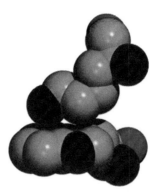

Color Insert 6 Representation of cation-π interactions in protein structures. An example between the residues Lys408 and Trp356 in sulfite reductase (1AOP) is shown.

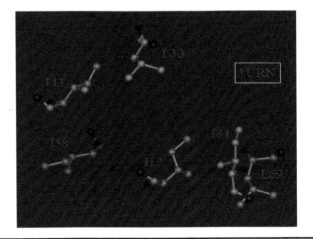

Color Insert 7 Hydrophobic clusters formed through long-range contacts in 1URN (Selvaraj and Gromiha, 2004).

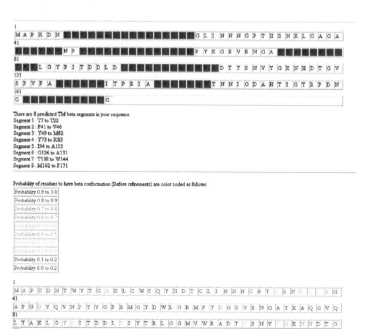

There are 8 predicted TM beta segments in your sequence.
Segment 1 : T7 to T22
Segment 2 : F41 to V46
Segment 3 : Y49 to M62
Segment 4 : Y73 to K83
Segment 5 : I94 to A105
Segment 6 : G126 to A131
Segment 7 : T138 to W144
Segment 8 : M162 to F171

Probability of residues to have beta conformation (Before refinements) are color coded as follows:

Probability 0.9 to 1.0
Probability 0.8 to 0.9
Probability 0.7 to 0.8
Probability 0.6 to 0.7
Probability 0.5 to 0.6
Probability 0.4 to 0.5
Probability 0.3 to 0.4
Probability 0.2 to 0.3
Probability 0.1 to 0.2
Probability 0.0 to 0.2

COLOR INSERT 8 Prediction of membrane spanning β-strand segments using TMBETA-NET. Output shows the stretch of amino acid residues in strand, list of segments, and probability of reach residue to be in β-strand.

GPCPRED - Results

Job identifier: edce4882e27948353df196157e780868

Viewing options: image zoom factor: [2 ▾] , number of contacts: [length/2 ▾] [refresh]

Predicted contact map:

Key to markings: small marks every 10 residues, large marks every 50.

Sequence (each predicted contact network is coloured differently):

```
  1   MNIFEMLRIDEGLRLKIYKDTEGYYTIGIGHLLTKSPSLNAAKSELDKAI     50
 51   GRNCNGVITKDEAEKLFNQDVDAAVRGILRNAKLKPVYDSLDAVRRCALI    100
101   NMVFQMGETGVAGFTNSLRMLQQKRWDEAAVNLAKSRWYNQTPNRAKRVI    150
151   TTFRTGIWDAYKNL                                        164
```

(b)

COLOR INSERT 9 Prediction of residue contacts in 2LZM using the GPCPRED server: (b) the contact map.

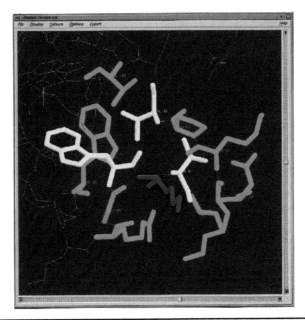

COLOR INSERT 10 Display of protein mutants and surrounding residues by RasMol. As an example, the surrounding residues of Lys 91 in Ribonuclease H (2RN2) are shown. The central residue, surrounding residues within 4 Å, and surrounding residues between 4 and 8 Å are shown in red, yellow, and green, respectively.

		Gly	Ala	Val	Leu	Ile	Cys	Met	Phe	Tyr	Trp	Pro	Ser	Thr	Asn	Gln	Asp	Glu	Lys	Arg	His
																To					
	Gly	---	230	57	5	0	12	0	10	2	6	14	55	2	5	26	28	24	6	19	17
	Ala	113	---	131	61	20	29	24	21	5	8	85	74	44	10	18	14	13	41	5	11
	Val	73	474	---	258	288	47	64	99	51	24	6	46	89	28	0	26	14	13	23	12
	Leu	34	350	114	---	88	55	40	86	2	4	18	16	14	4	4	13	32	12	23	5
	Ile	41	244	430	236	---	24	62	93	12	8	7	22	58	3	1	9	20	10	9	12
	Cys	4	137	17	26	1	---	2	2	2	1	1	105	86	0	0	0	4	0	0	0
	Met	18	65	50	124	45	1	---	17	2	0	0	0	6	1	0	6	6	18	21	0
	Phe	5	149	24	117	18	3	10	---	95	63	0	24	5	21	2	4	4	7	0	7
	Tyr	31	99	4	25	2	20	1	185	---	72	4	11	4	6	8	17	1	4	2	9
From	Trp	0	31	1	18	0	3	1	133	69	---	0	2	0	1	5	3	3	0	3	13
	Pro	66	187	7	17	2	9	0	4	2	4	---	68	10	4	4	6	5	5	8	3
	Ser	28	212	19	18	10	29	2	13	6	2	8	---	28	20	4	66	6	20	37	25
	Thr	37	204	126	32	103	38	11	24	28	2	3	118	---	20	27	18	92	7	40	21
	Asn	34	155	6	6	51	6	12	4	0	1	1	33	10	---	5	97	22	13	11	41
	Gln	42	74	3	23	5	12	3	3	4	0	9	5	1	14	---	10	34	33	14	6
	Asp	58	199	8	9	9	33	4	14	7	8	12	20	10	158	11	---	64	67	16	54
	Glu	50	299	67	48	6	10	16	25	29	17	8	29	12	13	103	32	---	134	30	13
	Lys	85	227	23	12	35	18	46	46	18	27	23	13	18	22	41	13	99	---	79	37
	Arg	40	161	14	13	0	19	18	3	0	1	1	8	2	0	20	1	71	26	---	67
	His	38	112	10	47	0	7	0	10	69	2	24	6	21	47	51	21	17	6	12	---

0-50:Blue ; 51-100:Magenta ; >100:Red

COLOR INSERT 11 Frequency of occurrence of mutants in ProTherm.

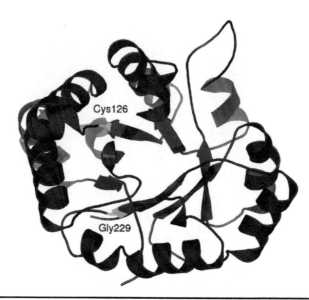

Color insert 12 Stabilizing residues in a typical TIM barrel protein, 1btm. The α-helices are shown as spiral ribbons and the β-strands are drawn as arrows from the amino end to the carboxyl end of the β-strand. The stabilizing segments are highlighted with red color. The two stabilizing residues Gly229 and Cys126 observed by thermodynamic, and kinetic experiments are also indicated. The picture was generated using PyMOL program (DeLano, 2002). Figure was adapted from Gromiha et al. (2004).

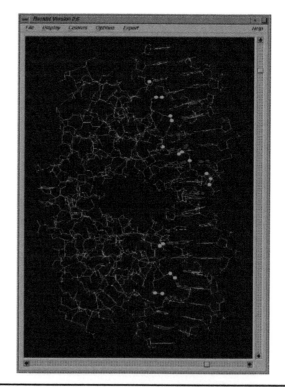

Color insert 13 RasMol view of λ repressor-operator complex (PDB code: 1LMB) showing contacting atoms of base (red) and amino acids (green) within a distance of 3.5 Å.

Submit a protein for prediction ::

Sequence Name: complex1

E-mail Address:

Sequence: (Type/paste FASTA format amino acid sequence)

```
>| SEQUENCE
MPRRRVIGQRKILPDPKFGSELLAKFVNILMVDGKKSTAESIVYSALETLAQRSGKSEL
EAFEVALENVRPTVEVKSRRV
GGSTYQVPVEVRPVRRNALAMRWIVEAARKRGDKSMALRLANELSDAAENKGTAVKKRE
DVHRMAEANKAFAHYRWLSLR
```

OR **Upload Sequence file** [   ] [Browse...]

Please submit only one sequence at a time

SVM threshold: [-0.2 ▼]

[Run Prediction] [Clear]

Prediction of RNA-interacting residues of complex1

Red: Interacting residues   Blue: Non-interacting residues

MPRRRVIGQRKILPDPKFGSELLAKFVNILMVDGKKSTAESIVYSALETLAQRSGKSELE
AFEVALENVRPTVEVKSRRVGGSTYQVPVEVRPVRRNALAMRWIVEAARKRGDKSMALRL
ANELSDAAENKGTAVKKREDVHRMAEANKAFAHYRWLSLRSFSHQAGASSKQPALGYLN

COLOR INSERT 14 Prediction of RNA-binding site residues using Pprint.

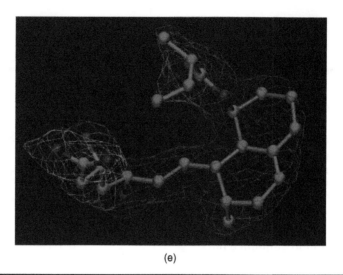

(e)

COLOR INSERT 15 Procedure for validating ligand structures in PDB: (e) ligand structure with electron density map.

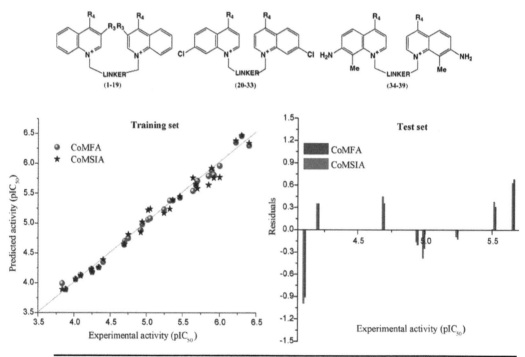

COLOR INSERT 16 Plot of actual versus predicted biological activities for training and test datasets obtained with CoMFA and CoMSIA models. The ligands were changed upon the substitution of atoms (groups) in R3 and R4. Figure was taken from Srivani and Sastry (2009) with permission from Elsevier.

Printed and bound by CPI Group (UK) Ltd, Croydon, CR0 4YY

09/10/2024

01042655-0001